A Lab for All Seasons

A Lab for All Seasons

*The Laboratory Revolution in
Modern Botany and the Rise of
Physiological Plant Ecology*

Sharon E. Kingsland

Yale

UNIVERSITY PRESS

New Haven & London

Published with assistance from the foundation established in memory of Calvin Chapin of
the Class of 1788, Yale College.

Yale University Press books may be purchased in quantity for educational, business, or
promotional use. For information, please e-mail sales.press@yale.edu (U.S. office) or
sales@yaleup.co.uk (U.K. office).

Set in Bulmer type by IDS Infotech Ltd.
Printed in the United States of America.

ISBN 978-0-300-26721-1 (paperback)
ISBN 978-0-300-26722-8 (hardcover)

Library of Congress Control Number: 2022946535
A catalogue record for this book is available from the British Library.

10 9 8 7 6 5 4 3 2 1

Contents

Preface

THE MOST IMPORTANT environmental challenges today are caused by two linked problems, population growth and climate change. Since we cannot stop either one, we have to adapt. We must figure out how to feed and house the world's population as well as mitigate the effects of climate change, and for these challenges we need science. This book is about one area of science, the science of plants, that seeks to unlock the mysteries of plant growth and development and to understand the ecological interactions that have created the earth's life-support systems. Investing in plant science is one of the most important ways to find solutions to our present and future environmental problems. The purpose of this book is to acquaint readers with the historical development of botanical science, especially since the 1940s, so that they may appreciate both the accomplishments and the challenges faced by scientists who have devoted their careers to understanding the lives of plants.

The history of botanical science opens up themes that are central to the development of modern biology in general. These include discipline formation, cross-disciplinary relationships and collaborations, innovation in laboratory designs, the synergistic relationship between laboratory and field science, and the process of scientific discovery. I examine these themes mainly through the history of two branches of botanical science, physiology and ecology, which are closely related. Early plant ecologists recognized the affinity between physiology and ecology; when ecology was still in its infancy, in the early twentieth century, they perceived the importance of uniting these approaches. Ecology was then a descriptive science just starting to develop experimental methods, while physiology tended to be narrowly focused on understanding how plants functioned rather than how the environment controlled those functions. Ecology needed to acquire rigor, and physiology needed to broaden its perspective.

Although scientists understood the need to synthesize plant ecology and plant physiology, these two fields matured without achieving a recognizable synthesis until the period following World War II, when physiological plant ecology emerged as a distinct field within ecology. I focus on one particularly important innovation that propelled the synthetic enterprise: the creation of new kinds of laboratories that made it possible to study whole organisms throughout their life cycles. These laboratories were invented after World War II, and they held out the promise for a revolutionary and global transformation of plant sciences. I use the term "revolution" not to suggest the overthrow of previous approaches or ambitions but rather to suggest that new technologies and laboratory designs, built at great expense, introduced a discontinuity into the gradual course of development in the fields of plant physiology and ecology. Although my first chapter discusses several lines of continuity between nineteenth- and twentieth-century biology, my emphasis will be on this postwar discontinuity, which had lasting effect.

These new laboratories, built between the late 1940s and the mid-1960s, were called "phytotrons," a term coined during their first generation to evoke the comparable Big Science instrument in particle physics, the cyclotron. The very first one, which opened at the California Institute of Technology in 1949, was the invention of Frits Warmolt Went and a team of talented engineers. The laboratories were weather factories, designed to produce any environment important for a plant's growth and development. They were truly labs for all seasons and for plants of all sizes, from tiny annuals small enough to grow in a test tube to whole trees. The goal was to create a rigorous, replicable experimental botany and to unravel the complicated interactions that occur between plants and the environment. These new designs brought a dramatically increased level of funding into plant sciences, boosted basic research as well as the application of science to practical problems, and brought scientists from different disciplines together. They remade botanical science worldwide and affected the course of the discipline of ecology as surely as particle accelerators affected the course of physics. Parts 1 and 2 of this book chart the invention of the phytotron and its worldwide dissemination, a development that I refer to as a "laboratory movement" because it entailed a concerted effort to spread and share knowledge among different scientific communities, to encourage increased patronage of such enterprises, and to create a social network that would sustain this movement.

Taking the material basis of science—laboratories and instruments—as my starting point has certain advantages. Instead of focusing on the history of an idea or discovery, or the development of a single discipline, exploring the laboratory as a place of inquiry means that attention must be paid not just to individuals but to communities of scientists from different disciplines who came together in the laboratory environment. Such laboratories were places for cross-disciplinary research that covered a wide range of basic and practical problems. The practical problems included agricultural questions about how to develop crops better adapted to environmental stresses as well as concerns about contemporary environmental problems such as smog and erosion. Although I am not writing about the history of agricultural practices, I am interested in the history of agricultural science and the role of such laboratory innovations in the development of scientific agriculture. One goal is to illustrate how the line between what we think of as "basic" versus "applied" research was consistently blurred in these laboratory settings, where fundamental research (or the study of plants with no commercial value) could be championed alongside research targeted at improving crops.

The rapid expansion of botanical research that occurred in the postwar period, aided in part by these new laboratory designs, also helped to invigorate such field sciences as ecology. The postwar rise of physiological plant ecology, which gained momentum from the changes occurring in botanical science more broadly, is the subject of Part 3. I present a few case studies that illustrate in different ways the relationship between laboratory-based experimental science and field studies. I also develop the idea that sciences like ecology, which are inherently interdisciplinary, must constantly grapple with the problem of synthesis, which can be understood as the fusion of different disciplines, the integration of laboratory and field research, and the effort to scale up from one level to another (as in scaling from the level of the leaf to the forest canopy to the ecosystem).

The case studies illustrate the challenges of synthesis that biologists have tackled and show how innovation in laboratory design and instrumentation has been a central feature in the ongoing quest for synthetic interpretations and in the emergence of physiological plant ecology as a vibrant subdiscipline within ecology. The underlying objective of the innovations that began what I have called the "laboratory revolution" in modern botany was to use science to improve the quantity and quality of food for the world's fast-growing population. That goal remains as central today as it was seventy years ago, but it is now made more urgent by the reality of climate change, which is proceeding much faster than has been expected. I have written this book in part to demonstrate how important plant sciences are to our health and well-being.

I have always regarded ecology as an inherently interdisciplinary science, but when I started studying physiological plant ecology I was struck by the regularity with which the literature sounded the theme of *synthesis,* or the pursuit of explanations that combined different viewpoints. I encountered it over and over in every problem that I studied. I realized too that this quest for synthesis, as I think of it, is a common scientific practice that has not often been directly analyzed in science and technology studies. I view the quest to achieve synthesis as a central scientific activity, at least in the fields I have studied, and therefore have tried to elevate its importance in my narrative. I wish to encourage greater attention to this aspect of scientific work. In the historical episodes I have selected I draw attention to texts that explicitly aim for synthetic interpretations, highlight strategies devised to reconcile conflicting interpretations, and describe efforts to bring together different communities for productive cross-disciplinary exchange. The quest for synthesis has sociological and cultural dimensions as well, because mechanisms are needed to bring people together and into communication with each other, to foster an ethos of collaboration, cooperation, and exchange. Creating a scientific culture that supports synthesis through thoughtful crossing of disciplinary boundaries has always required continuing effort and commitment.

When this book was about half written, I received an email from Robert E. Cook, an ecologist who had a question about an unrelated topic in the history of ecology. When I explained my present project on the history of phytotrons, he informed me he had been responsible for the construction of a modern incarnation of the phytotron when he was the director of the Arnold Arboretum of Harvard University. His doctoral research in plant population biology had been done under the direction of Arthur Galston at Yale University, and Galston had formerly been at Caltech, where he had worked in the world's first phytotron. Here was a direct link to Frits Went and the story I was researching. A few months later I was in Boston and Bob Cook arranged a visit to the Weld Hill Research Laboratory at the Arboretum, a complex of laboratories, growth chambers, and greenhouses that opened in 2011. Beautifully appointed and equipped with modern energy-saving systems, it was the direct descendant of the historical phytotrons that I was studying. Faye Rosin, the director of research facilitation, led us through the building, and Kea Woodruff showed me the growth chambers built by the company Conviron, one of the first companies to start manufacturing growth chambers for plant research in the 1960s. In the teaching laboratory, an experiment that Darwin had performed was set up for the next class of undergraduates. I discovered that Faye's early research had focused on the development of the potato tuber and maturation of the tomato, both classic themes going back to the first-generation phytotrons but updated using molecular biological methods. She had done postdoctoral research at the University of Wageningen in the Netherlands, which had built one of the earliest phytotrons in the 1950s, following Frits Went's lead. Each moment of the tour took me back to the historical themes of my project, and I realized that this history was highly relevant to today's science. This book's aim is to reveal the threads connecting past and present.

I am grateful to many people for supplying information and commentary that added important details to my central arguments. Snait Gissis invited me to Israel to lecture on the history of plant science as a guest of the Van Leer Jerusalem Institute in 2012. The scholars I met in Israel, especially Ayelet Shavit and Jaime Kigel, provided information that enabled me to develop the Israeli case study in chapter 4. Ditza Koller, the widow of Dov Koller, kindly read the draft of that chapter and corroborated my account. I thank Eugene Cittadino for comments on chapter 1. Elliot Meyerowitz at Caltech generously shared historical photographs of Caltech's phytotron, which he had obtained from Hans Kende. I am very grateful to Russell Monson for detailed information that strengthened the historical analysis in chapters 6 and 8. Barry Osmond supplied information on research connected to the Biosphere 2 project, discussed in chapter 6. On the history of research on mycorrhizal fungi I am indebted to Nellie Beetham Stark, Carl Jordan, Rafael Herrera, Ernesto Medina, Ariel Lugo, David J. Read, Michael Allen, and Edith B. Allen, whose suggestions and corrections greatly improved chapter 7. Jack Schultz sent comments relevant to the epilogue. I thank all these readers and take responsibility for any remaining errors. I am grateful to three anonymous referees for their helpful advice. Juliana Mulroy took keen interest in the history of physiological ecology, was a constant source of insight, and put me in contact with people who could clarify particular points.

Librarians and archivists provided crucial assistance, especially during the months of the pandemic when libraries were closed. The David M. Rubenstein Rare Book and Manuscript Library at Duke University supplied photocopies of archival materials free of charge. Archivists at the California Institute of Technology have provided much support over the years, and I particularly thank Loma Karklins for locating and digitizing images during the pandemic's lockdown period. Angela Todd at the Hunt Institute for Botanical Documentation at Carnegie Mellon University provided copies of Frits Went's diary and other materials from their collection. I thank the Rockefeller Archive Center for photocopies of correspondence relating to Rockefeller support of research at Caltech. Maggie Drain, digital archivist for the Carnegie Institution for Science, provided the photographs in chapter 8. Scholarly work during the pandemic would not have been possible without such electronic databases as JSTOR, as well as emergency digitization of monographs and journals by HathiTrust Digital Library.

At Yale University Press, Jean Thomson Black and Elizabeth Sylvia guided me through the proposal and publication process with great care. Paul Romney was a patient sounding board and editorial adviser during the long process of writing this book. The book is dedicated to Paul and to my son, David Romney, and daughter-in-law, Dana Levine; their creativity has long been an inspiration.

A Lab for All Seasons

PART ONE

Inventing the Phytotron

ONE

The Dream of Synthesis and the Laboratory Vision

OVER MILLIONS OF YEARS of evolution, plants have become adapted to live in some of the most extreme environments on earth. By studying these adaptations, scientists have uncovered the mysteries of how plants work and how natural selection has produced novel adaptive solutions to a wide range of environmental stresses. That knowledge can be put to use, showing us how to modify the species we depend on so that they benefit us more, by providing food, fiber, lumber, or medicines. Because the human population keeps growing, we need the plants that we have domesticated to keep pace with us. Historically, recognition that our population might outstrip food supply has stimulated greater investment in plant science; these investments have transformed botanical science over the past seventy years.

The human population has grown so large that our activities are altering the earth's climate and creating stresses for plants, including our most important crop species. Our actions may cause extinctions, and they are certainly causing shifts in the distribution and abundance of many plant species. Understanding how plants function is important for the study of global changes that are happening today, for plants do not just react to climate changes—they regulate climate. In the words of Christopher Field, an ecologist at Stanford University who studies global processes, plant physiology "is deeply entwined with climate change."[1] If we want to know what is happening to the earth's systems globally, we need to understand what plants are doing. This book is a historical study, told through selected episodes, of the science of plant adaptation and the scientists who work on these problems.

Today these studies fall within the province of ecology, especially physiological plant ecology, which emerged in the 1950s and 1960s when ecologists started to apply the interchangeable terms "ecophysiology" and "physiological ecology" to their field of research. However, as Hans Lambers, F. Stuart Chapin, and Thijs Pons rightly observe in their recent textbook on physiological plant ecology, the origins of this field go back to the nineteenth century, to treatises that linked physiological studies to broader problems of plant distribution and abundance.[2]

This chapter begins by considering these origins so that we can understand how botanical sciences and the subject that came to be called ecology were developing in the middle to late nineteenth century. Ecology is an inherently interdisciplinary field that concerns how organisms are adapted to their environments, both living and non-living. This involves the study of what Darwin metaphorically called the "struggle for existence," those complex and dynamic interactions that, as he argued, resulted in the formation of new species by means of natural selection. Ecology also studies interactions between populations and the patterns and processes occurring in larger associations of species that for convenience are described as making up communities and ecosystems. Ecology is therefore a hybrid enterprise, made up of many disciplines. Historically these were mainly taxonomy, morphology, physiology, and plant geography, but today one would also include genetics and molecular biology. A large part of the history of ecology involves figuring out how different disciplines or lines of research came together to support this broader type of inquiry. This book is mainly concerned with the relationship between plant physiology and plant ecology.

Ecology is an experimental science. It is studied not just in the field but also in laboratories. This book focuses on the evolution of laboratory designs for botanical research, because laboratory innovations were and still are a central feature of the development of ecology. Botanists trying to study plants indoors, in laboratories and under controlled conditions, faced many obstacles when technologies for controlling such basic things as light and temperature were relatively primitive. It took considerable ingenuity to find ways around these technological limitations, for many technologies that we take for granted did not exist until well into the twentieth century. Not many plants could be brought into the laboratory, and not many could be studied throughout their full life histories.

The narrative in Parts 1 and 2 is devoted largely to innovations in laboratory design that occurred in the postwar period, mainly in the 1950s to 1970s. These innovations made it possible to study more species of plants at all developmental stages under controlled conditions. Envisioning what such laboratories should be like and bringing them into existence was a major accomplishment of the mid-twentieth century. To set the stage, this chapter examines earlier efforts to envision and build facilities for experimental biology, to illustrate the kinds of difficulties biologists faced before World War II.

In the postwar period an influential visionary was Frits Warmolt Went, a plant physiologist and ecologist who built a controlled-environment laboratory, known as a "phytotron," at the California Institute of Technology in 1949. It was a weather factory, designed to produce and control all environmental conditions important for plant growth and development. Its high cost put it on a completely different level from most laboratories of the day. Went is central to this history both as an innovator in laboratory design and an eager exporter of these ideas to anyone willing to listen. His influence was felt around the world. The astonishing rapidity with which his invention of the phytotron spread worldwide can be understood once it is placed in the context of postwar discussions of food supply, population, and the future of arid lands for agricultural development. Technological innovations were not confined to large-scale enterprises like phytotrons. In Part 3, I also consider related developments that were significant for the synthesis of plant physiology and ecology that created physiological plant ecology. These include specially designed mobile laboratories for field research, development of "rhizotrons" for the study of soil environments, use of the Biosphere 2 Laboratory in Arizona for ecosystem-level experiments, and design of "ecotrons" for the study of artificial ecosystems.

The narrative that I develop here is international in scope. The question of how technological innovations travel and how they are picked up and adapted by research communities in different countries is a central theme of Part 2. Historians of science are interested in the global transformations occurring in the postwar period, but as a recent book shows, their attention has been given more to social and biomedical sciences and less to botanical sciences.[3] Historians of science have also long been interested in the history of field sciences but have paid less attention to the transformations that have taken place in these sciences since World War II. Unfortunately, recent attempts to remedy these deficiencies ignore plant sciences.[4] In the era of climate change, the botanical story demands our attention. This book shows how central plant sciences are to our understanding of the global transformations of science in the postwar period. I argue that the recent history of field sciences like ecology cannot be fully comprehended without simultaneously considering the revolutionary changes occurring in laboratory sciences.

The world's first phytotron at Caltech, rather remarkably, stimulated a worldwide laboratory movement. I use the term "movement" deliberately to suggest a concerted effort to join forces and campaign for these new types of laboratory. Scientists worked in concert and helped each other to achieve common goals for the improvement and modernization of plant science. How and why did this movement occur? What were the common elements that united projects in different countries? How did each country adapt these projects to suit its specific needs? Telling this story will take us around the world, from the United States to France, Australia, Israel, the Union of Soviet Socialist Republics, and Hungary. When we reach the Communist states, we will discover that these modernization efforts shed new light on our understanding of how scientists experienced the impact of Lysenkoism, the movement in the Soviet Union to suppress Mendelian genetics in favor of the doctrine of the inheritance of acquired characteristics.

After exploring the expansion of plant sciences broadly in Parts 1 and 2, the book culminates in Part 3 with several case studies that explore the development of physiological plant ecology through the study of different plant adaptations. The case studies involve the emission of volatile organic compounds by plants, the mutualistic relations between plants and soil fungi, the synthesis of genetics and ecology in the field known as "genecology," and the evolution of different biochemical pathways for photosynthesis. Each case study tackles a biological problem that has attracted a broad community of scientists. My examples focus on episodes in the history of physiological ecology that have intrinsic interest because they entail important physiological, ecological, and evolutionary problems.

A theme running throughout this book is the prevalence of *synthesis* as a scientific method and practice. Problems of adaptation, like other evolutionary problems, require multidisciplinary collaboration and the synthesis of knowledge from different fields. Historians of biology have paid attention to synthesis in certain specific contexts. Notable examples are studies of Darwin's evolutionary theory, Thomas Hunt Morgan's theory that genes were located on chromosomes, the creation of the modern evolutionary synthesis in the mid-twentieth century, and the formation of molecular biology following the discovery of the double helix structure of DNA. Historians of ecology have also recognized the interdisciplinary or synthetic nature of this field. But there is room for more analysis of the way scientists have deliberately sought synthetic interpretations and approaches, especially in disciplines that combine laboratory and field research. Biologists have pointed us in this direction. The evolutionary botanist G. Ledyard Stebbins, for instance, described the new cross-disciplinary linkages that were happening in several areas of biology in the early 1960s, and he was referring to more than the discovery of the genetic code and the rise of molecular biology. He perceived transformations in many fields, including taxonomy, morphology, and physiology. The synthetic disciplines, he argued, "are flourishing as never before."[5]

His comments about "synthetic disciplines" suggest that in some disciplines the need to synthesize is omnipresent but that certain conditions favor the synthetic enterprise. In doing research for this book, I was impressed by the number of times biologists made references to synthesis as a goal or essential step in advancing a field. Such comments should prompt us to examine what contexts and conditions promote synthesis, why scientists choose to seek synthetic interpretations, and what enables them to succeed. In what kinds of environment, in what kinds of institutions, and among what kinds of social groups is synthetic thinking encouraged and facilitated? These questions have historical, sociological, and philosophical dimensions. I examine this process especially in the case studies discussed in Part 3, but it is a theme of the entire book. One of the recurring justifications for building phytotrons was that they were laboratories meant for all of biology: they were places where scientists from different disciplines worked together.

In this book the challenge of synthesis should be understood in three ways. First, developing an explanation of adaptation can mean drawing on the skill set of scientists trained in different disciplines (such as biochemistry, physiology, taxonomy, genetics, and ecology). We must understand what helps to create opportunities for cross-disciplinary collaboration: new laboratories serving multiple disciplines are one part of the story. Second, and related to the problem of synthesizing disciplines, is the challenge of integrating laboratory-based and field-based research. Third is the challenge of scaling up, or going from the level of the leaf to whole plants to forest canopies, ecosystems, biomes, and eventually to the earth as a whole. All of these challenges will be evident in the examples I discuss. Understanding what promotes synthesis requires us to consider institutional contexts (the places of scientific inquiry, both in the laboratory and in the field), technological aspects (instruments, techniques, and methods), intellectual aspects (the brainwork of science), social and cultural aspects (the composition of social groups and what enables them to work productively) and broader social contexts that determine what kind of research is deemed relevant or worth supporting (how the science will be used or how it relates to problems of the day). This book encompasses all of these contexts and causes as we move up through time and as we travel around the world, visiting different countries to gain a comparative perspective on this complex history.

The biologists I discuss were, like Went, keen to cross the permeable border between the disciplines of physiology and ecology, and their research extended over a broad range of topics. As molecular biology grew dominant during the 1950s and 1960s, the biologists considered here continued to be interested in the whole organism and its relationship to the environment. They explored problems of adaptation and evolution that Darwin had studied earlier, and they appreciated the value of studying many species, not just one or two model organisms. The scientists at the center of the narratives in Parts 1 and 2 were generally of the same generation: they were born in the early twentieth century, started their careers in the 1930s, and shared a common vision of biology. To understand the traditions they were building on, we should first go back a few decades into the nineteenth century, to see how the subject of ecology came together.

Emerging Disciplines: The Nineteenth-Century Background

Botanical science was developing rapidly in the late nineteenth century, especially in Europe, while the scientific study of adaptation was advanced by Darwin's evolutionary theory, published in 1859. As scientists debated the merits of Darwin's theory, biological disciplines were taking shape, among them plant physiology, which scarcely existed as a coherent field in 1850 but was a dynamic and rigorous experimental discipline by the century's end. When botanists trained in laboratory sciences began to venture out into the field to study plants in their natural habitats, these fields of research started to link up, and from these various linkages the discipline of ecology was formed. Eugene Cittadino has provided the best discussion to date of European botanical research lying at the intersection of these fields in the late nineteenth century.[6] As he suggests, what we see emerging in the final decades of the nineteenth century is a body of research that we would identify as physiological plant ecology. At that time no one used this term, for "physiological ecology" or "ecophysiology" was not a distinct field until the 1960s, even though its origins lie in the nineteenth century.

The development of plant physiology from about 1850 on owed a great deal to the work of a highly talented group of European botanists: these included Wilhelm Hofmeister, Eduard Strasburger, Hermann Vöchting, and Nathaniel Pringsheim. As Cittadino has discussed, one man dominated this generation: Julius Sachs (1832–1897). Sachs contributed to many central inquiries of botany, including working out the basic process of photosynthesis (then called "assimilation"), and he was especially interested in plant "irritability" or reactions of plants to external conditions, which was connected to the study of "tropisms" or the directional growth of plants in relation to stimuli such as gravitation and light. He is recognized also because he published a handbook, *Handbuch der Experimental-Physiologie der Pflanzen* (1865), which set out an experimental approach to plant physiology and largely defined the field, as well as a textbook, *Lehrbuch der Botanik* (1868), which presented a masterful, exquisitely illustrated synthesis of the whole of botany, with detailed discussions of his own researches and those of other botanists. As Cittadino notes, in addition to Sachs's particular research advances, his outsized influence flowed from this comprehensive treatment of the subject of botany, told from the perspective of a physiologist. The German edition of the textbook in 1874 became the basis for the first English edition in 1875, and a second revised English edition followed in 1882.

A later commentator remarked that this textbook proved to be "the most potent factor in the development of botany in English-speaking countries." Sachs's emphasis on plant responsiveness and the "significance of protoplasm of the cell as the responsive material possessed with innate developmental and adaptive potencies" altered the way botanists approached anatomy, morphology, and taxonomy.[7] Sachs's approach was characterized by detailed microscopical examination, delicate experimental measurements, and close study of the life of the plant. The textbook was truly revolutionary for the way it moved through the whole of botany, covering topics in plant morphology (anatomy and development), physiology (including assimilation of nutrients, respiration, transport of water, and reactions to environmental conditions), and reproduction, ending with a full and sympathetic discussion of Darwin's theory of descent with modification. Although Sachs would later grow more skeptical of the role of natural selection in evolution, in his textbook he offered a balanced account of Darwin's argument, drawing attention to how a clear understanding of the struggle for existence provided insight into the origin of plant adaptations. As Cittadino comments, it was "as clear a statement of Darwin's view on evolution and natural selection as one can find in the botanical literature of the nineteenth century."[8] From this source, many botanists would be introduced to Darwin's evolutionary ideas.

After a series of short appointments, Sachs accepted a professorship at the University of Würzburg in 1869, remaining there until his death in 1897, and his laboratory attracted researchers from many countries. His students in turn became scientific leaders; among them were Hugo de Vries (the Netherlands), Jacques Loeb (United States), Wilhelm Pfeffer and Karl Goebel (Germany), and Francis Darwin, Frederick O. Bower, and Sydney H. Vines (England). Vines also worked on the second English edition of Sachs's textbook.

As Soraya de Chadarevian has shown, Sachs's laboratory students were given rigorous training in laboratory-based practices and a related set of skills that included being able to cultivate experimental plants, prepare microscopical specimens, and draw accurately.[9] Fritz Noll, one of Sachs's students at Würzburg, remembered that "whoever neglected his work, his apparatus, or his plant-cultures (upon plant-cultures Sachs laid a special emphasis) even once without a valid excuse might be perfectly confident that he would find his place occupied by another."[10] There was no shortage of prospective students. Sachs also developed a variety of laboratory instruments, the skillful use of which was central to sound experimental practice.

When Sachs was starting out, laboratory facilities could be primitive. One of his most fruitful research periods extended from 1861 to 1867, when he was at the Agricultural Academy in Poppelsdorf, about a mile from the University of Bonn. There, the "institute" where he worked "consisted of two very small rooms and an apartment in the basement, which also served the housemaster for a kitchen."[11] At Würzburg conditions were better; he had a large building for his botanical institute, where he assembled "rich collections, excellent contrivances for teaching, and instruments without number."[12] The instruments that Sachs developed were designed not just for precise measurements but also for achieving a degree of environmental control in the immediate vicinity of the plant. For physiological experiments temperature control was often desirable, and to achieve it the plant would be surrounded by a water-filled container heated by an oil lamp from below and capped by a small glass roof to keep the air warm, like a small house. That would serve for plants of a certain size, perhaps a few inches tall. But for microscopical studies it was necessary to encase the portion of the microscope holding the slide in a metal container filled with water, with a window to let in light. A lamp below would heat the water to the desired temperature, and if a cooler temperature was needed, ice cubes could be dropped into the box from an opening on top. These were among the kinds of contrivances that Sachs discussed in both his handbook and textbook.

Cittadino, in his analysis of the next generation of botanists whose careers started in the late 1870s and 1880s, argues that in this younger generation there came a shift from "the more analytical approach centered on the laboratory" toward "problems of a more synthetic nature."[13] It was that later generation, he suggests, who developed the field of research that, while not explicitly called ecology at the time, was in fact laying the foundation for what would later become ecology. His perceptive study establishes the background for the rest of this book.

Cittadino examined the careers of a small group of German-speaking European botanists who were born within a few years of each other in the mid-nineteenth century. Although he also considered a few lesser-known botanists, the central members of the group he studied were Gottlieb Haberlandt, Georg Volkens, Ernst Stahl, and Andreas F. W. Schimper. These men were trained in plant morphology and physiology by the professors of Sachs's generation and therefore received a thorough grounding in experimental and microscopical studies of plants—that is, in laboratory-based science. However, their education and early careers also fell in the 1870s and 1880s, when interest in Darwin's theory of natural selection was at its peak in Europe. As a result, their intellectual approach was strongly shaped by Darwinian ideas. Unlike most of their teachers, they became committed Darwinians, and their interest went well beyond paying lip service to Darwin's ideas. They wanted to develop new research programs focused on investigating problems in a Darwinian framework, which meant the study of plant adaptations. Their ability to advance this new Darwinian line of research in turn led them to forge new syntheses between different fields of science. Cittadino emphasizes the exceptional nature of this group, who represented a minority, but he also argues that their work had lasting impact.

I use just two of his examples, Haberlandt and Schimper, to illustrate his overall argument. First, however, we should review why the problem of adaptation would be central to an experimental Darwinian research program, as opposed to other lines of research that could be taken up, such as in phylogeny (the study of evolutionary trees or lines of descent, based on morphology). The problem of adaptation was in Darwin's time a subject of great interest to natural theologians, who argued that evidence of design or adaptation in nature was evidence of the divine creator's wisdom. Darwin's evolutionary ideas were opposed to this religious view, and to combat the religious interpretation he proposed alternative explanations of how adaptations arose through natural processes. Explaining adaptation was one of the central and most important problems that Darwin addressed in his work, including the *Origin of Species* (1859) and later books. For Darwin, the theory of natural selection was superior to other evolutionary theories precisely because it explained adaptation. Other evolutionary theories, which were far more speculative, had failed to provide satisfactory explanations of how organisms became adapted to their environments.

Darwin devoted a couple of chapters in the *Origin of Species* to examples of puzzling adaptations, but most of his research on the question of adaptation was published subsequently. In these later studies Darwin explored intriguing instances of adaptation in support of his argument that natural selection could account for the tremendous diversity of form in the world. Much of this research focused on botanical problems, which were the subject of books on fertilization in orchids (1862), variation in domesticated animals and plants (1868), insectivorous plants (1875), climbing plants (1875), cross- and self-fertilization in plants (1876), the different forms of flowers (1877), and the power of movement in plants (1880). As David Kohn has stressed, Darwin's research showed how his theory of evolution could be applied and supported through detailed study of botanical adaptations and thereby helped to create the field of evolutionary botany.[14] For Darwin himself, the problem of adaptation was a central concern of his later research program.

For the young botanists whom Cittadino studied, the decision to focus on problems of adaptation was a natural extension of their training in morphology and physiology. Adding the Darwinian framework, however, led them in new directions. Morphology, the study of structure or form, had been considered a subject that could be pursued completely independently of physiology, the study of function. Sachs's textbook clearly described how morphology could be approached in this way, without any reference to function at all. But for the younger Darwinian botanists, that separation made no sense. In a Darwinian world, where the struggle for existence led to the selection of new variations that improved a plant's function or made it fitter to live in a competitive environment, there could be no morphology independent of physiology. These two fields were necessarily united.

Haberlandt, whose research mostly dealt with photosynthesis, developed a new approach that he called "physiological plant anatomy," which insisted on the synthesis of morphology and physiology within a Darwinian framework. His treatise, *Physiologische Pflanzenanatomie,* was published in 1884. This new approach was based on the premise that every structure had a function and that all structures were adaptations. This assertion was radical and controversial for the time. It was, for a start, a teleological approach that did not meet with the approval of many established botanists (even though Sachs had sanctioned it in his textbook), and it could lead to circular arguments. It was often difficult to demonstrate the adaptive significance of a trait. Haberlandt did not have the knowledge to explain the adaptive functions for the traits he was observing; by insisting that they were adaptations, he was speculating. To give one example: he was the first to describe an unusual anatomical structure in the leaves of certain plants, which appeared to be wreath-shaped and therefore was given the name "Kranz anatomy," the German word for "wreath." In this case the correct function of Kranz anatomy was not discerned for another eighty years, when scientists found that the wreath-like structures were associated with a type of photosynthetic pathway called the "C_4 pathway," which indeed is an adaptation to hot, arid climates (discussed in chapter 8).

Haberlandt, professor at the University of Graz in Austria, stuck to his guns in the face of criticisms. But Graz was a small university with poor facilities, and he was not able to create a school centered on his new approach. Luckily, his approach found favor in the circle around Simon Schwendener, who had taught Haberlandt for a year and had in 1878 become professor at the University of Berlin, which had a well-equipped botanical institute and a large botanical garden and museum. Berlin was a vibrant center of botanical research, and Schwendener's students and associates enthusiastically developed the subject of physiological plant anatomy based on Haberlandt's lead.

Schwendener also appreciated the value of studying the relation of structure to function in natural environments, and he understood how this approach could be linked to the growing field of plant geography, a subject that Alexander von Humboldt had defined and that was gaining adherents in the second half of the century. Schwendener encouraged students to travel to exotic lands to study problems of adaptation in extreme environments, such as deserts, where adaptations might be more obvious than in European plants. Their ability to obtain funding for such excursions was helped by German colonial interests; Germans had by mid-century migrated to South American countries in large numbers, and later the German empire created protectorates, essentially colonies, in Africa and Micronesia. Colonial expansion created new opportunities for botanists, which they readily exploited to widen their inquiries in a variety of natural settings. Haberlandt himself got support from the Vienna Academy of Sciences to travel to Java and Malaya in 1891–1892, but his research remained laboratory-based. The Schwendener circle was important for extending Haberlandt's approach beyond the confines of the laboratory, broadening the scope of botany by linking physiology to plant geography and to problems of adaptation studied in exotic locations.

Outside the Schwendener circle, other botanists participated in this broadening, one of the most important being Andreas Schimper, an Alsatian. He had studied under Anton de Bary at Strasbourg, then under German rule. De Bary was an old-school plant pathologist and morphologist; Schimper's work was on the formation of starch grains in plants. His interest in Darwinism came later, after he went to Bonn in 1882 to work under Eduard Strasburger, botany professor at the university. Strasburger had come to Bonn from Jena, where he had been Ernst Haeckel's student and then colleague; like Haeckel he was a keen evolutionist. Haeckel was also responsible for coining the term "Oecologie" to denote the study of the conditions of existence, although he personally did not contribute to the development of the subject.

Before coming to Bonn, Schimper had traveled to the West Indies, where he became interested in epiphytes, plants that grow on other plants for support. Like the other botanists whom Cittadino profiled, he too became interested in the nature and origin of adaptation, an interest further stimulated by extensive travel to exotic lands. Some of his research in the 1880s was in Brazil, where he stayed with Fritz Müller, who along with his brother Hermann was an important early supporter of Darwin's theory of natural selection. Schimper made a total of four excursions, three to the Americas and the final one to Ceylon and Java, where he worked at the famed Dutch Botanical Garden at Buitenzorg (now Bogor) on Java. Frits Went, who later would also work on Java for five years, considered himself to be following in Schimper's footsteps.

Cittadino argues that Schimper's goals were synthetic. Where earlier generations had compiled anatomical and physiological facts without showing how they related to the problem of adaptation, Schimper's Darwinian perspective enabled him to bring these facts together to explain adaptation. Because of his extensive travels he also brought this research into line with recent work on plant geography. His final grand synthesis, *Pflanzengeographie auf Physiologischer Grundlage,* was published in 1898 and was translated into English as *Plant Geography on a Physiological Basis* in 1903. This, his best-known work, is considered one of the founding texts of the emerging discipline of ecology. Cittadino notes that Schimper considered the facilities at the Dutch Botanical Garden to be important for advancing the physiological branch of geographical botany. Not only was it well equipped for research and welcoming to foreign scientists, many of whom were German, but the tropical environment stimulated new research questions.

Once again, colonial expansion, this time by the Dutch in the East Indies, created opportunities for scientists that they gladly exploited. Cittadino makes the point that these scientists were, with a few exceptions, not connected in any way to the colonial missions of botanical gardens like this, which were directed more toward problems in agronomy and forestry. For botanists like Schimper, as for most of the others, the tropical locations were attractive for the opportunities they provided for research, but their research questions had to do with general problems of Darwinian theory, physiology, and morphology, not with any local needs. In fact, Schimper hoped that a comparable laboratory might be established in the Arctic, allowing for comparative study in regions with lower plant diversity. Botanists went to these locations, pursued their own research interests, and returned to their homelands without contributing anything to local science and without drawing on local knowledge.

Malcolm Nicolson, examining the spread of Humboldtian plant geography, noted that there were a lot of botanists working in plant geography especially in Scandinavia and in German-speaking countries in the second half of the nineteenth century.[15] Where other historians had depicted ecology as arising abruptly in the 1890s, Nicolson argued that treatises by Schimper and others were the culmination of a longer tradition of plant geography to which many people working in the Humboldtian tradition were contributing. Indeed, the research reported by Schimper was a summary of his past research and that of other people, including many plant geographers. Schimper recommended that readers consult earlier works of plant geography by Oscar Drude, Adolf Engler, August Grisebach, and Alphonse de Candolle, the latter being one of the founders of plant geography. Schimper's treatise was also modeled on an earlier textbook by the Danish botanist Eugenius Warming, published in 1895 as *Pflantesamfund,* meaning both "plant communities" and "plant societies." This text was translated into German in 1896 as *Lehrbuch der ökologischer Pflanzengeographie* and into English (1909) as *Oecology of Plants: An Introduction to the Study of Plant Communities.*

There is no question that a long line of work in plant geography created the foundation for what later became ecology. However, it is also important to note, as Cittadino does, the difference between Schimper's text and these earlier works, including Warming's text. Warming was not a Darwinian; while he accepted evolution, he adopted a Lamarckian stance, based on the inheritance of acquired characteristics. He did not ignore physiological work, but his approach leaned more toward physiognomic than physiological principles. In contrast to earlier plant geographers, Schimper was proposing a new physiological basis for plant geography, and in doing so he adopted Haeckel's term "oecology" to mean the science of biological adaptations. This would bring something new to plant geography, and Schimper was anxious to show that there was rigor in the new approach. As he wrote, "We may congratulate ourselves that botanists are turning more and more to oecological problems and are framing their theoretical opinions on the basis of accurately observed facts and critically conducted experiments."[16]

In Cittadino's view, this group of Darwinian botanists was distinctive. The hallmark of their approach was its firm Darwinism with its focus on understanding the effects of the struggle for existence and natural selection, processes that resulted in adaptive evolution. They were synthesizing morphology and physiology in order to explain adaptation from a Darwinian perspective. Those, like Schimper, who worked extensively in the field and developed the new ecological approach to research were also successful in synthesizing laboratory traditions and natural history traditions. These botanists were shifting what was viewed as a rigorous scientific approach from one environment (the laboratory) to another environment (nature). I will explore how later initiatives attempted similar kinds of synthesis.

Cittadino makes the point that for the most part his botanists were not trying to found a new discipline but rather were trying to expand the scope of botanical science in general. In doing so, they created a new area of research, which to us is recognizable as physiological plant ecology. Most were driven by general or theoretical goals that were not directly related to agricultural enterprises linked to colonialism. But there were exceptions and Cittadino does not neglect these.[17] The need for research related to colonial agriculture stimulated the creation of research stations in the colonies, journals focused on tropical botany and agriculture, and administrative agencies to guide colonial development. Botanists found research positions in these settings.

One such agency was the Botanische Zentralstelle für die Kolonien, the central botanical agency for German colonies located at the Berlin Botanical Garden and Museum, which was directed by Adolf Engler from 1889 to 1921. The Zentralstelle, created in 1891, served to study and assess the potential economic value of plants from the colonies. Botanical gardens in

general included experimental and acclimatization gardens for work relating to colonial agricultural development. Many were modeled on the Royal
Botanic Garden at Kew, outside of London, which was the center of a network of colonial botanical gardens in the British empire. I have elsewhere
discussed how the creation of the New York Botanical Garden along with its
museums and laboratories helped to further the study of the economic value
of plants in relation to territorial expansion of the United States and its western settlement.[18] The botanists in New York who were instrumental in creating the botanical garden looked not so much to Kew for inspiration as to
Berlin, because Engler was using scientific principles to guide his organization of the garden. The arrangements of the plants, the Americans believed,
should reflect scientific principles of plant distribution, floristics, and ecology, and for this scientifically guided approach Berlin was the model.

    The forces that were shaping the new science of ecology were coming
from two directions. One was the desire to expand the scope of botanical
science in general, to pursue new research questions within a Darwinian
framework, and to synthesize laboratory and natural history traditions.
Most of Cittadino's botanists were motivated by these general or theoretical
goals. However, the need for agricultural research in relation to colonial expansion was also prompting questions about crops and climate and what
type of agriculture would succeed in different environments. These are also
ecological problems. As these goals and questions came into sharper focus
in the 1890s, botanists also thought about what kinds of tools they needed
to reach their goals.

Finding the Tools for the Job

Ecologists strove to be good experimentalists, but in order to do so they
recognized that innovations were needed in two quite different settings:
they had to figure out how to do experiments in the field, but they also had
to figure out how to build better laboratories that would support the study
of the many species that were of interest. Robert Kohler took up this pre-
dicament in his study of field scientists (in which he included ecologists,
but also systematists and evolutionary biologists) and asked a number of

important questions about how scientists working in the field learned over time to perfect their methods, focus their queries, and develop their field techniques.[19] But this field-based learning curve is only half the story: the other half rests in the laboratory, where dramatic improvements were also needed to meet the requirements of modern field sciences. These improvements depended on technologies for indoor climate control.

The idea that botanical science needed not just new instruments but also a completely different concept of the laboratory was articulated quite early in Europe. Alphonse de Candolle, a French-Swiss botanist who was one of the founders of plant geography, stated the problem very clearly in 1866 in an address to fellow scientists.[20] He proposed that botanical gardens could be more useful if they carried out physiological research. Reminding his audience that he had already remarked on the deficiencies of existing laboratories in his treatise *Géographie Botanique Raisonnée* in 1855, he noted that Sachs in his later handbook on plant physiology had commented on the same deficiencies. It was particularly hard to assess the effects of environmental conditions like light and temperature on plants, because those conditions were constantly changing. In the open air, plants were also subject not just to constant changes in the atmosphere but to harmful gases given off by manufacturing activities. But bringing the plants indoors hardly improved matters: trying to control the environment of the plant, he pointed out, "the observer is led into error by growing plants in too contracted a space, either in tubes or bell glasses." Such plants, he argued, were no longer in a natural condition.

In his earlier treatise Candolle had raised the question of whether experimental greenhouses could be built in which the temperature was fully regulated, but that side remark had gone unnoticed. Now he raised the topic again: "I would like, were it possible, to have a greenhouse placed in some large horticultural establishment or botanic garden, under the direction of some ingenious and accurate physiologist and adapted to experiments on vegetable physiology."[21] Candolle imagined a building that would be partly below ground, like a vault, and well insulated to protect it from temperature variations. Glass windows facing different directions on top would let in light, but outside there would be a way to exclude light to create complete darkness. The building would be connected to another underground chamber containing heating and electrical apparatus, so that heat could be regulated using methods that engineers had already devised. Scientists would move into the experimental hothouse through a succession of doors, protecting the building from loss of heat. In such a laboratory, he continued, "the growth of plants could be followed from their germination to the ripening of their seeds, under the influence of a temperature and an amount of light perfectly definite in intensity."[22] In other words, he envisioned being able to study a plant under controlled conditions through its entire life cycle.

He ticked off several questions relating to the effects of environmental conditions on plants, especially the effects of different colors of light, that could be studied, resolving controversies that had continued for several years. "How interesting it would be," he suggested, "to make all these laboratory experiments on a large scale! Instead of looking into small cases or into a small apparatus held in the hand and in which the plants can not well be seen, the observer would himself be inside the apparatus and could arrange the plants as desired. He might observe several species at the same time—plants of all habits, climbing plants, sensitive plants, those with colored foliage, as well as ordinary plants. The experiment might be prolonged as long as desirable, and probably unlooked-for results would occur as to the form or color of the organs, particularly of the leaves." Experiments could involve changes in the gaseous composition of the atmosphere, and the effects of noxious gases could be studied. Or one could recreate an atmosphere higher in carbon dioxide (which he called carbonic acid gas) such as existed during earlier geological epochs, and test whether plants removed more carbon from the air or whether their existence was "inconvenienced" by it: "Then it might be ascertained what tribes of plants could bear this condition, and what other families could not have existed, supposing that the air had formerly had a very strong proportion of carbonic acid gas." He did not imagine that such increases might also occur in the future as a result of industrialization, but he realized that plants evolved as atmospheric conditions changed.[23]

Candolle's remarkable vision for experimental physiology vividly captured the handicap that botanists would overcome almost a century later. His visionary plan was quoted in a report published in 1905 by the chief meteorologist of the U.S. Weather Bureau, Cleveland Abbe, who is mainly known for his promotion of scientific methods for monitoring and forecasting the weather. In the 1890s the Weather Bureau, then part of the U.S. Department of Agriculture, commissioned Abbe to write a report on the relation between climate and crops. Abbe conducted an exhaustive survey of the literature, mostly European, on plant physiology and other sciences relevant to agriculture. In the course of that survey, he strongly endorsed the vision that Candolle had laid out decades earlier. He hoped that in such an experimental setting it would be possible to figure out how to maximize crop production or perhaps modify crops so that they could be adapted to any climate that a farmer might encounter. With laboratories of this kind the process of acclimating plants to different environments could be speeded up: "It is thus that we may hope to accelerate the natural course, which, on the one hand, has already produced grains adapted to the Russian steppes, and, on the other, will eventually evolve those adapted to the vicissitudes of our own arid regions and possibly our severe Alaskan climate."[24]

Abbe admitted that he had been so inspired by Candolle's vision that he had tried to build this type of "complete botanic laboratory." He spent his vacation in 1893 at the botanic garden and greenhouses of Harvard University and brought three hundred experimental plants of wheat and maize to Washington, D.C., where he was given space in the insectary of the Department of Agriculture. But he had woefully underestimated the task. "Unforeseen difficulties" arose, leaving him able to do no more than wish that "the idea of an experimental laboratory for botanic study may be carried out by abler hands."[25]

As Abbe illustrates, this impulse to reimagine the botanical laboratory arose from practical as well as theoretical concerns. In an agricultural setting, the desire to understand the relation of crops to climate and the frustration at not being able to subject this problem to experimental analysis prompted reflection on the need for a different kind of laboratory. The same period witnessed the first generation of self-described ecologists, many of them Americans with careers just starting in the early twentieth century, who were trying to define their new discipline, which they perceived as *scientific* natural history. The word "scientific" was meant to denote a certain level of rigor and to signal that the new science was not just descriptive but also experimental. These ecologists also started to reflect on the need for a new kind of laboratory suited to the requirements of their new discipline. Historically, ecology has developed in tandem with research in agriculture and horticulture, where the study of adaptation is also central and where similar challenges were experienced. These concerns intensified in the 1920s, in the wake of World War I, when the question of feeding a growing population prompted more investment in botanical research and agricultural improvement. Within the growing discipline of ecology, there was increasing sentiment that the science was being held back for want of better laboratory facilities.

The next two sections discuss two very different ways in which these desires and ambitions were expressed in American biology during the interwar period. In the first case, one of the wealthiest men in America, William Boyce Thompson, decided to invest massive amounts of money to build a botanical institute with state-of-the-art technologies for experimental botany at the service of agriculture. This ambitious undertaking showed that with enough money it was possible to design and build a new type of climate-controlled laboratory and greenhouse complex, but it also demanded a strong mission-oriented approach and an unwavering commitment to agricultural improvement. This was from the start a unique undertaking that could not be replicated anywhere else. It could not become a model for ordinary academic institutions and scientific research centers that were less well endowed. This example shows us what existed at the time that Frits Went began to plan his own laboratory, and, as we will see, he drew lessons from it that helped him plan a different kind of laboratory that was more easily adapted to other research settings.

In the second case, a leading ecologist, Victor Shelford, one of the founders of the new discipline in America, took to heart the idea that ecology demanded a close synergistic relationship between experimental science and field work. He realized this meant that the ecological laboratory had yet to be invented. He envisioned what this invention might entail but could not bring it into existence. I use Shelford's example to illustrate the interest that existed in the ecological community in pursuing the goal of designing new laboratories, for it helps us understand why, when Frits Went finally accomplished what Shelford could not, biologists from a wide range of disciplines were keen to try it out.

A Botanical Visionary: William Boyce Thompson and the Future of Agriculture

The type of laboratory that Cleveland Abbe conceived was in fact built within two decades of his remarks, but bringing it about required the imagination and money of one of the richest men in America, William Boyce Thompson (1869–1930). Although not trained in biology, Thompson was an extraordinary visionary who embraced the challenge of creating the resources needed to advance botanical science.[26] A product of the rough frontier culture of late-nineteenth-century Montana, he had studied mining at the Columbia School of Mines and turned his considerable talents toward making money by exploiting the west's rich mineral resources, especially copper, which electrification was making immensely profitable. He developed a series of highly lucrative mines in Nevada, Utah, and Arizona, which, coupled with shrewd speculation on Wall Street, made him a multimillionaire by the early twentieth century.

Thompson became interested in scientific philanthropy through the books of Robert Kennedy Duncan, a chemist and popularizer of science. At Duncan's suggestion, he established fellowships at the Mellon Institute in Pittsburgh to encourage research in various fields of industry and manufacturing. World War I shifted his focus to agricultural research. He accompanied an American Red Cross mission to Russia in 1917, just before the Bolshevik Revolution, and visited again in 1918.[27] Witnessing crop failure and starvation, he came away strongly impressed by the link between social stability and food supply.

The United States faced a potential hurdle in feeding its own population, which was then expanding relative to the country's agricultural production. It was expected that within fifty years the country would need to grow 75 percent more food than it had during the decade of 1911–1921. That

increase would have to come from more efficient agriculture, not just expansion of land under cultivation. Research was needed to make the land
more productive, improve breeds of plants, and control plant diseases.
Thompson decided to create a research institute for this purpose. His
model was the Rockefeller Institute for Medical Research in New York.
Where the Rockefeller Institute focused on human disease, the Boyce
Thompson Institute for Plant Research would focus on plant disease. In
planning the institute Thompson sought advice from John Merle Coulter,
the head of the botany department at the University of Chicago, Lewis R.
Jones, the head of the department of plant pathology at the University of
Wisconsin, and Raymond F. Bacon, the former director of the Mellon Institute of Industrial Research. Although a plant pathologist from Cornell University advised him to link his institute to a university, Thompson decided
instead to build it in Yonkers, New York, directly across from his mansion
so that he could be involved in its operations.[28]

Dedicated with much fanfare in September 1924, the Boyce Thompson Institute was a technological wonder, a completely new type of research
facility for botanical science. It was also a work in progress. Only the first
section, described as a "plant hospital and research laboratory," had been
constructed by 1924, at a cost of one million dollars.[29] The plan was to add
new sections to complete a quadrangle, for a total estimated cost of four
to five million dollars. One of its central aims was to understand plant-
environment relations, building on recent physiological discoveries and on
other research in botany. The institute was the best-equipped research
facility for botanical work in the country and included greenhouses, an arboretum, and grounds for growing experimental plants. Thompson died in
1930, just as ground was broken for an expansion that would double
the institute's size, but his legacy was not in danger, as he had set up a
ten-million-dollar endowment. The buildings were again expanded in the
1950s. The institute continued to operate at the Yonkers site into the 1970s,
when rising costs made it imperative to move out of the New York metro-
politan area; it made no sense to have fields growing vegetables on such
expensive real estate. In 1978 it moved to Cornell University in Ithaca,
where it continues to be an important center for research.[30]

The institute's laboratories and climate-controlled greenhouses exploited modern advances in air-conditioning technology. In the 1910s air-conditioning was being used in factories, especially in textile mills and food manufacturing, and by the 1920s it was proliferating in motion picture theaters. Engineers were developing expertise designing different systems for different places and functions, and in 1919 the American Society of Heating and Ventilation Engineers established its own research laboratory at the U.S. Bureau of Mines in Pittsburgh.[31] When Thompson decided to establish a plant research institute, the technology and expertise were just becoming available to allow an unprecedented level of control over the environment and to create a range of conditions for many kinds of experiments. In fact, Thompson's vision for his laboratory and greenhouse complex ran ahead of the engineering capabilities of the day. The laboratory drew on the best engineering skill available, but it was expected that future discoveries in plant science as well as improved engineering would require continual refinements of design and expansion of the facility.

When the institute opened in 1924 only some of the expected controls were actually installed. The basement contained ventilating, humidifying, and refrigeration equipment. There were constant-temperature light rooms and dark rooms, a carbon dioxide plant to allow for increased concentration of the gas in the laboratory atmosphere, and four greenhouses designed to provide different light conditions. A large crane equipped with lights could be rolled over the top of a greenhouse at night in order to extend the length of the day. There was no stinting on funds for equipment or labor. Scientists got the equipment they needed, a photographic and illustration division provided expertise that the scientists lacked, an engineering and mechanical division ran the machinery and built apparatus that could not be purchased, and a trained staff of workers was available to run the experiments. The institute published its research in two series of periodicals (*Contributions* and *Professional Papers*), and, although not yet formally connected to a university, it offered predoctoral and postdoctoral fellowships as well as fellowships for visiting senior scientists. The scientific staff represented the fields of biochemistry, morphology, pathology, physical chemistry, physiology, and microchemistry.

The institute's first director was William Crocker, a plant physiologist from the University of Chicago. In 1948 Crocker published a summary of the institute's research, which along with other essays enables us to get a sense of the intellectual approach in the early years as well as the scientific problems being studied.[32] The institute did not want to duplicate work done elsewhere in the northeastern United States, so it steered away from systematic botany and genetics and leaned toward plant physiology, pathology, and biochemistry. The idea was to do basic research but not to organize by department, as would be the case in a university. Instead the research was to be organized on the basis of projects, on which teams of investigators using different techniques could work.

The project method, as Crocker explained, involved identifying a problem and then tackling it from different angles, meaning from different disciplinary perspectives. The problem was always related to improving agriculture. Each project had several people assigned from various disciplines, working according to their expertise, but team members were supposed to cooperate and inspire each other, breaking through disciplinary barriers. Crocker saw this method as promoting synthesis of botanical disciplines. Botany had become specialized and divided into many branches, but Crocker argued in 1938 that "the next great advance in botany is a synthetic one, in which all the disciplines will develop in intimate relation with each other, resulting in a unified comprehensive botany of the future."[33] Crocker insisted that, in this type of research environment, the common distinction between "pure" and "applied" research was also invalid—indeed, the whole concept of pure research in botany was misleading. Work on practical problems had always contributed to basic knowledge in botany and would continue to do so.[34]

The Boyce Thompson Institute was impressive for both its size and focus on its agricultural mission. But it did not furnish a model that could easily be replicated or adapted to an academic setting, one where scientists might wish to pursue a broader set of questions. In the academic environment, the need for laboratory innovation was strongly felt within the field sciences, but it had to be on a smaller scale and ideally be replicable, not unique. Within the discipline of ecology, Victor Shelford became the primary advocate for laboratory innovation as a complement to field research. His discussions of the issue reveal his belief that without such laboratory improvement the science was being held back. His comments also reveal some of the obstacles biologists faced in bringing these kinds of laboratories into existence.

A Zoological Visionary: Victor Shelford and the Future of Ecology

Victor Ernest Shelford (1877–1968), a leading animal ecologist, was the first president of the Ecological Society of America in 1916. One of the first generation of American animal ecologists, he paid considerable attention to the laboratory environment as well as to field research. As Robert Kohler relates, starting in 1913 Shelford developed the vivarium at the University of Illinois into a physiological laboratory devoted to zoological problems. Here he worked for fifteen years, until in 1928 he left the vivarium and returned to field research. In discussing this shift from the lab to the field, Kohler identifies a common and persistent problem in ecology, which is to figure out how laboratory results can be extended to the more complex environment of the field. Kohler asks how field biologists like Shelford learned to do experiments in a way that was adapted to the field environment. Laboratory methods, with their exacting standards, could not simply be transferred into the field, as Shelford discovered.[35] While Kohler focused on Shelford's abandonment of the laboratory for the field, he overlooked Shelford's hope that in the future it might be possible to create a new kind of laboratory appropriate for ecological problems. In the 1920s he could envision what that laboratory might contain, but he could not build it because he could not find a suitable patron for such an expensive undertaking.

To grasp Shelford's dilemma, we must consider the arguments he made in *Laboratory and Field Ecology* (1929), which marked the end of his years of laboratory research. The book was largely devoted to studies of the relation of individuals to their environments, a field known in the early twentieth century as "autecology," which fundamentally involved physiological problems. Shelford rejected the term "autecology," insisting that ecology was the science of communities, or what was then called "synecol-ogy."[36] He seemed to be making the point that ecology was not limited to or reducible to autecology and that ecologists always had to relate their studies to entire communities of organisms. But he did not mean that ecology no longer included the subject matter of traditional autecology, for his book was chiefly devoted to problems that would have fallen into the category of autecology.

Shelford's discussion of how animals and plants responded to exter-nal conditions led to a detailed analysis of what kinds of facilities were needed to simulate different climate conditions indoors. The book was an appeal for a new approach to laboratory science and for a new kind of labo-ratory that would be suited to ecological investigations. In a key chapter dealing with buildings and equipment for climate simulation, Shelford showed how complicated the design of an ecological laboratory could be and how crucial it was to approach such design issues with care and fore-sight. One could not simply convert a commercial greenhouse into a labora-tory: ecological laboratories had to be designed from scratch and properly sited so that they met the requirements of experiments. Laboratory direc-tors had to collaborate with engineers and make sure that engineers under-stood that "the organism dictates the character of the equipment."[37] As he stressed, climate simulation work was not simple and the apparatus used could not be simple.

The designs that Shelford envisioned depended on technological in-novations in lighting, air-conditioning, and the means of modifying and monitoring environmental conditions. When Shelford was writing his book, such technologies were relatively new but were developing in a way that seemed promising for laboratory research. He commended the Boyce Thompson Institute for its use of innovative experimental designs, but he realized that it was nearly impossible to replicate what Thompson had built because of its cost. What he laid out in 1929 was an suggestion of what a laboratory and greenhouse complex could be, and what kinds of climate-control machinery and apparatus it might include. Describing his own work at the University of Illinois, he was at pains to point out not only what facilities he had at his disposal for experimental work but also what deficiencies hampered experimental work and what technologies should be abandoned or modified to get better results.

Shelford also complained about lack of funding for climatic and ecological research. Privately funded operations like the Boyce Thompson Institute could not easily be emulated elsewhere, and even that institute required external funds to supplement its endowment following Thompson's death in 1930. Shelford thought that such facilities could only be supported through public funds (for instance, through the Department of Agriculture), but he believed that it would take an economic crisis to motivate such expenditure and that public support would likely wane after three years if the work could not be brought to a close within that time period.[38] He spun a vision of a highly sophisticated laboratory-based ecological research program and was optimistic about success if facilities could be created, but he seemed to despair of this vision being realized. As Charles Kofoid remarked in his review of Shelford's book, "This is truly an imposing program. The author's estimate that 'a small plant' for this purpose could be built for $300,000 is another way of expressing this conclusion."[39] Another reviewer concluded that, given the "elaborate and expensive apparatus" that Shelford deemed necessary for ecological work, "it does not seem likely that his program will be followed up by very many experimental workers."[40] (A Carrier air-conditioning cabinet alone cost $2,000 in 1930.)

Shelford's ambitions for ecology, if unrealizable at that time, were not rejected as inappropriate in principle. The English ecologist Charles Elton pointed out that Shelford's book showed that the "progress of animal ecology has greatly depended upon the parallel development of physics and chemistry, and of technical inventions, without which it is impossible to make quantitative analysis of animal environments."[41] He continued, "Prof. Shelford has a vision of animal ecology as an exact quantitative science, and if the results of this type of ecological work are still in an early stage, and remain uncoordinated, no one can doubt their importance in the future." Shelford himself did not abandon this vision, even after leaving the vivarium to focus on field research. In 1953, when he was emeritus professor at the University of Illinois, he still lamented the imperfect conditions that he had worked under thirty-five years earlier, complaining that "about 5,000 teachers, investigators, and especially postgraduate students in the United States in the fields of plant ecology and taxonomy, wildlife, ornithology, animal ecology and ichthyology are deprived of opportunities to do experimental work because of obsolete campus and building plans."[42] Ever hopeful, he proposed new plans for an improved biological research facility.

Shelford's continuing interest in laboratory development can be better understood just by considering a subset of his research proposals, those connecting ecology to physiology. Lack of facilities could affect how quickly ecologists were able to exploit new physiological discoveries to expand the scope of ecology. For example, in 1920 two American agricultural scientists, Wightman Wells Garner and his assistant Harry Ardell Allard, published a discovery that was considered to be of far-reaching significance in physiology, ecology, and agricultural science. They found that certain plants appeared to be able to measure the length of the day and that in those species the key stages of plant development, such as flowering, were determined by day-length. After two years of painstaking experiments they concluded that plants fell into four basic types: short-day plants, long-day plants, plants intermediate between these groups, and indeterminate plants that seemed indifferent to length of day. Garner called the response of plants to day-length "photoperiodism," and the discovery was immediately hailed as important. Soon scientists found that other characteristics of plants, beyond flowering and fruiting, were apparently affected by the length of day. (Biologists in the 1930s later confirmed that the plants were actually measuring the length of night rather than length of day.)[43]

Barrington Moore, president of the Ecological Society of America in 1920, encouraged ecologists to examine this effect in the plants they studied in the field.[44] Shelford noted in 1929 that the physiological effect of light was a promising field for investigation and that photoperiodism appeared to be important as a stimulus for sexual reproduction in animals as well as plants.[45] Allard himself published an ecological study of photoperiodism in 1932, arguing that understanding the length-of-day requirements of plants allowed for a clearer interpretation of their natural distribution.[46] But the response by ecologists to this discovery was slow. As late as 1953 we find Shelford complaining that the effect of photoperiodism on wild plants had barely been touched, despite its importance for the kinds of problems that ecologists studied.[47] In part, he implied, this meager response reflected the lack of facilities for experimental analysis. He clearly believed that successful ecological science had to combine laboratory and field study. It was not simply a matter of perfecting the methods of field science or increasing the sophistication of field experiments, although improvements in field investigation were certainly important for the science of ecology. Shelford's dilemma was that he could not create the kind of laboratory that he believed was needed.

By this time, as Shelford was drawing up fresh plans late in his career, Frits Went had opened his new laboratory at Caltech, and Shelford cited a couple of articles by Went that described the possibilities of the world's first phytotron. Although Went was then known more for his physiological research than his ecological interests, he did accomplish what Shelford was unable to do, except that Went confined his work to botany, whereas Shelford had envisioned a combined laboratory complex that would serve both botany and zoology. Went was addressing the same question that had plagued Shelford for most of his career: how to create a rigorous science of the whole organism by improving methods on two fronts, the laboratory and the field. Went did not completely solve this problem, because moving from the controlled world of the laboratory to the complex environment of the field was (and still is) challenging. Went visited the Boyce Thompson Institute while preparing his own plans in 1946, but he thought that the institute had not solved the problem of integrating its equipment and research and was not taking full advantage of its technology. Went would try to solve the problems he perceived with a different approach that, while ambitious and very expensive for the time, was on a smaller scale than Thompson's institute. He invented and built a new kind of laboratory, dubbed a "phytotron," which despite its cost turned out to be highly replicable, stimulating worldwide interest. The Cold War concerns of the postwar period explain why Went succeeded where Shelford had failed.

An Atomic Age Laboratory

FRITS WARMOLT WENT (1903–1990) is best known for inventing a
new kind of climate-controlled laboratory called a "phytotron," which
opened at the California Institute of Technology in 1949. That laboratory
became famous: botanists from around the world visited Caltech, and many
of them left determined to build even better versions of this kind of laboratory in their home countries. The worldwide laboratory movement that resulted touched every continent except Antarctica, and it brought
unprecedented levels of funding into plant science by the 1960s. Went envisioned his laboratory as a place that would serve botanists from many fields,
especially fields emphasizing the whole organism and its responses to environmental conditions. This breadth of interest reflected Went's classical
European training in the interwar period. His education gave him an enduring commitment to physiology as the central discipline in botany, as well as
a commitment to the study of the whole organism and its relationship to the
environment. Went, like many others of his generation, had wide interests
that included physiological as well as ecological problems.

In the postwar environment new opportunities arose for an expanded
vision of plant science, and new philanthropic sources of funding made it
possible for Went to realize that vision. A large debate opened up in the
postwar years concerning population growth and food supply, and there
was a strong backlash against those who pressed the case for greater population control. Instead, improving the food supply was considered a more
humanitarian solution, and that alternative in turn motivated more investment in plant science. This chapter and the next discuss how the postwar
environment created these opportunities and how Went exploited them by

building a field of research that some would even claim achieved disciplinary status: "phytotronics."

Old World Traditions: A Botanist's Education

There is a direct intellectual line from Frits Went back to Julius Sachs, one of the founders of plant physiology in the nineteenth century and the teacher of a generation of plant physiologists who became scientific leaders in Europe. One of Sachs's brightest students had been Hugo de Vries, who had started with an interest in plant physiology before shifting to broader problems of heredity and evolution. De Vries became professor of botany at the University of Amsterdam, and among his early physiology students was Friedrich A. F. C. Went, Frits Went's father, who in turn became professor of botany at the University of Utrecht. Friedrich Went turned Utrecht into a renowned center for plant physiology, equipped with a botanical garden and modern laboratory.

Frits Went was born in 1903 in this scientific environment. The Went residence was located in the botanical garden, close to the botanical laboratory. Went later recalled the influence of many hours spent as a boy in the laboratory and greenhouses, where he was exposed to all kinds of exotic plants and to the superb collection of instruments being used to study them.[1] From the bright young botanists who visited his father he learned about a broad range of topics involving plant development, growth, and physiological processes. Went would also have seen the plants arriving from the Botanic Garden at Buitenzorg (now Bogor) on the island of Java in the Netherlands East Indies. Plants from Java, such as orchids, were grown in greenhouses in Utrecht to find out if they would behave as they did in their tropical homes. From an early age Went was exposed to a wide array of biological matters, including physiological research, but also ecological questions involving organism-environment relations and what happens when plants are moved from their native habitats. Writing later in life on how he had become a botanist, Went emphasized the importance of his immersion in botanical gardens, which helped to provide "an overall view of the plant kingdom" and also prevented him from becoming a narrow specialist.[2]

Went completed a doctoral dissertation in his father's laboratory at Utrecht in 1927. A predoctoral qualification required students to repeat work that someone else had done for a doctoral degree. Went repeated the work of one of his father's students, Anton H. Blaauw, who had studied the mechanism underlying heliotropism, the curvature of plants toward the light. Darwin had also been interested in plant movements, or tropisms, and the means by which a stimulus such as light could be "perceived" by the plant and transmitted to other parts of the plant, but his ideas were controversial. European research schools adopted different analytical approaches to these problems of growth and movement, with many different hypotheses developed over three or four decades. In the 1880s Julius Sachs postulated the existence of substances that moved throughout the plant and promoted development, substances that were influenced by light and gravity. Wilhelm Pfeffer at Leipzig imagined that the transmission of the stimulus was analogous to transmission of irritation in a nerve and involved a complex chain of reactions. Experiments in the early twentieth century increasingly suggested that a special growth-promoting substance, a plant hormone, was involved in producing the tropism response.[3]

Blaauw devised experiments to relate the curvature of plants more directly to the light energy falling on the plant. He concluded that light actually inhibited growth. A plant curved when light fell on one side because the dark side kept growing while the light side did not. By the time Went repeated these experiments in 1926, botanists had come to recognize the existence of a plant growth hormone, which was later called "auxin." Went's experiments, performed on oat seedlings, seemed consistent with observations from experiments on growth hormones, and this became the subject of his doctoral dissertation. He showed that if a gelatin block containing the growth hormone was applied to a plant, the shoot curved away from the hormone-soaked block because the hormone caused growth on the side of the shoot that it touched. These experiments helped to establish the principle that the hormone auxin was necessary for plant growth.[4]

The next phase of Went's career emphasized natural history and ecological studies. From 1928 to 1932 he worked at the Buitenzorg Botanic Garden in the East Indies, as his father had also done.[5] The Botanic Garden had facilities for general research as well as research related to agriculture, and the senior Went believed it important for university graduates—and not just graduates from agricultural colleges—to gain experience in the colonies.[6] Frits became the head of the Treub Laboratory, which had moved to a new modern building in 1914 and was named in honor of Melchior Treub, the former director of the Botanic Garden. In addition to the government's botanic garden, there were several experiment stations; Went found the station in the mountain rain forest at Tjibodas, West Java, to be exceptionally well suited to ecological studies.

The heat and humidity made it difficult to do experiments with complicated equipment, so Went adapted by turning to simpler experiments and ecological work on local flora. He got interested in the study of epiphytes that grew in the tree canopies of tropical forests. Tropical orchids, for example, often grow high up in trees. Different tree species harbored different communities of epiphytes, and orchids in particular were very specific in their hosts, allowing Went to identify the trees by the orchid communities that inhabited them.[7] He considered himself to be following in the path of Andreas Schimper, who had also worked at the Botanic Garden and had explored the origin of epiphytes in tropical rain forests. Schimper believed that the questions most central to ecological studies arose because botanists moved away from European laboratories to work in places with extreme climates. The highly managed and artificial European landscape simply did not stimulate the same questions about the nature of adaptation, whereas "in moist tropical forests, in the Sahara, and in the tundras, the close connexion between the character of the vegetation and the conditions of extreme climates is revealed by the most evident adaptations."[8] Went saw himself as working in this tradition, advancing ecological sciences as well as more specialized areas of plant physiology; his interest in these linked sciences remained strong throughout his career. His move to the United States opened new possibilities for developing these sciences, although he did not expect any major change in course when he arrived in California in 1933. The postwar environment, however, would present dramatic new opportunities, which he brilliantly exploited.

Encounters with the New World

As his tropical sojourn came to an end in 1932, Went prepared to return to Utrecht, where he was expected to succeed his father, then approaching retirement. But his father had just received a letter from Thomas Hunt Morgan in the United States, asking whether it would be possible for Frits to join the faculty at the California Institute of Technology. Morgan had come to Caltech from Columbia University in 1928 to create a new biology division. He and his collaborators had defined and advanced the science of genetics in the United States through their studies of Mendelian heredity in the fruit fly *Drosophila*.[9] Morgan brought to Caltech a strong group in genetics, including colleagues from Columbia. But his ambition was to develop other areas of biology that would complement genetics and reveal how the genes functioned in the organism. The separation of genetics from embryology and from physiology had long been seen as a problem, and his goal was to bring these fields back together.[10] Morgan intended to "organize groups of investigators in general physiology, genetics, biophysics, biochemistry, developmental mechanics, and perhaps later experimental psychology."[11] Lack of funds during the Great Depression forced him to drop physiological psychology and bacteriology and instead concentrate on the fields in which significant investments had already been made: general physiology, experimental embryology, biophysics, and biochemistry.[12]

A consequence of Morgan's strategy was that Caltech acquired a strong group of researchers in the plant sciences during the 1930s, including plant geneticists but also physiologists, biochemists, and biophysicists. Two early hires were Robert Emerson, a biophysicist who worked on photosynthesis, and Kenneth Thimann, a biochemist from the University of London who worked on plant hormones. Morgan also looked to the University of Utrecht, a center of plant physiology and biochemistry in the late 1920s, known especially for research on plant hormones.[13] He hired Herman Dolk, a student of Friedrich Went, to develop the field of hormone research in the United States. He built a small laboratory for Dolk, costing $10,000 and containing two insulated underground rooms, important for temperature and humidity control. From this laboratory came the first published work on plant hormones in the United States. Identifying the different classes of plant hormones and their chemical composition was one of the major new areas of plant physiology at the time.

Tragically, Dolk died in an automobile accident in 1932, so Morgan returned to Friedrich Went's laboratory to find his replacement. Morgan hoped to attract Went's son Frits, by then well known for his studies of plant growth hormones. Friedrich thought his son would benefit from experience in the New World, and Frits accepted the offer with enthusiasm. He told Morgan he was eager to return to the work on hormones that he and Dolk had done earlier, but he also looked forward to doing research on desert plants. He wrote to Morgan that he expected to take up problems in experimental morphology and the mechanics of development, and he even wanted to explore ideas about the physiological basis of heredity with Morgan's group.[14] After completing his term at the Treub Laboratory in 1932, he sailed to California in 1933 and settled into the Dolk laboratory.

Went explained to Morgan that the facilities of the Dolk laboratory would be completely adequate and that he would require only small additional investments in equipment. This turned out to be a massive understatement. In fact, Went built his American and international reputation by doing exactly the opposite, creating a new kind of laboratory for whole-organism study that required substantial additional investments. The goal was to gain total control over the experimental environment and hence over the experimental organism. It became Went's passion and his mission to reveal these possibilities to the rest of the world.

Connecting Worlds

Caltech boasted an exceptional concentration of talent in botany in the 1930s. With Went's move to Caltech in 1933, hormone studies continued to be a strong area of research, along with other projects in plant physiology and biochemistry. James Frederick Bonner became the first graduate student in plant physiology in 1931, after Dolk and Thimann got him interested in plant hormones.[15] Bonner finished his doctorate under Went's direction in 1934 and, after a postdoctoral tour of Europe, including time at Utrecht, he returned to Caltech as a research assistant in 1935, instructor in 1936, and assistant professor in 1937. Although Thimann left Caltech in 1935 for Harvard University and Emerson left Caltech right after the war, Morgan made two other strong hires. One was Arie Haagen-Smit, a bio-organic chemist from Utrecht, hired in 1937; he had worked on the chemical identification of the plant growth hormone, auxin. Johannes van Overbeek, a plant physiologist also from the Netherlands, joined the faculty in 1939. Ernest G. Anderson, one of Morgan's earliest hires, represented plant genetics. Even with the loss of Thimann and Emerson, Caltech was a center for plant science in the 1930s and 1940s.

By the late 1930s, after publishing a monograph co-authored with Thimann that surveyed the field of plant hormone research, Went grew tired of hormone work and began to look for other projects.[16] An opportunity arose through conversations with a retired physician, Henry Owen Eversole, who lived near Pasadena and interacted with Caltech's faculty. Eversole had retired after a remarkable career in medicine that involved humanitarian work overseas after World War I.[17] From 1923 to 1927 he served as director of the European office of the Rockefeller Foundation's division of medical education, reporting on medical education in eastern Europe. After his retirement from Rockefeller and his move to southern California in 1929, he went in a completely different direction and took up orchid culture, an interest he pursued obsessively. Eversole's enthusiasm for orchids in turn spurred a radical shift in Went's research and caused him to envision a far more ambitious research enterprise that culminated in the invention of the phytotron.

Went's ability to imagine, let alone build, his phytotron required adapting new technologies to laboratory and greenhouse spaces to make them ideal environments for growing plants. The inspiration for such innovation came from Eversole's idea that orchid culture could be commercially successful in California. The main domestic air-conditioning company, Carrier, had told him it was impossible to air-condition greenhouses, but he figured out how to do it by working with his own team of engineers. Eversole also conducted extensive experiments to discover what environmental conditions and nutrient supplements optimized the growth of his plants.[18] In the late 1930s he offered to help Went build such greenhouses at Caltech, and Went gladly accepted the offer, on the condition that they construct two greenhouses for comparative experiments. The funds needed, $16,000, came from Eversole's sister-in-law, Lucy Mason Clark, the daughter of Eli P. Clark, a businessman who had developed the electric inter-urban railway system in the Los Angeles region.

When the Clark greenhouses opened in 1939, Went decided to move in a direction different from the genetic and biochemical emphases of the other Caltech biologists. His idea was to look more closely at the environment and the effect of the environment on plant development. Went was impressed not only with the mechanical innovations of the air-conditioned greenhouse but also with Eversole's observational powers as he worked out what environmental conditions suited his plants. For instance, one of the first discoveries that Went made in the Clark greenhouses was what he called "thermoperiodicity"—the discovery that certain plants, such as tomatoes, required lower night temperatures for optimal growth. It had been Eversole's idea to set the night temperatures lower.[19]

Describing the Clark greenhouses in 1943, Went highlighted the improved efficiency that would result from designs that enabled scientists to control environmental conditions.[20] Experiments would be more reproducible, and conclusions would be more binding. He hoped other botanical institutes would construct similar facilities. Improved efficiency would become Went's recurring justification for this type of controlled-environment laboratory. He also believed it important to have better laboratories in order to make basic physiological research credible among agricultural and horticultural workers. Plants grown indoors had to be as healthy and vigorous as plants grown outdoors. If physiologists could not grow robust plants, practical cultivators would ignore their findings. Went had two related objectives: one was to connect the basic research of the physiological laboratory with the practical work of agriculture and horticulture. The other was to bring the results of modern physiological discoveries to bear on ecological problems of plant adaptation and species distribution, in the tradition of the nineteenth-century botanists discussed in chapter 1. In this way, field sciences could be strengthened.

Drawing more resources into basic plant physiology was also an important goal. In the 1930s scientists were making progress in understanding plant responses to their environment, but the biochemical mechanisms underlying these responses were still largely unknown. One of the most striking and unexpected of the interwar discoveries, as we have seen, was that certain plants were somehow able to measure the length of the day. This ability determined how the plant grew and developed over the course of the seasons. Two scientists employed by the U.S. Department of Agriculture, William Wight Garner and Harry A. Allard, demonstrated this effect in soybeans in 1920 and dubbed the adaptation "photoperiodism." Some plants were "short-day" plants (flowering in spring or fall), some were "long-day" plants (flowering in summer), and some were insensitive to the length of day.

This discovery helped make sense of transplantation experiments that had been conducted since the nineteenth century. These involved moving plants from one set of climatic and soil conditions to another to see how the plants responded to the new environment. In species that are sensitive to length of day, moving the plants north or south would affect their development. A plant that flowered well at one latitude might not flower at all if moved to another. Its appearance might also change, in one place being tall and straight and in another place bushy. A taxonomist observing such plants in different locations would judge them to be distinct species.

When changes occurred in plant development, they were often interpreted as supporting the theory, often attributed to the French biologist Jean-Baptiste Lamarck, that characters acquired during an organism's lifetime could be inherited.[21] Knowing about photoperiodism allowed for an explanation that did not involve the inheritance of acquired characters. Photoperiodism also helped make sense of plant distributions and explained why some plants were confined to certain ranges while others were widely distributed. Plants sensitive to photoperiod would have limited ranges, whereas plants that did not respond to day length were potentially worldwide travelers. Even closely related varieties could have markedly different photoperiods. Failure to recognize these differences had meant that field experiments could not be properly interpreted.[22]

In 1938, James Bonner and Karl Hamner, collaborating at the University of Chicago, made a critically important discovery about photoperiodism. They found that the plant they were studying, the cocklebur, a short-day plant, was actually responding to the length of the night and not to the length of day.[23] Further research showed that this was true of other short-day plants, so that strictly speaking "short-day" plants were actually "long-night" plants. This discovery reoriented thinking about the causes of photoperiodism and set off a search to find out what mechanism was responsible for this remarkable adaptation. It would take two decades to identify and isolate the pigment, named "phytochrome," that was responsible for this response, an honor that went to a group of talented biologists working at the Agricultural Research Center in Beltsville, Maryland.[24] As Went noted in 1941, knowledge of the plant's physiological responses to light was at that time "shockingly deficient."[25]

Better laboratory and greenhouse facilities were needed because by the 1930s scientists were aware that photoperiod and temperature interacted. Garner and Allard had used relatively primitive equipment when they first discovered this response. But teasing out the combined effects of day length, temperature, and other environmental conditions on plant growth and development called for a greater degree of control over the experimental environment. The Clark greenhouses represented a first attempt to demonstrate that climate-controlled facilities could be designed and built. Relatively modest in size and cost, they set the stage for the construction of a vastly more elaborate set of climate-controlled laboratories and greenhouses after the war.

Creating an Atomic Age Laboratory

The next step was to introduce a radically new conception of a plant research laboratory. What may have been a short-lived European precedent—a laboratory described as adopting the same concept as a phytotron—had been built in 1938 at the Kaiser Wilhelm Institute in Berlin, but it was destroyed during the war.[26] Once again Eversole's vision was crucial to the next stage of development, for Went credited him with perceiving what a more ambitiously designed facility might achieve.[27] The first task was to find the money for a greatly scaled-up enterprise. Robert Millikan, head of the university's board of trustees and Caltech's de facto president, recalled discussing the idea as early as 1940 with his close friend Harry Boyd Earhart, a Michigan-based oil baron who had established a philanthropic foundation in 1929.[28] At least by 1943 Millikan was sending out feelers to see whether the Earhart Foundation might pay for a new plant sciences laboratory, at an estimated cost of $200,000. The war prevented any immediate action, but the foundation provided modest funding for Caltech's wartime research on rubber production from the guayule shrub, which included work by Went in the Clark greenhouses.

In late 1944 or early 1945 Millikan again discussed the project with Earhart, and in May 1945 he conveyed a formal proposal from Went to the Earhart Foundation. He hoped to persuade the foundation to cover capital costs as well as operating expenses (estimated at $6,000 per year) for four years. Millikan's sales pitch painted a glorious picture of a laboratory that would transform agricultural practices. He compared the future impact of the Earhart laboratory to the great scientific centers in the world, such as the Cavendish Laboratory at Cambridge University, the Ryerson Laboratory at the University of Chicago, the Carnegie Observatory on Mount Wilson northeast of Pasadena, and the Pasteur Institute in Paris.[29]

These negotiations occurred during an interregnum in the university's biology division, which meant that Millikan and not the head of the division steered the plan through its final stages. Morgan had retired and the interim chair of biology during the war was Arie Haagen-Smit, who maintained the status quo in the division until Caltech hired George Beadle in July 1945. Beadle did not move to Pasadena right away, however, and thus it was Millikan, not Beadle, who negotiated the funding for the new laboratory later that year. In November Millikan wrote to Beadle to inform him of the plans for plant physiology, giving him the opportunity to veto the project. While visiting Caltech around Thanksgiving, Beadle must have appeared reserved about the project, but he later changed his mind. In early December he wrote back to Millikan giving his support. "I find my enthusiasm growing," he wrote, "and, as I recall our talk, I hope I didn't seem too unimpressed with the value of it. I hope the funds can be found for constructing and operating it on the scale he [Went] wants."[30] The initial plan was that the laboratory would support itself from external funds and would not require much direct support from Caltech.

Within days of receiving the nod from Beadle, Millikan secured Earhart's support for the project, offering assurances that the new laboratory would have far-reaching effects on agricultural practice, bringing scientific method to agriculture and enabling innovations in crop development. As he wrote to Earhart: "If it succeeds as I expect it to do, it should bring results of greater significance to the future of mankind that [sic] I can see even in any projects in the field of the utilization of atomic energy. For this Went project has its feet tied to the ground as the other products of over-stimulated imaginations do not. My honest judgment therefore is that this is Number 1 project in importance that is now on the horizon at the California Institute and I hope I am not over-sanguine in my estimates."[31] In January 1946, Millikan again reassured Earhart that the Caltech trustees supported the project "because of its altogether exceptional promise in their judgment in the whole field of agriculture" and also because it built on Caltech's strengths, not just in plant physiology but also in genetics and biochemistry. That month the foundation agreed to allocate $200,000 to build a new laboratory that would bear Earhart's name.[32]

In the spring of 1946, Eversole and Went embarked on a six-week cross-country tour to learn what was going on in botanical laboratories in the midwest and east. Went's report to Millikan was largely a litany of complaints that plant physiology was dismally supported overall.[33] At universities that had both basic science and agricultural departments, he judged the agricultural work to be of higher quality. This was his view of the University of California (Berkeley), Wisconsin, Rutgers, and Purdue. At Harvard, U.C. Berkeley, and the U.S. Department of Agriculture's main research facility in Beltsville, Maryland, physiologists were aggressive enough to get good working conditions and equipment. But some top universities, such as Johns Hopkins, Princeton, and Stanford, did not have plant physiologists on the faculty. At other places, such as the University of Chicago, the faculty members in plant physiology were, he thought, mediocre.

The agricultural research center in Beltsville impressed him by both the quality of the facilities and the quality of the researchers in plant physiology as well as other scientific fields. The shortcoming there was that government scientists did not train students and therefore were not reproducing themselves. Botanical gardens, often with large endowments, such as the Huntington and the Missouri Botanical Gardens, were too focused on recreation and were not able to devote enough of their resources to scientific research; at the New York Botanical Garden, however, Went thought the director, William J. Robbins, understood the scientific mission of the institution and might rejuvenate it. (Robbins developed microbiology at the garden during his tenure as director.)

The best-endowed facility for botanical research was the Boyce Thompson Institute in Yonkers, New York, with an annual budget of $300,000, of which two-thirds came from its endowment and one-third from industry. The institute was an early and highly ambitious example of a controlled-environment facility, but it had opened a full generation earlier, in 1924, when the technologies available for climate control were much less developed. When Went visited in 1946 he was disappointed. "A lot of expensive equipment and extensive laboratories and greenhouses are present," he wrote, "but they have in no way been integrated. Most of the special equipment was not in use, and seemed rather a drag than an asset."[34] Two air-conditioned rooms were not in use because of the high expense involved, and other air-conditioning equipment in the greenhouses had been removed due to inefficient engineering. Went returned to Pasadena persuaded that the Earhart laboratory would serve as an example of the highest quality engineering and demonstrate the feasibility of complete climate control in plant research. He must have already sensed that his laboratory could become an inspiration for the country and for the world: it would be the model that could be replicated.

After the end of the war, with construction costs soaring, Millikan and Went had to face the fact that the laboratory and greenhouse would run far over the $200,000 pledged by the Earhart Foundation. Went scaled back his plans, cutting the space by 25 percent but adding a room for radioactive isotope work and an elevator. The bids received for the new plans totaled

$407,000 and in 1948 Millikan pressed the Earhart Foundation to increase
its support to that level. Went and Millikan emphasized the important role
that such a facility could play in increasing agricultural productivity in the
light of rising world population. The justification for such an expensive
laboratory jibed with broader discussions about the postwar reconstruction of agriculture. Caltech's negotiations with the Earhart Foundation took
place against a backdrop of debates in the United States and other nations
about the postwar prospects for continued peace and security. Investment
in plant science was one way to meet those goals.

The Road to "Freedom from Want"

To understand how the Caltech phytotron came to be built we should keep
in view the global postwar context and the growing appreciation that peace
could not be maintained unless agriculture was reconstructed to meet the
food demands of present and future generations. Postwar debates about
feeding the world's growing population were polarized. Malthusians argued
that increasing the food supply would only promote population increase
and lead inexorably toward environmental catastrophe. Anti-Malthusians
countered that, given political will and commitment of resources, food
production could be increased enough to satisfy future needs. The Caltech
laboratory was built just as this debate was going into high gear, and Went
leaned decisively toward the anti-Malthusian argument that attention should
be given to improvements in agriculture and nutrition to ensure world
peace. In his view, this investment had to focus on the design of new kinds
of laboratories that were better equipped to answer questions about what
controlled plant growth and development.

 Central to the huge task of postwar reconstruction was provision of
adequate food for war-torn populations. These discussions built on the
prewar international food movement, which had been taken up by the
League of Nations in the mid-1930s. The League formed a committee
known as the Mixed Committee, which issued a report in 1937 on the relation of nutrition to health, agriculture, and economic policy. This report
urged the formation of national nutrition committees, envisioned as inter-

disciplinary groups that would approach nutrition from different angles and would both advise governments and influence public opinion. From 1937 to 1939 twenty-five governments acted on the recommendation, but the outbreak of war ended most of these activities.[35]

During and after the war, discussion of food and nutrition in relation to postwar reconstruction revived. The League of Nations published a series of reports dealing with food relief, the world economy, and population trends in Europe and the Soviet Union. In 1943, the United Nations convened a conference on food and nutrition in Hot Springs, Virginia, which took a bold, visionary approach to the need for the postwar expansion of economies in order to eliminate poverty, world hunger, and dietary deficiency.[36] This conference laid the foundation for the creation of the United Nations Food and Agriculture Organization (FAO) in 1945, an international agency for research and advice on all matters of food and agricultural policy. The FAO's organizer and first director, the Scottish nutrition expert John Boyd Orr, received the Nobel Peace Prize in 1949 for his role in promoting peace through improvement in human diet.

American social scientists were also drawn into debates over postwar reconstruction in relation to food, population, and advances in agriculture and nutrition science. The Harvard economist John D. Black emphasized the importance of avoiding the chaotic aftermath of World War I, when "hungry crowds of central Europe wandered about for months pillaging for food and fighting every effort, good or bad, to restore order." When the current war ended, he warned, "the suffering people must be fed, and the disorderly, irresponsible elements held in check."[37] At the University of Chicago the Norman Wait Harris Foundation convened a conference in 1944, Food in International Relations, which brought together experts in the fields of nutrition, agricultural economics, population, and international relations. The final paper by Karl Brandt, agricultural economist at the Food Research Institute of Stanford University, expressed the forward-looking and cautiously optimistic mood of the conference as a whole. Brandt recognized that freeing the world from hunger and malnutrition demanded exceptional leadership and international cooperation, but he also believed that there had never been a greater opportunity to make progress than existed at the end of the war. In *The Reconstruction of World Agriculture* (1945), Brandt similarly argued that postwar reconstruction had to build on a foundation of agricultural reconstruction.[38]

Hitting a different note from these discussions, arguments that cast a harsher light on human misdeeds in the past also surfaced after the war. These arguments drew on many of the same studies that supported the debates about postwar reconstruction, but they sounded an alarm about the impending crisis posed by increasing population in a world of limited resources. Two popular best sellers, both published in the spring of 1948, were meant to be wake-up calls to Americans, warning them of the danger of exploitation of resources and the dire consequences of our unbalanced relationship with nature. The milder of the two was Fairfield Osborn's *Our Plundered Planet,* which made the case that humans had to work with nature rather than against it to conserve the earth's species and its natural resources. Osborn argued that the human impulse to dominate was destroying the very basis of human well-being and that a more ecologically sensitive understanding of the natural world would reveal the necessity of adopting sound conservation policies. Scientific solutions that were not attentive to nature's ways were misguided, he warned.[39]

William Vogt's book *Road to Survival* was in the same vein but was more strident in its call for population control.[40] In fact, Vogt criticized Osborn's book on the grounds that he did not include suggestions about checking population increases, leading readers to infer that "improved land-use would take care of mankind, even after we were piled three deep."[41] Vogt, an ornithologist and prominent conservationist, condemned human misdeeds and warned of impending famine in breathless prose. Much of his argument about population was taken from an earlier book by Guy Irving Burch and Elmer Pendell, *Population Roads to Peace and War,* published in 1945, with a revised Penguin edition titled *Human Breeding and Survival* appearing in 1947. Their arguments, laden with eugenic concerns about deterioration of populations, forcefully made the case for population control, advocating the sterilization of "biologically or socially" inadequate people as well as the creation of incentives to encourage voluntary sterilization of "normal persons" who have "had their share of children."[42] Vogt similarly harped on the dangers of overpopulation, accepting the Malthusian argument that population had already outstripped resources and that it was urgent to adopt measures to control it through vigorous birth control campaigns and by making contraception available. Given that America was in a privileged position and able to offer aid to other countries, he also insisted that aid be tied to policies that promoted population decline, though he emphasized that this stabilization should occur through the "voluntary" action of people. Referring to President Franklin D. Roosevelt's speech to Congress in 1941 upholding a vision of a world founded on four freedoms, namely, freedom of speech, freedom of worship, freedom from want, and freedom from fear, Vogt added another: "Quite as important as the Four Freedoms, which we have made a shibboleth, is a Fifth Freedom—from excessive numbers of children. Far more than much of the world realizes, even the partial achievement of the first four is dependent upon this last."[43]

Many experts rejected the Malthusian logic behind the crisis scenarios that these books envisioned, opening up a large debate about how to deal with population growth and food supply that continues today.[44] No one doubted the seriousness of the problem, but the policy recommendations from the anti-Malthusian side focused less on population control than on economic and social reforms, including food and nutrition policies to improve standards of living and health worldwide.[45] Boyd Orr, for example, emphasized the importance of mounting an international effort coordinated through the U.N.'s agencies to solve the food problem.[46] Karl Brandt viewed Vogt's book as highly distorted, the work of "a naturalist who broke into the social sciences and went berserk." Anyone who knew about farming, land use, or resource economics, he suggested, would be "moved to anger by it and may want to throw it into a wastebasket."[47] Other critical reviews cited the authority of agricultural scientists who were projecting more optimistic scenarios.[48]

The arguments that Went made in appealing for funds for his laboratory dovetailed with these views that the food shortage could be resolved; his appeal also coincided with growing interest in the agricultural development of semi-arid regions, for if such development were to proceed, then basic scientific research was also needed. As Went stressed to Millikan, he did not believe that the world was headed for inevitable famine. He thought productivity could be increased two- or threefold if scientific research could be directed to the problem of making photosynthesis more efficient. Such agricultural improvements, he asserted, could be the answer to the world's overpopulation.[49] Achieving that goal in turn required nothing short of the new facilities that Went envisioned at Caltech, and he stressed that there was not a moment to lose. As Went explained, "we should start right away, for the basic research we are planning does not yield results immediately, and it takes considerable time to adapt new principles to practical problems."[50] Millikan conveyed Went's arguments to the Earhart Foundation and urged a quick decision so that Caltech could respond to the price bids. The foundation obliged by approving the funding in early June. Millikan effusively thanked Harry Earhart, telling him that the laboratory would equal and possibly exceed the scientific value of Caltech's $6.5-million telescope at Mount Palomar, which had just been dedicated.[51]

For Millikan and Earhart, this project perfectly illustrated the way that private philanthropy and not government patronage should operate to promote scientific advances that would have economic benefits. Both detested the idea of a planned economy and the policies of the New Deal, which Earhart believed had produced gross corruption, as illustrated precisely by postwar inflation. Millikan viewed Roosevelt's patronage system as corrupt and strongly opposed efforts after the war to establish a national system for the support of science, which he believed would be equally susceptible to pork-barrel politics.[52] Earhart's guru was the economist Friedrich Hayek, whose book *The Road to Serfdom* appeared in the United States in 1944 to great acclaim.[53] Hayek warned that all forms of socialist planning, whether of the left or the right, set up conditions that enabled totalitarian forces to gain the upper hand, even if socialist movements did not intend to create totalitarian states. His argument against government control matched Earhart's views of the importance of free enterprise and the dangers of a planned economy. Millikan assured Earhart that the laboratory was "an object lesson of what a free enterprise system, created for us by the founding fathers, can do in stimulating human progress and in enriching the life of man on earth."[54]

Construction began forthwith and the Earhart Laboratory for Plant Research was dedicated just one year later, on June 7, 1949. Putting "plant research" in the laboratory's title implied a broader research program than the words "botany" or "botanical" would have conveyed, but the botanists wondered whether a snappier name would better express the bold vision behind this building. After all, they had dared to do what was not uncommon in physics but was unusual for botany: build a very large instrument at very great expense. Went and Millikan happened to visit the cyclotron at U.C. Berkeley about this time, and Millikan had remarked that botany deserved a facility equally impressive.[55] Hearing this comment, James Bonner and postdoctoral fellow Samuel Wildman quickly coined the word "phytotron" for the Earhart lab, with Bonner later explaining that he intended to convey the idea of a big complicated machine.[56] In fact, the Earhart laboratory's control panel resembled that of a cyclotron. Bonner recalled that Went was originally annoyed at the joke, but Millikan loved the term. Went's annoyance passed quickly: he must have realized how much capital could be gained by drawing on the authority of both Millikan and atomic physics. Went gave Millikan credit for coining the word "phytotron" and happily declared that the laboratory was destined to play the same role in plant physiology and applied plant sciences that the cyclotron was already playing in physics.[57] The birth announcements went out, and Went invited the world to visit Caltech's fabulous atomic age laboratory.

Big Science in a Small Pond

FRITS WENT HOPED THAT HIS laboratory would serve all of botanical science and not just one or two fields of research. He insisted that the phytotron concept was something novel, not simply a reiteration of earlier ideas about improving climate controls in greenhouses and laboratories, because it provided an unprecedented level of control of many environmental variables. He also thought that the kind of experimental control that the phytotron offered would advance field sciences by providing a place where hypotheses could be rigorously tested. The science done in the phytotron was meant to help end long-running controversies.

Because so many conditions could be controlled at once, the phytotron was also designed to increase the efficiency of research, so that answers could be obtained faster. The comparison of phytotron to cyclotron had a double meaning: these were not only big, complicated machines but also *accelerators*— the phytotron's great selling point was its ability to accelerate the research process. Finally, it was meant to serve both basic and applied science; potential applications included the study of crops and other plants of economic value and environmental concerns such as erosion and photochemical smog.

In this chapter I consider the phytotron concept in greater detail and examine a few of the scientific problems studied in Caltech's laboratory. Two areas of research are explored in more detail: the biological clock as an example of a basic research question and the damaging effects of photochemical smog as an example of an environmental problem. The phytotron's use in connection with California's emerging smog problem illustrated how the authority of precise, laboratory-based science could come to the aid of a controversial theory about smog formation. The study of the biological clock was an emerging field of research in the 1950s. The phytotron turned out to be a valuable instrument for biologists who were trying to prove that an organism's "clock" mechanism was in-built, an idea that was controversial at the time. In both of these cases, experiments in the phytotron helped to resolve controversies, which Went viewed as one of the chief advantages of this type of laboratory.

In the end, however, inflexibility in the laboratory's design could be a handicap to scientific creativity, and some were critical of it for this reason. In addition, Went found it hard to raise funds to sustain what amounted to a Big Science program, and the burden of paying for the laboratory's operation became too great for a relatively small university like Caltech. As a result the Earhart laboratory was eventually closed and razed many years after Went left Caltech. Its demise did not signal the end of the phytotron story, for by this time it had ignited a worldwide movement.

A New Kind of Biology

The speed with which Caltech's phytotron was built was a testament to Eversole's experience and vision and to the skill of the engineers, architects, and contractors who worked with Went and Eversole. Local architect Henry Palmer Sabin designed the laboratory building. He was known for his Spanish-revival style of house design, but he also had experience in hospital architecture. Sabin's role was to create a simple, attractive building without "sacrificing a single wish of the scientists."[1] The outside design was meant to be unobtrusive and aesthetically pleasing, resembling a one-story house (figure 1). The building's design had to "satisfy the donor, the scientist, the administration, the neighbors, the City Planning Commission, the building inspectors, and many other interested parties."[2] The modest exterior design scarcely hinted at its unusual high-tech interior except for one thing: there was a telephone placed by the front door, which visitors used to gain admission, for the door was kept locked.

Going through the front door, one entered a different world. A popular article published on the Earhart in 1952 dubbed it an "atomic age greenhouse" equipped with a "futuristic bewilderment" of knobs, switches,

Figure 1. Exterior of the Earhart Laboratory, designed to resemble a house and blend in with the neighborhood. (From the photographs and images collection of the California Institute of Technology Archives)

buttons, dials, flashing lights, and recording devices.[3] The article quoted Arthur Galston, an associate professor at Caltech and a former Navy lieutenant, as joking that it all resembled the Combat Information Center on a warship; the writer added that the basement, with its "blowers, boilers, conduits, and general uproar, could be the engine room of Dr. Galston's warship."[4] The greenhouses were "filled with plants on which a shimmering light plays," the shimmer caused by water sprayed on the glass roof to absorb heat.[5] The air, which was completely replaced twice each minute, was fresh and pleasant, not muggy as in a normal greenhouse. In keeping with the ship analogy, Went viewed the laboratory as an integrated construction, not just a set of rooms serving various experimental purposes. He emphasized that the term "phytotron" did not apply to single controlled-temperature growing chambers; instead it designated a comprehensive set of laboratory spaces and growing rooms designed to achieve an unprecedented level of control over experiments.[6] Greenhouses with some level of temperature control and

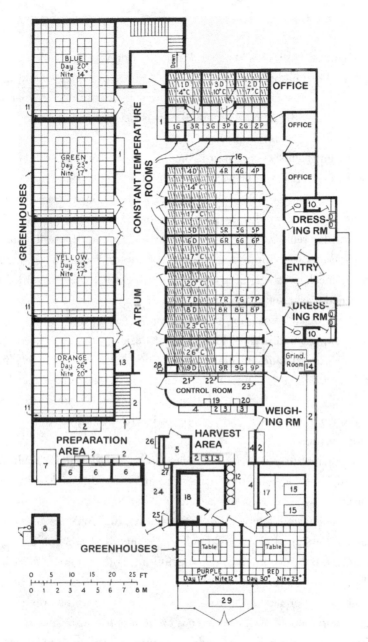

Figure 2. A plan of the upper floor of the Earhart laboratory, showing how green-house spaces and controlled-environment laboratories were integrated into a single complex. Air was circulated from the basement. The main entrance (*right side*) led to washrooms and changing rooms, not to an open foyer as in most buildings. Greenhouses are on the left side and bottom, and controlled-temperature rooms are in the interior. (Schematic adapted from F. W. Went, *The Experimental Control of Plant Growth, Chronica Botanica,* v. 17, 1957, p. 14; courtesy of the California Institute of Technology Archives)

artificially lit rooms were old news. The phytotron was on a different scale altogether. The building as a whole was like a complex instrument designed to control the plant's environment, and in this sense it can be compared to an instrument of Big Science (figure 2). It was also big in that it was expensive to run and to maintain.

In Went's effusive description, the construction of the Earhart laboratory was a marvel of team effort and coordination. He credited the contractor and the foreman for their skill in ensuring that controversies with the architect, engineers, and scientists were avoided. Eversole also got credit for the way his remarkable intuition got "crystallized into tons of concrete, miles of ducts and piping and scores of motors" (figure 3).[7] Arthur Hess, the mechanical engineer who had worked with Eversole on his airconditioned greenhouse, accomplished the "superhuman feat" of designing systems of air ducts, hot and cold water piping, circulation pumps, air conditioners, and heaters. James Taylor designed the equally complex layout of electrical conduits, motor circuits, and lights. Lighting came from incandescent lamps and fluorescent tubes, a new technology that developed largely after the war. All the lamps, motors, and other machinery throughout the laboratory were connected with the power supply in the transformer room.

The laboratory complex included six air-conditioned greenhouses with daylight as the light source and thirteen air-conditioned laboratories with artificial light.[8] Eleven rooms were kept dark, nine general laboratories were kept at constant temperature and humidity, and small air-conditioned offices were assigned to researchers from the nearby Kerckhoff biological laboratories so they could remain in the building and not have to move back and forth. The building not only allowed for control of air temperature and humidity but operated as a "weather factory" for the production of wind, rain, and fog; it could even alter the gas content of the air in special rooms. A wind tunnel in the basement could produce winds of up to 30 kilometers per hour, although in the end it did not see much use. Photographic equipment enabled scientists to do routine photographing of all plants growing in the greenhouses at regular intervals. The complex contained a total of fifty-four separate rooms where plants grew, and plants were moved around

Figure 3. Frits Went inside the phytotron: "miles of ducts and piping." (From the photographs and images collection of the California Institute of Technology Archives)

on wheeled tables for maximum flexibility, because it was easier to keep rooms at different constant temperatures and just shift the plants, rather than change room temperatures (figure 4). As Went explained, "an almost infinite number of artificial climates, with occasional wind or rain storms or frost periods, can be maintained, limited only by the size of rooms and greenhouses."[9] In all, about one-third of the space was allocated for growing plants; about one-third for handling plants, laboratories, offices, and quarantine space; and about one-third for the machinery needed to run the whole operation.

The control room, the "brains of the building," allowed scientists to monitor different rooms and greenhouses simultaneously (figure 5). Any apparatus in the greenhouses or darkrooms could be made to register in the

Figure 4. Laboratory staff shifting plants on carts in the phytotron. Plants were shifted twice a day, which took about twenty minutes for the entire building, while rooms were kept under constant conditions. (Courtesy of Elliot Meyerowitz and the California Institute of Technology Archives)

control room. The superintendent, George Pret Keyes, who operated the control room along with a team of technicians, had an extremely important role, for he had to manage the intricate shifting, care, and feeding of the plants and in general make sure the building kept humming—and it did hum constantly, with all the equipment operating. The building as a whole was like a complex instrument and all systems had to be monitored, rules on decontamination had to be enforced, and photographic recording methods managed. Keyes and his crew were on call seven days a week.

The phytotron represented a new kind of experimental mini-world designed to reproduce "the whole range of climates under which plants do or can grow."[10] One approached this experimental world with the same precautions as a surgeon entering an operating room. Diseases and pests that might complicate experiments had to be excluded. Fumigation with poisons killed animals in the building, and air, water, sand, plants, and humans

Figure 5. The "brains" of the building: the control room, where the superintendent monitored conditions throughout the laboratory and greenhouses. (Courtesy
of Elliot Meyerowitz)

were made as germ-free as possible. When entering through the front door,
one stepped not into an open foyer, as in most buildings, but rather into
washrooms (one each for men and women) that led to changing rooms with
lockers. Immediately on entering, one washed hands, combed hair, and ex-
changed outer clothing for sterilized laboratory clothes resembling the
"freshly laundered hospital gowns and caps or the regular doctor's uni-
forms used in operating rooms."[11] The floor was washed regularly with insecticides, cigarettes were sterilized in autoclaves, and even laboratory
notebooks were fumigated. These elaborate quarantine precautions did not
completely prevent disease and pest infestations, which occurred about
twice yearly, but they kept the infestations to a manageable level.

One of Went's main concerns was to resolve long-standing controversies. Many of the conflicting theories and arguments rife in plant science, he
argued, were the result of growing plants under uncontrolled or poorly controlled conditions. As a consequence, experiments could not be reproduced,

evidence was conflicting, and unknown variables were affecting the outcome of experiments. With better control of the environmental variables, experiments could be more precise. Went believed the phytotron would assist in creating a "theoretical biology" that was comparable to "theoretical physics," and although he did not specify what he meant by theoretical biology, the context of the remark makes it clear that he had in mind a set of universally accepted principles, or in other words scientific consensus on fundamental problems.[12] Compared with the exact sciences of physics and chemistry, botany lacked agreement on many basic problems, and Went saw the need for consensus as essential to scientific progress.

Went also argued that better control, and hence greater reproducibility, would mean that fewer experiments could be run to substantiate a given interpretation, "which makes the research worker more efficient."[13] His appeal to efficiency was an attempt to address the problem of the relatively high cost of running the laboratory. The Earhart Foundation provided $6,000 annually for operating expenses at first, but costs were much higher, increasing from $41,000 in 1949–1950 to amounts ranging from about $72,000 to $97,000 by the mid-1950s. These costs excluded salaries of research scientists or costs of individual projects. The laboratory relied on outside entities to cover about two-thirds of its costs: these included various industry sources, such as the Sugar Beet Development Foundation, Riker Chemical Company, and Campbell Soup Company. Private research institutions such as the IBEC Research Institute, the Rockefeller Foundation, and the Carnegie Institution of Washington, as well as public universities, contributed support for research projects.[14] Since photochemical smog was just becoming a problem in the region, the Los Angeles County Air Pollution Control District provided funding for an analysis of the components of smog and its effect on crops.[15] Other government support came from the U.S. Department of Agriculture, U.S. Public Health Service, and eventually the National Science Foundation, which made a large three-year grant in 1955 to help with operating costs. Despite Millikan's and Earhart's initial opposition to governmental subsidy of science, private philanthropy and corporate interests alone could not cover all the costs of such an expensive undertaking.

The expense could be justified in part by the economic benefits that came from research in the phytotron. The laboratory came to the rescue in 1955, for instance, when the tomato crop failed in the United States due to unusually warm temperatures. For the Campbell Soup Company, which was marketing its condensed tomato soup as a key ingredient of American comfort food, this failure was a catastrophe. The company's research director sent a plant breeder to the phytotron to breed a tomato that would grow in warmer temperatures, and within a couple of years a new tomato strain was produced by crossing an American and a Philippine breed.[16] This project relied on traditional approaches to selective breeding, but the phytotron made it possible to speed up the breeding process fourfold compared with working in the field. The phytotron, in analogy to the cyclotron, came to be seen as a "plant accelerator" by scientists around the world who built their own phytotrons.

Another baffling case concerned tuber production in potatoes, where the evidence was conflicting: some scientists believed tuber formation occurred during short days, but in places like Holland, Maine, and Ireland there was extensive potato production during the long days of summer. Studies in the phytotron showed that at higher temperatures potatoes were short-day plants for tuber formation, but at lower temperatures they were indeterminate. As Went emphasized in discussing these and other cases at a symposium convened by the Campbell Soup Company in 1962, puzzles relating to optimizing productivity of crops could only be solved in facilities like the phytotron.[17]

Practical work involved not only domesticated species but also wild species. A serious problem in southern California in the 1950s was control of fire and erosion, which was being jointly studied by the U.S. Forest Service, Caltech, and the Los Angeles County Flood Control District. Authorities hoped to improve the erosion control methods used to stabilize the soil in an environment where rainfall was erratic. On slopes denuded by fire, the short-term goal was to create a temporary vegetation cover that would last long enough to allow the native shrubs to regrow over five to ten years. Black mustard (*Brassica nigra*) had been the preferred cover, based on field trials of different species, but large-scale field sowings of black mustard had produced almost no cover in some cases. Mixtures of grasses, clover, and alfalfa also failed to produce a good cover crop. The scientists decided to bring the problem to the Earhart laboratory. The phytotron made possible speedy and efficient assessments of which cover plants would survive best in the California landscape.[18] Tests of different species in the phytotron allowed scientists quickly to narrow the range of suitable cover species to just two: Italian ryegrass (*Lolium multiflorum*) and black mustard, which grew well together and did well in field trials. The Earhart lab provided accurate and quick tests of species for their adaptability to field use.

A second, longer-term approach was to identify shrubs that would hold the soil in place on rapidly eroding but unburned slopes. Exotic plants were sought to replace or supplement the native shrubs, which did not have the desired form and rooting characteristics. Botanists from Mediterranean climates around the world recommended up to a hundred promising exotics. Because field testing and supplementary nursery-growing would be very costly and could take up to a decade, the plants were first screened in the Earhart laboratory to assess how well they grew under normal temperature ranges. Both native and exotic species were screened and compared, with the idea that a blend of native and exotic forms would be selected for their ability to control erosion. The value of the laboratory tests was to show what species would not be able to grow in the mountains and why.[19] The laboratory tests accomplished in less than one year what field trials would have accomplished in five to ten years, and the number of species that had to be field-tested for drought resistance was reduced by more than 80 percent. In this example, speed, efficiency, and reduced costs justified the expense of constructing a phytotron.

Bridging Disciplines

With growing confidence as the phytotron's fame spread, Went began to project a larger role for it as a central instrument in interdisciplinary fields of research. In an article in *Scientific American*, "Climate and Agriculture," Went included an extended discussion of the phytotron's applications to agricultural problems and promoted the idea that in places where interdisciplinary teams already existed, the phytotron should be considered a necessary instrument. "Now that the need for such a laboratory has been recognized," he wrote in 1957, "we can expect that in the near future a number will be built, not only for basic research but also for the solution of many problems in agriculture, horticulture and forestry."[20] He returned to the earlier theme that had featured in his first proposals for the phytotron: the need to improve the efficiency of photosynthesis in higher plants.

Went envisioned a transformation in the relationship between various branches of plant science, in which the phytotron served as a key instrument linking different groups in the plant sciences. In an agricultural experiment station, for example, researchers in agriculture, horticulture, plant nutrition, plant breeding, pathology, and climatology all would require a range of environmental conditions that the phytotron could supply. Similarly, in a large botanical institute, physiologists, morphologists, ecologists, geneticists, taxonomists, and biochemists would find uses for a phytotron. Moreover, the phytotron would push the descriptive sciences such as ecology, morphology, anatomy, and evolution toward new and fertile experimental studies.[21] Ideally it would reduce the bewildering complexity of field experiments through better control of variables, and it would allow greenhouse experiments to be extrapolated to the field.

Finally, the phytotron was an instrument of internationalism, bringing together investigators from all continents and through this interaction fostering international understanding.[22] In this respect Went was continuing in the tradition of his father, who along with other Dutch scientists had worked to reunite the scientific community after World War I.[23] Friedrich Went had believed that international organizations, such as the International Union of Biological Sciences, of which he was president in 1928, were important for facilitating activities that required cooperation among nations, for example, the procurement of living material that had to be sent across borders or the preservation of nature.[24] Frits Went's interest in internationalism and cooperation sustained his father's legacy.

The aim of the work was to get an overall picture of basic plant behavior. Over fifteen years in the Clark and Earhart labs, about 150 species were investigated. In a review of these laboratories' research published in 1957,

Went listed five main criteria for the choice of different organisms. The first was to study problems connected to agriculture, horticulture, and medicine. The second was to select plants suited to the study of specific problems, such as photoperiodism, germination, tuber formation, adaptation to cold, and the effect of smog on plants. The third standard was based on recognition that certain plants exhibited unusual behaviors that were biologically interesting, such as *Mimosa pudica,* a sensitive plant that collapsed when touched, or insectivorous plants, whose adaptations had fascinated Darwin. The purpose of this grouping was to attract interest to the study of these plants. The fourth criterion was to study the effect of the environment on gene expression and possibly on the genetic composition of the plant itself. The fifth was to select organisms suitable for the study of ecological problems, especially as they related to the distribution of plant species. One central goal of phytotronics was to create better links between field and laboratory research, especially in relation to the problem of distinguishing between genetic and environmental effects.[25]

Given the amount of machinery that made up a phytotron, one might assume it was a forbidding environment of instrumentation. But despite all the buzzing technology and modern look, Went conceived of the phytotron as a place to reconnect the experimentalist with the experimental organism and, through the process of the experiment, to get a feel for the reactions and responses of the organism. The apparatus and machinery for the experiments were separate from the growing rooms and mostly installed in the basement, leaving the first floor for growing and handling plants. As Went explained, "Anyone carrying out an experiment in the Earhart Laboratory is continuously in the closest contact with thousands of other living plants, all being tested for their responses, and naturally one gets an appreciation of the general validity of the response obtained with one's own experimental plants."[26]

The plants studied had to be suited to the kinds of research questions being asked but also to the space itself. Many of the plants, such as tomato and coffee, were economically important and were chosen because financial support could be obtained from public and private organizations. But for some studies, limitations on space required practical decisions about the size of plants. Early on, Went had chosen the tomato as a good organism for studying the relationship between climate (especially temperature) and growth. But the pea (*Pisum*) was smaller and more convenient, allowing for a ninefold increase in the number of plants, so the trend was toward growing smaller plants. Went also searched for even smaller annuals, which he called "belly plants" because you had to crawl low to the ground to study them. Along with other botanists in the 1940s, Went began deliberately searching for "botanical *Drosophilas*": small, widely distributed, and easily grown annuals that could be raised in large quantities. German botanist Friedrich Laibach believed he had found the ideal plant in *Arabidopsis thaliana* (a weed in the mustard family, native to Europe), whereas Went preferred flowering plants native to southern California.[27] (Laibach's choice did indeed come to fulfill the role of the "botanical *Drosophila*" and was the first plant to have its genome sequenced in 2000.)[28]

Two lines of research deserve a closer look because they demonstrate how laboratory results could offer authoritative evidence that helped to settle controversies, which was one of Went's chief goals. The next two sections consider both a practical and a basic scientific problem in more detail. The practical problem, photochemical smog, illustrates how the expense of the phytotron could be justified by its contribution to research on an increasingly urgent environmental problem. Research in the phytotron on smog's biological effects provided compelling evidence in support of a controversial theory of smog formation, helping to turn the tide in the theory's favor. The second, more general problem concerned the study of the biological "clock," a subject of increasing attention in the 1950s but also a contested zone. Experiments in the phytotron provided evidence supporting the claim that the biological clock was in-built or endogenous, and in doing so they also gave momentum to the rising discipline of chronobiology.

The Authority of Laboratory Experiments: Understanding Photochemical Smog

One of the most serious environmental problems to emerge during and after World War II was photochemical smog, an irritating haze produced by the chemical breakdown of hydrocarbons in the atmosphere under the action of sunlight. The problem was noticed first in southern California, where severe smog attacks began to occur as a result of the boom in wartime industry. By 1947 a systematic control effort began in the hope that it would be possible to locate the main sources of pollution and put an end to the nuisance. To everyone's dismay the situation worsened after the first pollution control measures were implemented, and intense arguments over the causes and sources of smog continued for many years.

The term "smog," coined in 1905, designated a mixture of smoke and fog that affected industrial regions in Britain and the eastern United States. The Los Angeles County Air Pollution Control District (APCD), which was formed in October 1947, initially focused on sulfur dioxide, a known pollutant in other regions. The head of the APCD, Louis McCabe, was under pressure to fix the problem quickly, but Arnold Beckman at Caltech and Robert Vivian at the University of Southern California urged him not to cave in to public demands for quick fixes but rather to do more research into the causes of the pollution.[29] As the study of Los Angeles smog grew more scientific, it became apparent that this smog was different from smog found in other regions, for it could occur on bright sunny days.

Arie Haagen-Smit, a bio-organic chemist at Caltech, proposed a novel explanation in 1950, arguing that this form of pollution was caused by chemical reactions in the atmosphere that transformed otherwise harmless molecules into irritants under the action of sunlight: hence the name "photochemical smog." It was not obvious that the main cause of Los Angeles smog was this new form of photochemical smog, for there were other sources of air pollution, including sulfur dioxide, carbon monoxide, nitrogen oxides, metal oxides, mineral dust, and oil and soot from burning. As the controversies over the cause of California smog intensified, Caltech's scientists figured at the center of the disagreement. The phytotron also had a starring role, for the laboratory opened in 1949 just as the debate was reaching a high pitch.

The story of photochemical smog has been told several times, and these accounts focus on the disputes between scientists in the pay of industry and scientists in academia who worked with government agencies that were trying to solve the smog problem. Haagen-Smit is the hero of these stories, and scientists in the pay of industry who opposed him are seen as biased because of their connections to industry. Ralph Nader used this clash as an example of outright fraud or "junk science." In his view, this dispute illustrated the evil effects of the unholy alliance between science and industry, which would inevitably produce corrupt science.[30]

These interpretations do not take account of the complexity of identifying both the nature of air pollution and its main sources. Haagen-Smit's initial published announcement in 1950 was not in a peer-reviewed journal but in Caltech's general magazine, *Engineering and Science.* Some scientists accepted his ideas right away, but there were legitimate grounds for challenging them. Making a case that was persuasive enough to achieve broader scientific consensus required additional research, as well as marshalling evidence from different kinds of research. Work in the phytotron provided some of that essential support.

Haagen-Smit began by asking what kind of chemical reactions produced the typical effects of Los Angeles smog, especially the eye-watering haze, distinctive smell, and characteristic damage to vegetation. He got the idea of extracting the contents of smoggy air by passing the air through the cold traps that he normally used to collect the vapors given off by fruits. He meant to analyze the "bouquet" of smog just as he would have done for fruits or wines: "I wanted to know more about the nature of the stuff I had collected, where it came from, and what happened to it," he remembered.[31] The smog had strong oxidizing action, and he postulated that the distinctive smell was caused by oxidized hydrocarbons originating from gasoline, not sulfur compounds.[32] Sulfur dioxide was a reducing agent and could not explain the oxidizing property of smog.

The key advance was to realize that Los Angeles smog was not something dumped into the atmosphere by industrial and other activities. It was actually created in the atmosphere by the breakdown of saturated hydrocarbons in the presence of oxygen or ozone, all driven by the sun's light. These hydrocarbons formed peroxides, which decomposed to form smaller aldehydes and organic acids, which in turn were broken down into carbon dioxide. Formation of these organic peroxides resulted in aerosols, or the haze of smog. The trouble-causing organic materials, it appeared, came

Figure 6. Arie Haagen-Smit came up with the idea of manufacturing smog in the laboratory, to demonstrate that it had the same damaging effects on plants as "L.A. smog." (Courtesy of the Archives, California Institute of Technology)

from gasoline released into the air from petroleum refineries, oil fields, automobiles, and gasoline filling stations.

Having made a plausible deduction, Haagen-Smit was able to synthesize "artificial smog" in his chemistry laboratory and then test whether it had the characteristic effects of Los Angeles smog (figure 6). This approach made it possible for him to identify the probable chemical cause of smog as early as 1950 and also to identify the sources of the hydrocarbons producing the smog, chiefly petroleum refineries and automobiles. However, there were still unanswered questions: how was ozone formed in polluted atmospheres, and what was its role in smog formation?[33] Haagen-Smit and co-workers modified the initial theory, arguing that nitrogen dioxide (produced by combustion) also played an important role and that the photochemical reaction of nitrogen dioxide with hydrocarbons resulted in the formation of ozone. This idea was controversial.

Industrial scientists did their research at the Stanford Research Institute, a private non-profit research institute in Palo Alto that also had laboratories in Los Angeles and Pasadena. Those scientists took a different approach. They disagreed with Haagen-Smit and argued that smog had other causes, for instance, sulfur dioxide emissions, which deflected blame away from the automobile industry. Haagen-Smit did not think sulfur dioxide emissions were very important, because anti-smog actions were targeting these and effectively reducing pollution of this kind without eliminating the irritating smog.[34]

The analysis of photochemical smog involved the use of Caltech's phytotron; this was one of the earliest projects that justified the expense of the phytotron. Went directed the portion of the research that concerned the Earhart laboratory, and this research provided a counterweight to the industry-sponsored research. The Earhart laboratory proved to be a crucial resource in aid of Haagen-Smit's controversial hypothesis, because in the phytotron it was possible to connect hypotheses about what smog *was* to observations about what smog *did*. At a national air pollution symposium held in the fall of 1949, the first of its kind, Went reported on experiments at the Earhart laboratory to determine what constituents of smog were harming plants in the region.[35] Went had noticed in 1945 that lettuce leaves were damaged by smog, and John Middleton, a plant pathologist at the Riverside Citrus Experiment Station, had observed the same damage as early as 1944. Clearly something in the air, although not visible, could have marked effects on plant development and health. More controlled experiments in the phytotron could possibly nail down what these effects were.

The ability to use plants as indicators of the cause of smog was possible because plants showed different reactions to different forms of smog. Went argued that it made sense to attack the smog problem from the botanical side and use the unique facilities of the phytotron, which were ideally suited to this kind of analysis (figure 7). When the phytotron first opened, in fact, smog was already affecting the plants inside, and one of the first things the engineers had to do was install activated charcoal filters to

Figure 7. Researchers testing the effects of smog on plants,
one of the first environmental projects undertaken in the
phytotron. (From *Engineering and Science* 14, no. 3 [1950]:
12; resolver.caltech.edu/CaltechES:14.3.0)

purify the air and prevent these effects. It was then possible to expose plants
to different air compositions in the phytotron and examine the difference
between plants grown in smog-free air and smog-infested air. Certain plants
showed characteristic damage when exposed to smog from southern California. Spinach, sugar beets, and endive developed a metallic sheen on the
underside of their leaves. Alfalfa and oat leaves got bleached. The kind of

damage produced by photochemical smog was different from damage done by inorganic pollutants such as sulfur dioxide, chlorine, or hydrogen fluoride, which was confined to the immediate neighborhood of the emissions. Photochemical smog, because it was being produced by chemical reactions in the air, caused more widespread damage.

The phytotron experiments were immediately seen as central to resolving this controversy. In 1950 the *Los Angeles Times,* which inexplicably depicted the phytotron as like a "mystic Oriental temple" concluded, overly optimistically, that "smog is nearing its Waterloo."[36] Robert Millikan was excited by the work and reported to Harry Earhart in May 1951 what an important role the phytotron was playing in the great smog war: "The Earhart Laboratory has been a godsend during all the rumpus that we have had this year in Southern California arising from the increasing chorus of protest against the smog."[37] The laboratory, Millikan assured its patron, was the chief instrument by which Haagen-Smit was able to make his "extraordinary analysis of all the elements that enter into the smog" and solve "the whole problem of what causes the evil effects," which was somewhat of an exaggeration. The phytotron experimenters also enlisted the help of about a dozen Los Angeles County high schools, where students grew plants that were subject to smog effects and reported back on the damage.[38] At a National Air Pollution Symposium in 1952, research from Caltech's phytotron was well represented.[39] The great advantage to the Earhart laboratory was that it was possible to grow smog-free plants there and therefore know exactly what smog was doing to plants.

The initial plan had been to expose plants to different compounds that might be in the air and check for smog damage, but tests on about fifty compounds showed no results. Haagen-Smit then got the idea of manufacturing artificial smog, dubbed "Haagen-smog," using ozone combined with hydrocarbons, which he then employed in a variety of experimental demonstrations to illustrate the effects of smog. In further experiments he combined vapors from cracked gasoline with nitrogen dioxide, which could oxidize a number of compounds in sunlight. In both kinds of experiments plants showed typical smog damage when exposed to these airs.[40] These experiments were described fully in January 1952 in the journal *Plant Physiology,* where Haagen-Smit and four co-authors concluded that their investigations showed "for the first time that hydrocarbons, normally harmless air pollutants of organic nature, can cause severe damage through their reaction with substances known to be in the air."[41] The artificial smog had caused typical smog damage to five different plants, with different symptoms in different species. The ability to link artificial smog to specific damage to plants was taken as strong confirmation of Haagen-Smit's theory of smog formation.

Edgar Stephens, a chemist who took up the problem of Los Angeles smog in 1953, remarked later that this 1952 article became a landmark in the field and was "very convincing."[42] The combined evidence from chemistry and plant physiology drew other researchers into the field and stimulated closer study of these photochemical reactions. In 1953 the American Petroleum Institute farmed out two projects on photochemical smog to the Franklin Institute Laboratories for Research and Development in Philadelphia. Stephens was involved in these projects, which were designed to duplicate and test Haagen-Smit's theories. The Franklin Institute group drove their trailer-laboratory cross-country and were based at the Stanford Research Institute's amenities in South Pasadena. They used a new long-path infrared spectroscope, which had just been introduced by Perkin-Elmer Corporation, to try to detect ozone while duplicating Haagen-Smit's experiment. The instrument, known as Silent Sam, the Smog Detective, allowed the scientists to measure materials in the atmosphere at very small concentrations.[43] They verified Haagen-Smit's conclusions and presented these results in 1955 at a meeting of the American Chemical Society.[44] As Stephens recalled, Haagen-Smit recognized that their work completely vindicated his results.[45]

Many pieces fell into place at this meeting, where paper after paper confirmed and extended the "brilliant theoretical analysis and sound experimental evidence" of Haagen-Smit and his associates.[46] Further confirmation came from the Stanford Research Institute in February 1956 at a conference on chemical reactions in the urban atmosphere organized by the Air Pollution Foundation.[47] The Air Pollution Foundation, in its final report of 1961, made sure to give the Stanford Research Institute scientists credit for solving part of the puzzle of smog formation.[48] Haagen-Smit viewed these two conferences in 1955 and 1956 as marking the end to the controversy; thenceforth "the photochemical origin of Los Angeles smog through the action of sunlight on a mixture of oxides of nitrogen and hydrocarbons was no longer in dispute."[49] That did not mean that all of the mysteries of photochemical smog had been resolved. Stephens and an interdisciplinary group of collaborators in southern California continued to study smog and its effects on vegetation through the 1950s, aiming to clarify the chemical composition of its damaging components.[50] But effectively the years of polarized debate and controversy had come to an end. The phytotron experiments, by offering clear, authoritative, and non-controversial results, were an important means toward this end.

Making a scientific discovery is not a discrete event but a step in a social process that takes time. The discovery of the cause of photochemical smog should not be narrowly restricted to the article where Haagen-Smit first announced his theory in 1950. The subsequent social processes of arguing, testing, validating, and consensus building were all part of the discovery of the cause of photochemical smog. The smog problem was complex and controversial, and evidence from plant physiology documenting the toxic effects of smog proved to be a key contribution to the debate. The phytotron's role was to provide evidence that seemed clear and incontestable, demonstrating that laboratory-made smog affected plants in distinctive ways that matched field evidence of damage. That evidence was authoritative and persuasive, and because it was also linked to Haagen-Smit's theory of smog, it stimulated further research to evaluate Haagen-Smit's theory. That research helped to sway scientific opinion in Haagen-Smit's favor, with agreement emerging by the mid-1950s. We should understand the "discovery" of the cause of photochemical smog as comprising this entire social process over half a decade.

The experimental work published in 1952 on damage to vegetation did not offer a full understanding of what was occurring in the atmosphere and what exactly was producing this damage. That story turned out to be more complicated and required more research to reach a conclusion. But the phytotron-based experiments were decisive in shifting the focus of research in a way that resolved the controversy in Haagen-Smit's favor. The remarkable thing is that the consensus formed fairly rapidly after the land-mark article of 1952. Three to four years is a very short time to resolve a controversy of this kind where the two sides have major disagreements. The Earhart Plant Research Laboratory played a pivotal role in this dispute because of its ability to produce results that no one disputed.

Support for an Emerging Discipline: The Biological Clock

Research programs at the Earhart lab were more likely to be successful when there was already a sizable group of scientists interested in a particular question that the phytotron seemed capable of answering. In this section I consider an example where a new field—potentially a new discipline—was engaging the attention of a wide range of biologists. This was the study of the biological clock, or the mechanisms underlying rhythmic behavior in plants and animals. The phytotron turned out to be well designed for research on this problem.

In plants, rhythmic behavior operating on a roughly 24-hour or "circadian" cycle had long been observed and was one of the subjects that Darwin studied. Scientists had known for a couple of centuries that rhythmic behaviors could continue even in the absence of an environmental stimulus. A plant that moved its leaves down at night and up during the day, for example, would continue these movements even if kept totally in the dark. A new idea gaining credence in the twentieth century was that organisms showed these rhythmic behaviors because they possessed time-measuring mechanisms, analogous to clocks. In other words, organisms' ability to respond to daily changes in the environment appeared to be endogenous, the consequence of an in-built mechanism, and therefore these behaviors continued even in the absence of environmental stimuli.

The German botanist Erwin Bünning was credited with discovering the endogenous nature of daily rhythms in plants in the 1930s. In 1958 he published a summary of his research in a book titled *The Physiological Clock*. Colin Pittendrigh, a British zoologist working at Princeton University, also promoted the theory of an endogenous clock mechanism. As the idea of an endogenous clock gained support in the 1950s, the debate in-volved figuring out what the nature of this clock mechanism was, or how to think about the clock analogy. Bünning thought the mechanism entailed an oscillation, or an alternation between a state of tension and relaxation. However, some biologists contested this idea and asserted instead that the rhythmic behaviors were not produced by an in-built clock-like mecha-nism. Those arguing for an endogenous clock were working hard to estab-lish their claims and quell any opposition. This field grew into a major area of research, and by about 1960 some scientists viewed it as a new discipline, "chronobiology."[51]

A biological clock is an adaptation to a varying environment, in this case the variation caused by the earth's rotation. The implication is that or-ganisms with such clocks require these environmental variations to function well, so that if placed in a constant environment they will be sick. The realiza-tion that a constant environment makes a plant sick focuses attention on the plant's circadian rhythms and on the possibility that it possesses a clock-like mechanism. The phytotron was therefore suited to the study of the biologi-cal clock in many ways. One could examine adaptations to variable environ-ments, such as photoperiodism (the ability of plants to measure length of day/night), in order to test specific hypotheses concerning an endogenous clock mechanism. One could investigate what happened when the period of the environmental rhythm departed from the normal 24-hour cycle. One could explore the relationship between periodic behavior and specific envi-ronmental variables, such as temperature. Finally, one could study the effect of growing plants in constant environments to which they were not adapted. All these ways of tackling the biological clock problem benefited from the phytotron's climate-controlled laboratories. For instance, Karl Hamner, who had moved to the University of California, Los Angeles, in 1944, used the phytotron to explore the relationship between photoperiodism and endog-enous circadian rhythms.[52] Hamner's work supported the existence of an endogenous 24-hour rhythm in the photoperiodic response of soybeans and provided evidence that backed Bünning's controversial hypotheses.

Went drew attention to one problem involving circadian rhythms that had not been suspected before the advent of the phytotron.[53] This

concerned the role played by circadian rhythms in the temperature response of plants. For instance, tropical plants do not grow well in cool climates; they gradually die even if the temperatures are not freezing. Similarly,
plants accustomed to cool climates would die in the tropics, even though
they are able to stand occasional high temperatures. There was no obvious
biochemical explanation, because there was no enzyme system that failed or
became toxic when the temperatures changed.

Experiments in the phytotron revealed that the underlying problem
was the circadian rhythm. If the environmental rhythm was shifted, for instance using an 18-hour light-dark rhythm or a 32-hour light-dark rhythm
instead of a 24-hour rhythm, the plants could be made to thrive. An African
violet that would die if kept at 10 degrees centigrade would develop normally if given 16 hours of light followed by 16 hours of darkness. As Went
explained, it was not that the plants were unable to tolerate higher or lower
temperatures but that at these temperatures the circadian rhythm was
wrong. This observation had ecological significance, helping to explain
plant distributions. When a tropical plant died if moved to a temperate region, it was not because of a biochemical abnormality but because its clock
was desynchronized with the environment.

The phytotron was especially well suited to experiments involving
circadian rhythms and temperature. John Woodland (Woody) Hastings
took up the study of circadian rhythms in relation to bioluminescence in the
1950s, and he found the phytotron valuable in his work on the biolumines-
cent marine dinoflagellate *Gonyaulax polyedra*.[54] The luminescence of
these single-celled organisms was rhythmic: cells became brighter and dim-
mer at roughly the same time every day, and the daily rhythm persisted even
when they were taken out of their natural environment and kept under constant light and temperature. Hastings spent a year during 1955 and 1956 at
the Scripps Institution of Oceanography in San Diego, working with Beatrice Sweeney on this problem. Wanting to test how the dinoflagellate's
rhythm behaved at different temperatures, Hastings and Sweeney brought
their cultures to Caltech, where Went made several rooms of the phytotron
available. They found that the flashing rhythm was different at different
temperatures.

Bünning had reported a similar effect in the bean *Phaseolus,* whose up and down leaf movements had been studied extensively. He argued that the clock ran faster at higher temperatures, which is what one would expect. But Hastings and Sweeney found to the contrary that the clock ran slower at higher temperatures. They proposed in 1957 that a pair of compensating reactions, both temperature dependent, were controlling the clock's speed. The two reactions were like positive and negative elements in a feedback loop. Their experiments suggested that the biological clock was independent of temperature as a consequence of these temperature-compensating mechanisms. This research addressed the observation made by Pittendrigh and others that many biological rhythms appeared to be independent of temperature. The key point was that temperature independence supported the idea of an in-built clock, or a biological timing device or mechanism. As Pittendrigh argued, "the near-independence of temperature must surely be a prerequisite for a useful organic clock."[55] Hastings and Sweeney's work also suggested that the term "temperature independence" was misleading. Possibly one could understand these rhythms in terms of oscillating chemical systems and feedback loops.

Harry Highkin at Caltech used the phytotron to study the effects of constant environments, which produced decreased vigor in plants.[56] Those decreases could be reversed by returning plants to a fluctuating environment. He also found that these effects carried over into the next generations of plants, that is, the inhibition of growth caused by a constant environment became more pronounced in successive generations of plants. Therefore the constant environment was not simply inhibiting growth but was having a more profound effect that accumulated for about four generations. That meant, in turn, that a plant's environment consisted not only of its present environment but also the environment in which the parent plants were grown.

These and other results, all supporting the idea of an endogenous biological clock, were presented at a Cold Spring Harbor Symposium organized by Pittendrigh in 1960.[57] Went, Sweeney and Hastings, Hamner, and Highkin all gave papers. Went had by this time left Caltech to take up the directorship of the Missouri Botanical Garden, but his interests in the biological clock continued, and research on the subject at the Caltech phytotron also continued. As new research questions like the biological clock rose to prominence in the 1950s, laboratories like the phytotron, although not constructed for this purpose, appeared to be ideally suited to this type of complex problem. As the phytotron model was copied elsewhere, part of the justification for pursuing phytotronics, as it came to be called, was that it could help to elucidate the nature of the biological clock. With marvelous serendipity, the biologists who favored the idea of an endogenous clock were now able to work in a laboratory uniquely equipped to help them find evidence in support of their central hypothesis. (In 2017 the Nobel Prize in Physiology or Medicine was awarded to three biologists who succeeded in elucidating the mechanism by isolating the gene controlling these biological rhythms and studying the processes under this genetic control.)

Challenges to the Supreme Commander

Went's atomic age greenhouse would trigger a worldwide movement, but its time at Caltech was limited. Although the Earhart laboratory had been approved during the tenure of Robert Millikan and at a time when the biology division lacked a strong leader, by the time the laboratory opened in 1949, George Beadle had been divisional head for over three years and Caltech had a new president, Lee Dubridge. The 1950s brought new challenges just as Went's laboratory was opening for business. The difficulties of sustaining a costly Big Science approach to biology were evident even in the very early days of the Earhart laboratory. Went was enthusiastic, but he did not approach his Big Science laboratory with the skills of a Big Science administrator. The phytotron was never self-supporting and therefore required financing from Caltech's general funds. This might not have become a stumbling block had Went's approach been in keeping with the general trends at Caltech, which emphasized research at the intersection of biological, physical, and chemical sciences. As James Bonner remarked, Caltech was the home of modern biochemical plant biology, and it would also become the home of modern molecular biology.[58]

George Beadle, head of the biology division, had turned to biochemical genetics in the late 1930s and had shifted from research on the genetics of maize and *Drosophila* to the fungus *Neurospora* (bread mold). Together with the biochemist Edward L. Tatum, he made the important discovery that genes controlled specific biochemical processes, which led to their sharing the Nobel Prize in Physiology or Medicine in 1958. In 1947 Beadle hired Max Delbrück, a physicist who had turned his attention to biological problems in the 1930s and who would contribute to the development of molecular biology through his work on bacteriophage, viruses that infect bacteria. Beadle was also close to Linus Pauling, a physical chemist who was working on the structure of macromolecules and who would in the early 1950s become interested in the structure of DNA. Beadle and Pauling had been collaborating since 1946, and in 1954 the university received a grant from the Rockefeller Foundation of $1.5 million for three years, contingent on raising matching funds, for the support of chemical biology.[50] Caltech was already becoming a center of molecular biology by the time James Watson and Francis Crick deduced the double helix structure of DNA in 1953. Went's promotion of phytotronics, and his insistence that physiology be considered the central discipline of biology, bucked the trends at Caltech from the start.

Went's inability to make his phytotron self-supporting meant that his leadership of the Earhart laboratory in the post-Millikan era never had full support from Caltech's administration. The difficulty of balancing Went's needs with those of the other biologists at Caltech was well illustrated by Went's negotiations with the Rockefeller Foundation's Division of Natural Sciences and Agriculture, from which he hoped to get modest support in the early 1950s. Warren Weaver, director of the Division of Natural Sciences, heard from a visiting foreign scientist in 1951 that Went desperately needed funding for basic research. This news prompted Weaver to check into the matter with Earnest C. Watson, Caltech's dean of the faculty. Watson was annoyed that Went had given this impression and informed Weaver that the real problem was that Went was not using the funding he had received in the most efficient way. In 1951, the operating costs for the Earhart exceeded $66,000, of which three-quarters came from Caltech's general funds. Went, Watson suggested, was simply out of control, continually going ahead with unauthorized commitments and then needing a bailout from

Caltech. As Watson explained, "What he [Went] should do, instead of telling everyone what he could be doing if only he had a little more support, is to make the most efficient use of the fairly adequate funds he now has. Once this is done, there should be no difficulty in getting him additional support."[60] Watson also doubted that they were getting as much value out of the Earhart money as they were from the rest of the biology division.

At that time, Went hoped to get Rockefeller funding for a project on the water relations of plants, a subject that his recent visits to Israel, southern France, and Spain had convinced him was important. J. G. Harrar, the Rockefeller officer who was negotiating with Went on this grant, visited Pasadena in November 1952 to examine the Earhart laboratory's work. He judged Went to be a very capable botanist, "industrious and intelligent," with "great curiosity" and a willingness to approach problems in an unorthodox way.[61] But he also perceived that Went, being classically trained, retained "some of the spirit of the naturalist" and was "not in entire accord with the direct chemical and biophysical approach to biological problems without reference to the approach from the viewpoint of the ecologist and physiologist as well." Went, Harrar observed, was the type of individualist who "works best as supreme commander and has some difficulty in delegating responsibility or in sharing it with scientific associates."[62]

The point at issue here did not so much concern responsibility for day-to-day operations in the Earhart, but rather the overall direction and emphasis of the biological work and how this direction was translated into grant proposals. This problem would come up explicitly in the next round of Rockefeller funding. Toward the close of the granting period in 1956, Caltech's biologists prepared another application for support of work in the Earhart laboratory, this time hoping to obtain about $20,000 to $25,000 annually for five years. But the proposal for a project entitled "The Climatic Responses of Plants" came from James Bonner, who was interested more in the biochemical side of this problem. Beadle also thought their chances were better if they emphasized the biochemical rather than physiological aspects of the science.

The Rockefeller officers favored this project, but Went was alarmed because it seemed to shut out the research he had been doing on the water relations of plants, which was based solidly on physiological methods. As Harrar explained to Beadle in discussing the case, Went worried that he might be prevented from carrying out the work he had been doing under the previous grant.[63] Went perceived this dispute to involve the conflicting approaches of biochemistry and physiology. As he explained to Harrar: "I am delighted with the interest Dr. Bonner and his group have taken in the work which can be carried out in the Earhart Laboratory and that a biochemical analysis of the climatic response thus becomes possible on a much larger scale than had been done before. At the same time I feel that a physiological analysis of these processes is essential, particularly because through the general trend towards biochemistry in the biological sciences the physiological approach tends to become neglected."[64] Clearly Went felt himself to be competing with the rest of the biology division and perceived the growing emphasis on biochemistry as a threat to his physiological approach.

Beadle decided to submit a revised and expanded proposal, only to have Rockefeller officers reject it. In the end Caltech submitted a proposal with Bonner and Went both listed as principal investigators, but the proposal was very similar to Bonner's original one, with a similar budget. Lee Dubridge, Caltech's president, made it clear in submitting the proposal to Rockefeller that Bonner was really in charge of the research but that Went and his associates would cooperate in the work. By the fall of 1956, Went's ability to determine what lines of research should be eligible for funding was being challenged. His departure from Caltech was foreshadowed by his failure to raise the funds needed to make the Earhart lab truly self-supporting. After a year's leave of absence in 1958–1959, he formally resigned from Caltech in August 1959 to become director of the Missouri Botanical Garden. His departure did not put an end to phytotronics at Caltech, but the days of the laboratory were numbered.

The scientists working in the phytotron were also aware that it had limitations caused by its inflexible design. As a scientific instrument comparable to an ultracentrifuge or computer, the phytotron determined what kind of work could be done. In a sense there was too much design and too little room to improvise. The recollections of Anton Lang, Went's successor as director of the Earhart lab, suggest that the phytotron was not always responsive to the needs of individual researchers: "Being designed with the idea of pro-viding a large variety of environmental conditions for a large number of inves-tigators, the facility proved insufficiently flexible; on quite a number of occasions, when special needs arose in my own work or that of an associate, it became necessary to improvise or go outside the phytotron."[65] Lang thought that the Caltech phytotron had served a useful purpose, but his advice for future phytotron projects was to allow for uncommitted space "to provide for new needs as these may arise."[66] This deficiency was addressed in later de-signs, where equipment could be modified easily and different facilities could be added as research goals changed.[67]

High costs and the inflexibility of the instrument were not the only hurdles phytotronists faced, for the growth of molecular biology in the 1950s steered scientific attention, and with it funding, to the exciting ques-tions generated by the new discoveries surrounding DNA and RNA. James Bonner had no difficulty shifting his research interests to the study of RNA in the mid-1950s: he was happy to follow the new biological directions.[68] But Went was highly resistant to these trends. In 1962, the same year that Watson, Crick, and Maurice Wilkins received the Nobel Prize for the dis-covery of the double helix structure, Went proclaimed: "Statements sug-gesting that if we knew all about DNA and RNA, we would understand the secrets of life, or that molecular biology holds the solution to the problems of biology can only be made by immature minds. Such minds are carried away by the glitter of modern physics and chemistry, and ignore the com-plexity of nature."[69] Went wanted the phytotron to serve "all of biology, clas-sical as well as molecular" by bringing biologists together "instead of pulling them apart, as the establishment of a separate department of molecular biol-ogy tends to do."[70] Went did believe that biology should be more like phys-ics in certain respects, but not if that meant an end to the classical biological concern with whole organisms.

After Went left Caltech, the other biologists wanted to close down the laboratory and focus on other kinds of research. However, the plant sci-ences at Caltech recovered from this period of crisis and were even able to expand with new funding sources. Rockefeller support helped to finance refurbishment of the laboratory, and a new laboratory-greenhouse building,

Figure 8. Caltech's plant sciences complex in the early 1960s after the addition of the Campbell Laboratory (*left foreground*), which was connected to the Earhart lab (*behind the Campbell lab*) and the original Dolk lab (*right side*). (Courtesy of Elliot Meyerowitz and the California Institute of Technology Archives)

the Campbell Laboratory (funded partly by the Campbell Soup Company and dubbed the "soup kitchen"), was added in 1960. In combination with the Earhart and the original Dolk laboratory, this addition created a large plant science complex (figure 8). However, as Lang recalled, by the time he left in 1965 he was the only faculty member at Caltech "still interested in plants as distinct organisms," and there was no one outside his own group with whom to discuss problems in depth.[71] The Earhart closed in 1969, partly because of age and obsolescence, and it was razed in 1972 to make room for a new laboratory devoted to behavioral biology and neuroscience. In its last days it was used for artists' studios. The Campbell Laboratory remained a bit longer and was used to house animals in support of the behavioral biology program, but it too eventually was torn down.

In contrast to the Caltech story, elsewhere in the country and indeed around the world, phytotron development was moving into high gear in the

1960s, and Went was active in promoting these other enterprises long after leaving Caltech. Went's ideas about unifying biology and connecting worlds did not fall on deaf ears. They proved to be persuasive arguments that helped to bring new resources into botanical science. They represented not just resistance to molecular biology but also recognition that support of whole-organism biology was needed to solve the world's food problem. Just as genetics, biochemistry, biophysics, and microbiology were being re-made into molecular biology in the postwar period, so also among the sciences of the whole organism, new interdisciplinary linkages were forming, new conceptions of the organism were emerging, and environmental and agricultural sciences were being remade. The mini-world of the phytotron gives us insight into these postwar changes in biology. I consider this inter-national story in Part 2.

PART TWO

Phytotrons Around the World

FOUR

Crossing Borders

IN ADDITION TO DOING "phytotronics," as it came to be called, Frits Went was keen to replicate this laboratory model in other locations, and he worked hard to raise its international profile. He was always willing to collaborate with scientists interested in building a version of the phyto- tron, and on his frequent travels in the 1950s he ceaselessly promoted his laboratory's unique features. His efforts helped to stimulate a worldwide laboratory movement that extended through the 1970s and tapered off in the 1980s in favor of designs that were more flexible. Some of the laborato- ries inspired by the Caltech phytotron were relatively modest in size, whereas others were more elaborate and required several years of cam- paigning and planning because of their complex design. Getting a phyto- tron was hard work, often extending over several years.

Many of the scientists who built their own versions of the phytotron had visited or worked in the Caltech laboratory. In translating Caltech's model to other settings, scientists also adapted its purposes to serve particular national needs, which might include practical agricultural concerns such as wheat or rice improvement as well as research on general botanical problems that were not directly linked to economic applications. Phytotrons came to be seen as symbols of scientific achievement and embodiments of the national goals of the governments supporting these enterprises. The need to address practical problems of crop improvement was always central to the work of these labo- ratories, and some, such as the phytotron at the International Rice Research Institute, were instruments of the "Green Revolution." But the increase in experimental rigor could also improve the authority of the "softer" sciences that were devoted to the study of organism-environment relations. For Went and other phytotronists, this included the science of ecology. The pursuit of ecological problems in a rigorous way was a recurring justification for building phytotrons, or for building ecological spin-offs to phytotrons that became known as "ecotrons." The development of the phytotron plays an important role in the growing sophistication of ecology as an experimental science.

However, before narrowing the focus to ecology, I first need to lay the groundwork by looking broadly at the phytotron movement as a worldwide laboratory movement, which brings to light many interesting features of postwar experimental biology. This chapter therefore takes a general look at the laboratory movement, with case studies chosen to highlight different contexts and objectives that gave the movement its momentum. In referring to these projects as a "laboratory movement" I mean to suggest not simply the replication of one laboratory model in other locations but a whole range of activities that involved mounting campaigns, selling the idea, building networks of scientists, and coordinating efforts nationally and around the world. Biologists worked together and tried to share knowledge, so that the process of building a phytotron could become streamlined and less dependent on the need for custom design as time went on.

Among biologists, the desire for improvement and modernization of botanical sciences, especially in view of the need for better food security, already existed before their acquaintance with the Caltech prototype. Went's laboratory immediately caught their interest because, in the words of a British botanist, it "showed the great possibilities of providing a wide range of different controlled environments within a single installation."[1] It became a model to emulate and expand upon. After a brief survey of the international phytotron movement, I turn to a selected group of phytotron campaigns and projects that illustrate in different ways the ambitious visions for science that were associated with these laboratories.

I explore how phytotron projects developed in six national settings: the United States, France, Australia, Israel, the Union of Soviet Socialist Republics, and Hungary. These examples illustrate the diverse ways in which phytotrons were promoted and the blend of internationalism and nationalism that characterized this movement. The cases are not treated symmetrically; rather, each one brings out something different about the nature of the phytotron movement as an international trend, while highlighting the national goals that motivated specific campaigns. Each case offers its own story.

In the American case, I focus on the emergence of Big Science in biology and the idea that phytotrons (as well as labs combining botany and zoology called biotrons) might serve as regional laboratories much like the national laboratories created after World War II. The next three examples—France, Australia, and Israel—were directly influenced by Frits Went and by experience in Caltech's phytotron. The French example illustrates the importance of creating a worldwide social network through support of international conferences, publications, and newsletters that would enable scientists and technicians to learn from each other. The French leader, Pierre Chouard, believed that phytotronics represented a new strategy for the pursuit of biology, and he spoke of phytotronics as though it was a new discipline. The Australian case is another example of a Big Science project; it reflects the emerging leadership of Australia's Commonwealth Scientific and Industrial Research Organisation, based in Canberra, in promoting basic and applied biology related to national and regional problems. In this case, the idea that phytotronics was a separate discipline was emphatically rejected. The Canberra phytotron, which has been periodically renovated, is still in operation and it exemplifies a successful adaptation to changes in biology over more than half a century. The Israeli case study demonstrates how phytotron development was a response to the broader question of the future of agriculture in arid regions, an issue of particular relevance to a new nation experiencing rapid population growth after the war. The last two cases—the Soviet Union and Hungary, discussed in chapter 5—are included in part to show that these innovations were of equal interest to Communist regimes and in part to show how these particular examples suggest new perspectives on the history of the doctrine known as "Lysenkoism," on which much has been written.

Spreading the Word

Several phytotrons built after 1949 involved some form of collaboration with Went or were modeled directly on the Earhart laboratory. Many foreign scientists visited Caltech. As Went remarked, with his customary boosterism,

"A truly international spirit reigns in the laboratory, with people of all parts of the world sharing space and experiences with each other, and in many cases collaborating in research problems."[2] These experiences stimulated efforts to replicate the lab, but it was not simply a matter of exporting a brand new idea to Europe and elsewhere. Went's travels showed him that European leaders were already working aggressively to develop their plant research facilities. Although contact with Went provided some new ideas, they needed little persuasion to move in that direction. On his travels through Europe in 1950, Went discovered that there was already a strong commitment to putting resources into botanical and agricultural sciences, but he was still proud that the Earhart lab was holding its own and developing an international reputation. At Stockholm for the International Botanical Congress, Went gave a talk to an estimated 700 or 800 people, showed a film of the Earhart lab, and described the work being done there. "The interest was very great," he reported to Harry Earhart, adding that he was now sure that the laboratory was "the best known laboratory in the world."[3]

In 1952 Went promoted the phytotron energetically throughout the year, spreading the word and inviting young researchers to Pasadena. While in Jerusalem to attend an international conference on arid zones, he spoke about his phytotron and about desert ecology.[4] This was the first international conference held in the new state of Israel, and Went noted that "the interest of the local scientists for these meetings was so great, that the smaller halls were much too small, and people would stand for hours, even outside the windows."[5] Went lectured in Madrid, Spain, where botanists were already campaigning actively for a phytotron. In England that September he lectured on the phytotron and showed his film to various audiences in different venues, including the International Horticultural Congress in London and the Rothamsted Experimental Station. After moving on to the Netherlands for more lectures in his hometown of Utrecht, he traveled to Sweden, Denmark, and Germany. That year Went gave a total of twenty-seven talks in eight countries (and in four languages) in Europe and Britain, five talks in Asia, and "innumerable" talks in America.[6]

The result was a small boom in phytotrons, many modeled on the Earhart laboratory. The first, a smaller and less expensive laboratory than the Earhart, was at the University of Liège, Belgium, built by Went's friend Raymond Bouillenne, director of the university's Botanical Institute and Garden. Bouillenne had collaborated with Went in the 1930s at the Dutch botanical garden in Java, and like Went he had broad interests in physiology, ecology, and biogeography. In 1950, just before the laboratory was completed, Went visited Bouillenne and showed the group his film about the Earhart laboratory so they could see how it worked. The Liège model demonstrated what was possible with a relatively small investment of money.[7]

Another early version opened in 1953 to serve the Institute of Horticultural Plant Breeding at Wageningen, the Netherlands. The institute's director had visited Went in Pasadena during an American tour in 1945–1946 and benefited from Went's advice. The institute was associated with the State Agricultural College in Wageningen, a technical school that was distinct from a university. As Harro Maat describes, Wageningen was unusual in having relatively close connection between its research institutes and the professors of the college. It aspired to become a full-fledged university, which eventually happened in 1986. Building a new phytotron at the Institute might have helped the agricultural college to behave more like a university by strengthening its research capacity.[8]

Phytotrons were also built in the 1950s and early 1960s in Sweden (Stockholm and Uppsala) and in the Austrian Alps.[9] The pace of development stepped up in the 1960s and by the early 1970s there were phytotrons all over the world, with relatively large units in Australia, New Zealand, France, Japan, Sweden, the Netherlands, the Soviet Union, and the United States. A survey in 1980 of phytotrons and research institutes with controlled-environment installations listed fifty-three facilities worldwide, of which about half were full phytotrons or biotrons. Another survey, also published in 1980, listed thirty-nine phytotrons or biotrons in nineteen countries.[10]

The Japanese building spree was especially noteworthy. The first Japanese phytotron, modeled on the one at Caltech but smaller, was built at the new National Institute of Genetics (NIG) in Mishima and opened in 1952. The goals at NIG were to study environmental effects on gene expression as well as to breed new crops. Other phytotron projects followed at

Kyoto University, Tokyo University, and Kyushu University. The Japanese also began early to manufacture domestic machines and apparatus, so that they did not require foreign imports. A report in 1972 stated that most Japanese universities with agricultural departments either had or were in the process of getting a phytotron or its zoological equivalent (biotron, zootron, or aquatron).[11]

China too, although cut off from the West after 1949 until the late 1970s, built a phytotron in 1969, during the Cultural Revolution. When an American plant physiologist visited China for a month in 1979 he was surprised to find a phytotron at the Institute of Plant Physiology of the Academia Sinica in Shanghai. Built entirely with local resources, it had not been described in the West. It had modern equipment and its staff members were well informed about Western science.[12] There was also a connection to Caltech. The director of the institute, Hung-Chang Yin, had received his Ph.D. at Caltech in 1937 for research on the growth hormone auxin and on leaf movements. It would be valuable to have more scholarship on both the Japanese and Chinese initiatives in connection with the development of botanical sciences in these countries.

By the 1970s there was less need for custom design compared to Went's laboratory. Commercial companies, such as the Canadian company Conviron and the Japanese company Koito, started manufacturing growth chambers, so that by the 1970s it was only necessary to construct a building and install the chambers, as is done today. Whereas the first-generation phytotrons were custom-designed, later versions had greater flexibility and took less time to build. Conviron, now a multinational company, remains today one of the leading companies in the manufacture of controlled-environment facilities, and as its website describes, it offers a large variety of products for any kind of controlled-environment research involving plants. The fact that the Earhart laboratory was seen as a facility that could be replicated was important in making commercial development possible, which in turn guaranteed continuation of the original laboratory movement. Without this commercialization, which started in the 1960s, the movement might have plateaued at the first generation, instead of growing and continually adapting to new research needs.

The United States: Creating Regional Laboratories for Biology

Momentum to create more laboratories like the phytotron increased in the United States during the 1950s as biologists became more enthused about Big Science projects in biology. Interest in improving research capability was growing in the U.S. Department of Agriculture (USDA) and federal agencies such as the newly formed National Science Foundation (NSF), which started funding operating costs at the Earhart lab in the mid-1950s. As more people learned about the Earhart laboratory, more grant applications went to the NSF for similar facilities. In 1957 the NSF asked the American Institute of Biological Sciences to form a committee to evaluate how biologists viewed such facilities and to suggest ways to judge the merits of the grant applications. Went was a member of the committee and Kenneth Thimann from Harvard University was its chair. That committee confirmed the need for large facilities comparable to the Earhart but also for smaller individual growth chambers.[13]

Coinciding with these efforts to gain resources for research, the Department of Agriculture was also trying to encourage basic research, especially at its main center in Beltsville, Maryland. Beltsville had grown into a national research center under the visionary leadership of Henry A. Wallace, secretary of USDA from 1933 to 1940.[14] For botanical research a new building was erected in 1936, and by 1937 two controlled-environment laboratories for plant physiology were operating. Physiologists Harry Borthwick and Marion Wesley Parker took up the problem of photoperiodism there at a time when few people were working on it. Sterling Hendricks, a physical chemist working on mineral nutrition at Beltsville, joined the photoperiodism team after the war. When Went and Eversole made their cross-country tour in 1946, this group impressed them greatly. The Beltsville team, after years of research, succeeded in isolating the blue pigment responsible for photoperiodism, which they named "phytochrome" in 1960.

In 1957, USDA administrator Byron T. Shaw, long a champion of research in agricultural institutions, authorized the designation of certain laboratories or research groups as Pioneering Research Laboratories.[15]

This designation meant exceptional freedom to do basic research. Pioneering research groups were not required to limit investigations to specific crops or animals, and they were free to do research without having to fulfill many routine administrative duties or justify the practicality of their research. Borthwick's group in plant physiology and Hendricks's group in mineral nutrition of plants were among the first to be designated as pioneering laboratories. This was an opportune time for those in agricultural research to join forces with academic biologists to advocate for better facilities. Even with the backing of the Department of Agriculture, however, it could take time to acquire those facilities. A full phytotron, the Phyto-Engineering Laboratory, finally opened at Beltsville in 1966.

To explore university needs, NSF funded a second committee, the Committee on Feasibility of Biotron Construction, to find out what laboratory facilities were most essential to both botany and zoology. A biotron was comparable to a phytotron but included animals. The committee members included Went and Sterling Hendricks (representing the USDA) along with Arthur Hess (engineer for the Caltech phytotron), botanists Paul J. Kramer and Albert J. Riker, and zoologists Colin S. Pittendrigh and Clifford Ladd Prosser. The committee met with groups of biologists across the country, and in 1958 Hendricks and Went published an initial report of its activities in *Science* in order to gauge scientific interest. They noted that botanists were strongly committed to having controlled conditions, whereas zoologists felt less urgency about the matter. They argued that the phytotron had been accepted as an experimental tool "comparable to telescopes, particle accelerators, fossil collections, and other tools of science," and they pointed to the increased funding from the National Science Foundation as proof of interest in these tools.[16]

These discussions took place during a period of general soul searching about how American scientific research was supported, which often took the form of appeals for greater support of "basic" over "applied" research. A major symposium on this topic in 1959, attended by President Dwight D. Eisenhower, brought together leaders from academia, industry, government agencies, and philanthropic foundations to assess support for basic research. It was understood that there was no sharp distinction between basic and applied research and indeed that it was often difficult to define what these terms meant. In general, the term "basic" applied to research where scientists had relative freedom to define research questions, and it usually referred to research done in academic settings. "Applied" research was directed toward specific goals or products. There was general agreement that basic research was not nearly as well supported as applied research, or as Eisenhower put it, "only about four percent of our scientists and engineers are engaged in basic research."[17]

Philanthropic foundations had underwritten basic research in biology for a long time, but this debate was directed more at government agencies that were subsidizing science and engineering. Since the National Science Foundation was the main funding agency with a mandate to support basic research, one outcome of these discussions was to expand NSF's level of support for capital projects. NSF accepted responsibility for funding nuclear reactors, accelerators, and computers, and financing for controlled-environment facilities for biological research also fell within this expanded mission. The agency preferred to fund specialized research facilities that were not found in every college and university; by the late 1950s phytotrons and biotrons were added to the list of facilities that NSF believed worthy of support.[18]

NSF viewed such facilities as national or regional laboratories. Although associated with universities, these labs would give access to investigators from other institutions.[19] With that in mind, the biotron committee had specifically focused on large-scale facilities and not on individual controlled-environment chambers. One of the most ambitious projects that came from these discussions was the Biotron Laboratory built at the University of Wisconsin, and its development illustrated the typical course taken by such Big Science projects. With partial support from NSF, the project broke ground in Madison in 1964. As the budget escalated beyond the initial projections there was a scramble to find new funding sources. Initial goals for achieving climate control that were too ambitious had to be scaled back. Finally, more time was needed to resolve the inevitable engineering and equipment complications. Although the estimated completion date for the Wisconsin biotron was the fall of 1966, the first research project did not begin until May 1967, and the official dedication was in September 1970, over a decade after the project was first proposed in 1959. Bringing such complex and expensive projects to fruition was enormously challenging, but once created they were considered invaluable in their contributions to a wide range of scientific projects.[20]

In 1968 a pair of phytotrons, the Southeastern Plant Environment Laboratories (SEPEL), opened as a cooperative venture between Duke University in Durham and North Carolina State University in Raleigh.[21] Although they were meant for the use of plant scientists in the southeastern United States, the SEPEL facilities were available to scientists from around the world. Paul Jackson Kramer of Duke University was the main force behind the Duke phytotron. He had worked at Caltech during a sabbatical semester in 1955 and found the experience so profitable that it prompted him to campaign for a phytotron in the eastern United States, starting in 1956.[22] He also served on NSF's Biotron Committee. The director of the Duke phytotron, Henry Hellmers, had also worked in Caltech's phytotron.[23]

With these advanced laboratory facilities, Duke University, already strong in ecological research, solidified its place as a premier research and graduate training center for physiological ecology. But because Kramer focused on woody plants, his research also had direct bearing on forestry, and about half of his graduate students were foresters. The director of the North Carolina State phytotron, Robert Jack Downs, had been at the USDA's Beltsville research center, and the phytotron was in part supported by funds from the tobacco industry for research on tobacco. Other crop plants were also studied in the North Carolina State laboratory. These laboratories were good examples of how academic and agricultural interests were combining in this period to gain resources for botanical science. These phytotrons, in common with other early phytotrons, brought basic and applied sciences together.

Phytotron leaders tended to be from the same generation: they were born in the early twentieth century, were plant physiologists with strong ecological interests, and were committed to the idea that basic and applied problems should be studied together. They also believed that the problems of biology had to be solved at the whole-organism level rather than the molecular level. At Duke, Kramer embodied this vision and approach to biology.[24] Harold Mooney, an ecologist trained at Duke who worked with Kramer, commented on why Kramer stood out: "He wanted his science to be useful to practical people. That he never saw a boundary between cutting-edge science and the application of science to solve practical problems left a deep impression on me. At that time in most universities there was a deep schism between so-called 'basic' and applied sciences. To his great merit he never let these artificial boundaries keep him from what he thought was right and from seeking solutions to problems of interest."[25]

Mooney also noted Kramer's interest in linking plant physiology and ecology: "Kramer developed this interaction to a high degree and thus greatly appealed to the large number of ecological graduate students who passed through Duke University, of which I was one. Kramer, though, considered himself a plant physiologist, but one especially interested in whole plant physiological integration."[26]

This common breadth of vision was at the core of the phytotron movement, which was not simply about controlling the environment through advanced technologies. Fundamentally it was about bringing scientific fields together, promoting multidisciplinary interactions, and breaking down the artificial distinction between basic and applied sciences, all the while remaining focused on the whole organism. Kramer expressed his ideas about the importance of synthesizing research findings in reflections on the state of plant physiology published in 1973, the year before his retirement from Duke. Complaining about the tendency toward overspecialization in plant physiology, he remarked that "there is a great need for scientists with training and interests broad enough to enable them to collect information from various areas, correlate and organize it, and use it to explain plant behavior."[27] He also noted what he deemed a "healthy renewal of interest in the physiology of whole plants" after the earlier burst of enthusiasm for molecular biology: "Some of the scientists who rather arrogantly claimed in the late 1950s and early 1960s that the only worthwhile approach to biological problems was through molecular biology have now changed their views. I think it has become apparent to even the most devoted supporters of molecular biology that an organism is much more than a collection of molecules or even of cells and one cannot solve all the problems of organisms solely at the molecular level."[28] These sentiments about biology were shared by Frits Went and many other phytotronists of that generation.

However, the cost of phytotronics was skyrocketing. In the academic setting, questions of cost and size, as well as the need to keep abreast of changes in biology, were important. The two units at Duke and North Carolina State cost nearly $4.5 million, with annual operating costs just over $400,000 in 1970–1971.[29] The more ambitious phytotrons and biotrons of the 1960s were far more expensive to build and operate than Caltech's prototype, could take years to complete, and in the end did not always function as the regional or national centers they were intended to be.[30] The enthusiasm for large-scale, expensive phytotrons and biotrons tapered off at the end of the 1970s, although the use of controlled-environmental cabinets and smaller eco-cells continued to be important for physiological and ecological research. Went's own career followed this shift from larger to smaller facilities.

The idea of the phytotron as a place where disciplines met was still seen as a worthy objective in the ones that remained in operation. Robert Downs put it this way in 1980: "In a phytotron the gap between physiology and microclimatology can be narrowed, ecology can become an experimental science, the genetics of response to climate can be explored and the biochemical and physiological mechanisms affected by climate stress can be identified and understood."[31] One can think of such laboratories as cultural forces for scientific synthesis or interdisciplinarity, and it would be valuable to explore further how well disciplinary barriers were broken down in such workplaces. We would need to recover the individual stories of scientists who worked in these kinds of laboratories and to explore further how their ideas or attitudes changed as a result of interactions in these environments.

The French Phytotron: Advancing a New Strategy, Creating a Social Network

Went's European and Middle Eastern travels in 1952 took him to France, where he met Pierre Chouard (1903–1983), a physiologist who was then working on photoperiodism and hoped to come to the Earhart lab to work out a few problems. Second only to Went, Chouard would become a world leader in promoting what he called "phytotronics" in the 1960s and 1970s, even to the point of arguing that work in phytotrons represented a distinctive and new approach to biology, although it was not always clear what he meant by this claim. He was passionately committed to showing that biologists needed phytotrons, and the one he built near Paris during the 1950s was one of the largest and most elaborate of that time. The French case is also of interest because French scientists perceived quite early the need for an ecological version of the phytotron, dubbed the "ecotron," although their plans did not materialize for a long time.

Chouard's research involved the physiology of plant growth and development in relation to the environment. He was also interested in deserts and was a member of the international advisory committee for arid zone research of the United Nations Educational, Scientific, and Cultural Organization (UNESCO). Although he specialized in plant physiology, Chouard's activities ranged over many disciplines and embraced problems in agriculture and horticulture, plant geography, morphogenesis, ecology, and conservation.[32] In 1930 he backed the founding of an international station for Mediterranean and alpine plant geography at Montpellier, which was an important center for ecological studies in France. From 1935 to 1937 he was Maître de Conférences in plant biology at the University of Bordeaux, and from 1938 to 1954 he held the chair of agriculture at the Conservatoire National des Arts et Métiers in Paris.[33] He helped to create the Néouvielle Natural Reserve in the central Pyrenees, a site of exceptional natural beauty and biodiversity, and in 1947 became a member of the National Council for the Protection of Nature. As a professor of plant physiology at the Sorbonne (University of Paris) starting in 1953, he became the force behind the building of a French phytotron. Known simply as "Le Phytotron," it was located outside Paris at Gif-sur-Yvette as part of the Centre National de la Recherche Scientifique (CNRS). While serving as director of the phytotron from 1958 he also held the chair of plant physiology at the University of Paris.

The French phytotron was relatively large. Construction of the laboratories at Gif started in 1953, but the phytotron itself opened in 1958 and was fully operational only in 1962. "Le Phytotron" was meant to be a prototype

for phytotron development, that is, the goal was to push the limits of the
phytotron by increasing the complexity of the factors controlled and the ac-
curacy of their control. In other words, its purpose was in part to show the
potential of such facilities. Its deputy director between 1958 and 1962, Jean
P. Nitsch, had done his doctoral work on the role of plant hormones in fruit
development at Caltech and had continued to work in the Earhart lab while
on the faculty at Cornell University in the 1950s.[34]

Chouard coined the term "phytotronics" to describe the approach to
science in these laboratories. He gave credit for this innovation to Went and
his Dutch predecessor Anton Blaauw, who in the 1920s had experimented
with temperature-controlled rooms and cabinets in the Agricultural Physiological Laboratory at Wageningen in the Netherlands. But Chouard also recognized French precursors of the concept of phytotronics. Two were of
special significance. Gaston Bonnier, early in the twentieth century, had observed how plants changed in different climates, and he used the cellars of
Les Halles, the central fresh-food market in Paris, to recreate these climates,
albeit crudely. Bonnier's side of the phytotron's family tree represented the
sciences of systematics, biogeography, ecology, and physiology. Chouard also
gave the eminent nineteenth-century physiologist Claude Bernard the credit
for setting the stage by formulating the principle that physiological experiments should aim for control of all environmental factors, varying only one at
a time. "If we follow Claude Bernard's dictum," he explained, "innumerable
investigations will require the use of a phytotron or phytotron-like device."[35]
Thus the idea of phytotronics was given a respectable French pedigree.

An article on the Gif phytotron published in 1963 emphasized the
importance of creating ties between the phytotron's team of researchers and
those of other laboratories at the same CNRS site: laboratories for genetics,
photosynthesis, enzymology, biochemistry, hydrobiology, and so forth created a workforce of nearly seven hundred researchers and technicians.[36]
This critical mass in turn justified the high expense of a phytotron. The
question of funding such enterprises would become important during periods of belt-tightening in French science, but the 1960s were good years in
France for the support of science. By the early 1970s the phytotron accommodated a research staff of about forty, with some twenty guest scientists.

Chouard conceived of phytotronics as a technology and a method of scientific investigation, a way of attacking scientific problems involving plants and their environment. Improving the science meant on one hand gaining technical knowledge and improving the reliability of instruments. These were matters of engineering and design. But "pure phytotronics" was also a style of doing science, one meant to produce innovative hypotheses that could be tested in the phytotron. One could adopt a Baconian methodology, analyzing all the phenomena "with no preconceived ideas, by a 'multi-factor grid' of climatic combinations."[37] Or one could be guided by ecological principles, whether they dealt with individual ecology (autecology) or community ecology (synecology). Chouard stressed the value of multidisciplinarity and of working on different levels: one could study whole organisms, organs, tissues, and cells, and investigate problems through physiology, biochemistry, biophysics, or other disciplinary approaches. The point in bringing them all together was to be able to relate discoveries at the lower levels to processes at the higher level. Explanations just at the molecular level were not the end goal: the goal was to extend such explanations to the higher levels of complexity of cells, organs, and whole plants. He also stressed the practical benefits of phytotrons in solving horticultural and agricultural problems, most importantly to discover ways to increase productivity in plants. In Chouard's description, the phytotron was clearly intended to foster synthesis in many different senses.

Chouard considered the phytotron to be a tool for developing a new research strategy in plant science.[38] It is difficult to tease out what claims he was making for the unique opportunities the phytotron afforded, for his writing on this subject is by no means clear. By "strategy" he meant a plan of action or an approach to a large number of problems in plant biology that revolved around the question of how plants developed during their life cycles, starting with germination of the seed. The strategy entailed a comprehensive, multi-level attack on this problem. In this way one could build up a full physiological and ontogenetic "portrait" of each species and each stock. This knowledge could be used to understand how plants behaved under cultivation as well as in their natural environment. Chouard considered phytotronics to apply not just to physiological science but also to ecology, botanical geography, and mycology. A large array of problems could be studied within the phytotron.

Chouard was completely convinced of the importance of the phytotron as a scientific tool and a way of organizing research, and he and his colleagues at Gif (especially J. P. Nitsch, Nicolas de Bilderling, and Roger Jacques) became active promoters of phytotronics. By the early 1970s the French group concluded that a reasonable expectation would be to have a few large multipurpose phytotrons operating on each continent, with a staff of about fifty investigators carrying out about a hundred simultaneous experiments involving various climatic combinations.[39] They reasoned that although the cost of such an installation might be an order of magnitude greater than conventional biological research, it was still an order of magnitude less than similar units for nuclear physics. The justification for such large budgets remained the same as it had been for the very first phytotron: to relieve hunger by making nature more bountiful.

A sense of Chouard's comprehensive vision can be gleaned from a report of the 14th General Assembly of the International Union of Biological Sciences, which met in Amsterdam in 1961.[40] Chouard spoke at the meeting about phytotrons and their uses, for both basic and applied research. By basic research he meant the study of physiological and ecological reactions important for growth and development in plants (such as dormancy and photoperiodism) or study of the internal factors controlling growth and development in conjunction with external factors. Applied research problems might include selection of new varieties of cultivated plants, acclimatization of wild plants, and determining what factors affected maximum productivity under different conditions of climate and soil.

Many of the discussions concerning phytotrons harped on the same theme: that both basic and applied research had to be supported. Chouard, like Went and other phytotron enthusiasts, reiterated this point in order to counter the belief that basic research on biological processes was not relevant to achieving practical improvements in agriculture and horticulture, or the belief that targeted applied research was sufficient for such improvements. Chouard was at pains to hammer home the message that certain kinds of innovations could not have been achieved without "fundamental research."[41] The phytotron therefore helped to blur the boundary between what was considered "basic" and "applied" research at the time, a distinction that reflected the different kinds of institutions where basic and applied research was carried out. The overall message of phytotronics, driving home both its novelty and its importance, was that in these laboratories these two approaches to science would work together synergistically. At the 1961 conference Chouard also argued for rational siting of phytotrons, with the larger installations being used internationally and the smaller ones used as regional or national laboratories. He envisioned phytotrons sited strategically in different environments—in arid zones, in the humid tropics, in arctic climates, and at high altitudes—and hoped that nations would cooperate in using them for studies of international concern.

Chouard promoted phytotronics in the 1960s and 1970s, much as Went had done in the 1950s, but on an expanded scale. He also took the lead in creating and maintaining a social network of phytotron researchers and technicians, an action that helped give this international movement greater momentum. He realized that the success of phytotronics depended on collaboration and exchange of information among all interested groups. Attention had to be paid to bringing groups together and ensuring that lines of communication were open. The French phytotronists, under his leadership, took over the important social role of facilitating communication across groups of researchers as the laboratory movement grew. The more people they could bring into the conversation the better.

In 1969 Chouard directed a half-hour film to display the laboratory's architecture, climate-control systems, and research programs.[42] He organized a series of roundtables and symposia in London (1964), Tel Aviv (1972), Warsaw (1974), and Leningrad (1975).[43] In 1971 he also started a *Phytotronic Newsletter* (with French and English editions) to facilitate exchange of information about technologies and work strategies. Biologists had suggested creating such a newsletter in 1964 at a conference on phytotronics held in London.[44] The hope was that UNESCO, which had sponsored the conference, would pay for the newsletter, but the idea went nowhere. When Chouard took responsibility later on, he had scant resources and depended on donations to keep the newsletter going through twenty-one issues, ending in 1980. His idea was to promote exchange between physiologists who used phytotrons, engineers and technicians who constructed equipment, and anyone involved in agricultural and horticultural work who might reap the practical benefits of the phytotron. The French contribution was not only to test the capabilities of such an instrument but also to create the social networks needed to promote innovation by sharing information about what was going on around the world. By the early 1970s this network consisted of about five hundred specialists who were able to keep in contact with each other as a result of Chouard's organizational work.

The *Phytotronic Newsletter* reported regularly on conferences, reprinted essays from phytotron researchers worldwide, and informed readers of new phytotron developments. It serves as an excellent source for tracking the growth of this laboratory movement worldwide during the 1970s. In November 1973 it reported that at the next three international meetings of plant science, for the first time either a part or the entire program would be devoted to phytotronics. The meetings were the 19th International Horticultural Congress, held in Warsaw in 1974, the 12th International Botanical Congress held in Leningrad in 1975, and the International Symposium on Biotrons and Biotronics held in Japan in 1976. In 1974 the newsletter opened its pages to Eastern Europe and the Soviet Union with contributions from Hungary and Kiev.

However, the high cost of running the French phytotron would become a problem as it began increasingly to compete with molecular biology. One estimate was that the consumption of electricity by the Gif phytotron equaled that of a town of 28,000 inhabitants.[45] The oil crisis of 1973–1974 suddenly made such an expense seem extravagant. Chouard retired at the end of 1975, and Pierre Champagnat, professor of plant physiology at the University of Clermont-Ferrand, became the new director. He was followed by Roger Jacques, who was the last Phytotron director. State support for science declined in the 1970s but picked up after 1981 with the election of François Mitterrand as France's president. But after five years of expansion under a socialist government, the government of Jacques Chirac, elected in 1986, reversed that trend by slashing the budget for science and research, while shifting power to the universities and away from the CNRS.[46] The

Phytotron did not survive these cuts and was closed in 1986. By this time molecular biology was ascendant and attention was being given to DNA and to protein synthesis. The central unifying science of phytotronics had been physiology, which by the 1980s was considered to be outmoded.

Although the Gif phytotron was the only one built in France in these early years, it was not the only one conceived at that time. Another plan was afoot for a facility to serve the needs of ecological sciences specifically. The University of Montpellier was the center of ecological research in France, and in the early 1960s the CNRS was supporting a large-scale vegetation mapping project there, in the Botanical Institute. This was a massive enterprise of data-collecting, involving the recording of up to 180 characteristics of vegetation and habitat for each site studied. By 1961 eight hundred sites had been studied. Since studies of vegetation were linked to physiological questions, the Plant Physiology Department at Montpellier was hoping for funds for what was dubbed an "ecotron," described as "a sort of super-phytotron, similar to the famous Cal-Tech instrument for controlled environment studies but much more flexible."[47]

The designer and force behind this project, Frode Eckardt, was a Danish plant ecophysiologist (a term he coined) who was then head of the plant physiology department at Montpellier. He had also worked in Pasadena in 1956–1957. Eckardt understood both the advantages and the shortcomings of Caltech's prototype. He hoped to build a better laboratory that would preserve the advantages while avoiding the shortcomings. In a UNESCO news article published in 1960, the ecotron concept (erroneously attributed to Louis Emberger, director of the Institute of Botany at Montpellier) was said to promise the opening of a new chapter in the history of Montpellier, which for centuries had been the center of French botany. One feature of the proposed ecotron was to allow observation of plant reactions without anyone entering the room, in order to eliminate the possibility that humans were modifying experimental conditions in uncontrolled ways.[48]

In 1963 Eckardt presented a plan to Montpellier University to build an ecotron specifically for studies in ecophysiology (as opposed to the usual focus of phytotrons, which was the physiology of development).[49] This ecotron was not built, but Eckardt became a leading ecophysiologist and used climate-controlled chambers in his research, while developing new techniques for ecophysiological research. A full history of physiological plant ecology would grant him a prominent role in the development of this discipline, in particular in extending the science beyond the study of individual plants to the study of ecological communities and ecosystems. He helped to organize and chaired an international symposium on plant ecophysiology in 1962 at Montpellier under the sponsorship of UNESCO and the International Union of Biological Sciences. The symposium, published in 1965, was intended to broaden the compass of ecophysiology by focusing on the development of new methodologies.[50]

Eckardt's particular concern was to bridge the laboratory-field divide, which involved taking instruments into the field for better measurement of environmental conditions as experienced by plants outside. His multidisciplinary approach inspired the volume *Terrestrial Global Productivity*, published in 2001 and edited by Jacques Roy, Bernard Saugier, and Harold A. Mooney. Both Saugier and Mooney had worked with Eckardt. Mooney recalled encountering Eckardt's unusual multidisciplinary approach to plant ecology in the mid-1960s: "The process was unprecedented for the times, with physicists, engineers, and biologists working together as a team. The program was comprehensive and pathbreaking. Utilizing a battery of approaches, some of which would be considered innovative even today, they were doing pioneering work on scaling from leaf-level to whole-ecosystem measurements."[51] The experience of working in Caltech's phytotron likely helped to transform Eckardt's ideas of ecological research. Even without his ecotron, he creatively adapted the multidisciplinary, multi-level approach that the phytotron represented to ecological research. Perseverance in the development of ecophysiology at Montpellier paid off much later: two scientific generations down the road the CNRS did finally build an ecotron there, which opened in June 2010 at the Centre d'Ecologie Fonctionelle et Evolutive.[52]

The Australian Phytotron: A Big Project for Biology

The largest Australian phytotron, built in Canberra and opened in 1962, was also stimulated by the Caltech model but aimed to improve on its designs. Great care and attention to detail went into the new facility's design and building, which took a few years. It too was seen as an innovative Big Science project, comparable to projects such as the Parkes Radio Telescope, which was being built at the same time (in that case with American philanthropic support). Still operating today (after major renovation), Canberra's phytotron has adapted successfully to changes in biology over the course of a half century.

The scientists who visited the Caltech phytotron for research purposes typically stayed for about a year, and in some cases that experience transformed their expectations of science. The building of the Canberra phytotron was an example of this effect. Lloyd Evans (1927–2015), a New Zealander who was studying at Oxford, heard Went give a talk about the Earhart laboratory in 1952, just as Evans's three-year stint at Oxford was ending. He obtained a postdoctoral fellowship to work at Caltech, where he stayed for eighteen months to study photoperiodism. In the phytotron, he recalled, "Frits Went was the unchallenged authority," but he also remembered being surprised by the egalitarian atmosphere that allowed postdocs to disagree with faculty members at Caltech, something that was not the custom at Oxford.[53]

While Evans was at Caltech he had a visit from Otto Frankel, chief of the division of plant industry at the Commonwealth Scientific and Industrial Research Organisation (CSIRO) in Canberra. Evans got Frankel interested in the idea of building an equivalent laboratory in Canberra. Frankel lost no time getting the ball rolling and in 1956 recruited Evans as part of this effort. As Evans recalled, part of the attraction was the "challenge of a big project."[54] Plans were under way for the Parkes Observatory, and Frankel thought it a good idea to try for a high-priced biological equivalent. He recruited Went to the cause. In July 1955, at the invitation of the CSIRO, Went made an extended visit to Australia to discuss ways of developing plant physiology and ecology.[55] After a short visit to New Zealand, he spent about three months in Australia, visiting Canberra, inspecting laboratories elsewhere, and making various excursions into the field. Canberra, he concluded, was the ideal location for a phytotron, which would make it the botanical center for Australia and the region.

In addition to recommending a phytotron to support agricultural research, Went's report encouraged the Australians to develop ecological research. The motive for developing ecology was to make full use of Australia's natural resources, and the motive for building the phytotron was to rid ecology of "sloppy work and thinking." As Went explained, "With the facilities of a phytotron, ecology enters in a new, largely experimental, stage, where the causal relationships between plant and environment can become established."[56] Went was pleased to see that already in Australia there were controlled-temperature cabinets in practically every laboratory or research station, but he thought too much work was still being done in the field or in uncontrolled greenhouses and cold frames. For practical crop ecology, as well as for detailed study of native flora, he strongly recommended a phytotron about the same size as the Earhart laboratory. This was a larger space than the CSIRO thought was needed, but Went argued that they needed a larger facility to accommodate scientists visiting from other universities. He offered his assistance, recommending that whoever was in charge of an Australian phytotron should visit Caltech and become thoroughly acquainted with the Earhart laboratory.

In fact, Frankel had already arranged for the officer in charge of CSIRO's engineering section, Roger N. Morse, to visit Caltech while Evans was still there.[57] Morse also visited other controlled-environment laboratories in Europe and picked up some new ideas from the Rothamsted Experimental Station in England. The Institute of Plant Physiology of the Imperial College of Science and Technology had a research laboratory at Rothamsted, north of London. Scientists there were using controlled-environment cabinets and small glasshouses in studies of photoperiodism. The Australians saw how to adapt those facilities for the larger-scale phytotron. The key idea was to use controlled-environment cabinets in addition to controlled-environment rooms, so it would not be necessary to shift plants to different rooms twice a day, as was done in the Earhart lab. Such cabinets could be factory-built at a high standard and would allow for more rigorous control of environmental conditions. Inside the Canberra glasshouses, these cabinets would occupy more than half of the space, providing the highest level of control. Morse and Evans took charge of the planning, design, and testing of the equipment for the phytotron, while Frankel searched for funding. In the end he succeeded in raising A$1,350,000 for the project.[58]

After years of preparation, Canberra's phytotron, the Controlled Environment Research Laboratory, appropriately shortened to CERES (the Roman goddess of agriculture), opened in 1962, a year after the Parkes Radio Telescope.[59] The building, designed by Roy Grounds, was two stories high and about 65 by 24 meters on the sides; it is considered an example of postwar modernist architecture because of its smooth wall surfaces and cubiform patterning in the sunhoods, as described on its webpage.[60] It was a credit to the "demonic energy" of Frankel as well as to the dedicated efforts of Morse and Evans and their colleagues.[61] The relatively large facility had fifteen glasshouses with shuttered cabinets that could provide a range of day-lengths and temperatures. In the other half of the building, cabinets enabled scientists to combine these conditions with variations in light intensity, humidity, and other environmental conditions. By drawing on some of the innovative features of other laboratories, especially Rothamsted's use of self-contained growth chambers, the Australians were able to improve on the Earhart's designs.

Frits Went was on hand to give the opening remarks at the international symposium that celebrated the dedication of CERES.[62] At the symposium Evans commented on the still vexing problem of relating experimental results obtained in controlled environments to the vastly more complex conditions found in the field.[63] In the discussion provoked by his paper, speakers drew attention to the computer-like nature of the phytotron, able to produce any environmental complex. As with the computer, it could provide "reliable, intelligible answers only when presented with intelligent, logical questions."[64] Asking the right question would be the main challenge of phytotronics.

One research problem that Evans had started earlier and then brought to the Canberra phytotron for refinement involved the physiology of flowering in a grass, darnel ryegrass (*Lolium temulentum*). The grass grew in wheat fields, but its seeds could carry a fungal disease that was toxic if eaten, so people had learned very early to separate the wheat from this noxious weed. Flowering in this plant could be induced by exposure of just one leaf blade to one long day. In Australia, such crops as wheat, barley, and rye were long-day plants, so although *Lolium* was a weed it provided a good model for other species. Evans had to figure out what the mechanism of flowering was: did the leaf exposed to a long day produce a mobile stimulus, a hormone for flowering? Or were the plants growing on short days exporting an inhibitor of flowering? The prevailing view was that the plant in short days did not export an inhibitor, but Evans concluded that both things were occurring: the long-day leaves exported a stimulus to flowering and the short-day leaves exported a mobile inhibitor of flowering. These two interacted at the shoot apex. The phytotron enabled Evans to refine his experiments on this problem. Although Anton Lang of Caltech disputed his explanation at the time the Canberra phytotron opened, fifteen years later he admitted that Evans had been right.[65]

The *Lolium* research was considered basic science without direct practical consequences, but Evans also worked on crop plants such as wheat and the problem of improving yields. As he wrote in 1977, "Scratch a plant physiologist, as many a sceptical review committee has done, and you find an agronomist, plant breeder or horticulturist."[66] This comment was made to address growing public skepticism about science in the 1970s, including the view that support of topics deemed to be "basic science" was not important to solving agricultural problems. The value of phytotrons for agricultural research was, in Evans's view, a compelling reason to support their construction. By the 1970s several large phytotrons were engaged explicitly with agricultural problems: they were located in Wageningen in the Netherlands, Canberra and Brisbane in Australia, Raleigh in North Carolina, and Palmerston North in New Zealand. Evans considered these laboratories especially useful for the study of various problems that had agricultural applications. Among these were crop diseases, clarifying the bottlenecks in plant adaptation (environmental conditions that prevented plant development), manipulating life cycles of plants so that wider crosses could be done, and extending the adaptive range of plants to different environments.[67] In all cases the ultimate goal was to increase the yield, quality, and stability of crops for the world's growing populations.

But basic research, including inquiries on seemingly useless plants, still had to be explicitly defended, since its value was not self-evident to people who were not scientists. Evans's arguments in the 1970s were not

meant to place basic science above practical applications but rather to demonstrate that the fields of basic physiology and agriculture were inextricably
intertwined. He also stressed that field and laboratory studies were complementary. Data obtained in the phytotron could help to explain the field performance of plants.[68] As Went had done earlier, Evans promoted the
phytotron as an instrument useful for multiple disciplines, including physiology, biochemistry, genetics, plant nutrition, ecology, forestry, horticulture,
plant pathology, soil microbiology, and anatomy.

The papers delivered at the symposium to celebrate the opening of
CERES included many of the central subjects of plant physiology, including the climatic control of plant growth and development, the biological
clock, plant adaptations to extreme environments such as drought and cold,
and the extrapolation of results from controlled environments to the field.
But the Canberra phytotron also included an unusual and innovative line of
research in a program focused on a small weedy plant of Eurasian origins
called *Arabidopsis thaliana,* mouse-ear cress or thale cress, which within
about fifteen years would become an important model organism for flowering plants. Friedrich Laibach, a German botanist, had been working with
Arabidopsis since 1907 and in 1943 had recommended it as a model organism for genetic, developmental, and evolutionary studies, essentially as the
botanical equivalent to *Drosophila,* but relatively few people had heeded
his call. The plant was convenient for experimental research because it had
the fewest number of chromosomes of any flowering plant, was small, had a
short life cycle, produced abundant seeds, and could self-fertilize. Its value
as a model organism was much better appreciated by the mid-1980s, and it
was the first plant to have its genome sequenced in 2000.

Well before *Arabidopsis* took off in the 1980s, however, the Canberra
botanists realized that the phytotron facilities were ideal for culturing this
plant. John Langridge and his collaborators directed this line of research.
Langridge had been looking for a plant suited to the kinds of experiments
that George Beadle and Edward Tatum had conducted on the bread mold
Neurospora during the 1940s to try to elucidate the function of genes. They
produced mutants in *Neurospora* that were unable to synthesize key vita-
mins such as thiamin, and their experiments gave rise to the idea that the

function of the gene was to produce enzymes that catalyzed various metabolic pathways in the cell. Langridge discovered Laibach's recommendation of *Arabidopsis,* obtained seeds from Laibach, and in his doctoral dissertation at the University of Adelaide (1955) he also produced a mutant that was not able to synthesize thiamin. But even more importantly he developed a method of culturing *Arabidopsis* on a sterile medium in a test tube for its entire life cycle. This was a crucial step in making the plant an attractive experimental subject.

Langridge's innovation in culture method was ideally suited to the kinds of controlled experiments that could be done in a phytotron. Bruce Griffing and Randall Scholl later explained how the sterile culture method accelerated research: "First, test tube culture permits many plants (at least 1000) to be grown in a single growth cabinet. Second, rapid plant growth results in rapid experiments (15–20 days), which ensures economy of cabinet use and minimizes the probability of cabinet breakdown during an experiment."[69] The method eliminated competition between plants and minimized environmental variation, while the control provided by the growth cabinets meant that plant responses to many environmental variables were reproducible. Because of these culture methods it became possible to make inroads into the study of variables that were controlled by many genes at different loci and that were strongly influenced by the environment. This line of research is an important part of the early history of the use of *Arabidopsis* as a model organism: with the phytotron, scientists could easily scale up this research so that experiments could be run on thousands of plants. The decision to make greater use of growth chambers at Canberra created the opportunity for expansion of this line of research. As Evans remarked in 2003, "it was at this level that most space for future expansion and improvement was left, and that is where *Arabidopsis* now reigns supreme."[70]

Like the French with their phytotron, Australians celebrated their achievement in building a more complex and better-designed facility than the Earhart. CERES was a source of national pride, illustrating Australia's maturity as a scientific leader and CSIRO's role in that maturation process. David Munns has developed this point in his analysis of the Australian phytotron.[71] That sense of maturity was achieved in part by emphasizing basic science, which meant bringing in experts from several disciplines so that more scientific fields were represented. Munns pointed out that, before planning the phytotron, Frankel had instituted important reforms in CSIRO's plant industry division during the 1950s, bringing in people from basic science disciplines (including chemistry, physiology, genetics, and physics) and shifting away from the previous focus on particular projects or crops. In 1967 the Canberra phytotron, like its Big Science counterpart the Parkes Radio Telescope, was included in the first global live television link-up, called "Our World."[72]

Evans disagreed with Chouard's claim that phytotronics constituted a new discipline, and he was not comfortable even using the word "phytotronics." Such language seemed damaging in its elitism. In the same way that he perceived the common interests of physiologists and agronomists, he proposed that if one were to "scratch a phytotronist" one would "find a plant physiologist or an ecologist or a geneticist." He continued, "The full potential of phytotrons will be realized only when all these and many other scientists have learned to exploit controlled environment facilities for their own purposes. By calling ourselves phytotronists we may give the impression that we are a separate breed, and may prevent the fullest use of phytotron facilities."[73] Where Chouard had tried to create an independent identity for phytotronists, which included reinforcing their sense of community through conferences and newsletters, Evans did not want to lose their identification with established disciplines.

However, the established disciplines at this time were being confronted with a major cultural change in biology, the change wrought by the rise of molecular biology. It became important therefore not just to maintain ties to these disciplines but also to explain why these disciplines mattered. CERES opened in the same year that James Watson, Francis Crick, and Maurice Wilkins shared the Nobel Prize in Physiology or Medicine for their contributions to the discovery of the double helix structure of DNA in 1953. That momentous discovery ushered in a new era that saw complete reinterpretation of the concept of the gene as a code represented by the sequence of bases that made up the DNA molecule. Biology swiftly turned toward the study of how the RNA molecule read the genetic code and guided protein synthesis within the cell. What alarmed biologists in classical disciplines was the reductionist agenda of molecular biologists, who seemed bent on converting all of biology into molecular biology. That meant killing off the disciplines that could not easily be reduced to the molecular level. Hence Edward O. Wilson, recalling Watson's disparagement of ecology after he came to Harvard, referred to Watson in his autobiography as the "Caligula" of biology.[74]

Many biologists were in no doubt about the reductionist ambitions of molecular biologists. Hans Mohr, a plant physiologist at the University of Freiburg, Germany, stated it bluntly in 1989: "The program of molecular biology has been strictly reductionist from the very beginning until the present day. The goal has always been the unity of biological sciences: to explain the major elements of the classical biological disciplines in terms belonging to the most fundamental biological science, molecular biology."[75] This sense of biology being under attack produced defensive reactions and, as discussed earlier, Went and others shared these reactions.

At the symposium celebrating the opening of CERES, the question of how to respond to the advances of molecular biology was in the air. Frederick C. Steward's address acknowledged that current thought regarded metabolism as genetically determined, but Steward pointed out that there was abundant experimental evidence showing that environmental conditions affected the metabolism of plants. Genetics was undoubtedly important, but a large range of nutritional and environmental factors could "intervene to control, or modulate, the genetically feasible events."[76] Those factors also had to be recognized and explained. In the discussion he commented that this response to the environment was a feature of the whole organization of the plant, arguing that the whole was greater than the sum of the parts, which indirectly was a rebuttal of the reductionist agenda. He was referring to change in the way genes were expressed, a subject not well understood at that time but one that would develop in subsequent decades into the field of epigenetics.[77]

Frankel's closing comments at the symposium, building in part on Steward's discussion, carefully laid out the challenge for the future, which was to use the advances of molecular biology to achieve a new level of integration of the sciences. "As yet our minds are conditioned to regard 'genetic' and 'environmental' pathways as distinct and, in a mechanistic sense, unrelated," he asserted. The future course, as he saw it, was to integrate the current lines of research to create a "unified theory of differentiation in plants."[78] That quest for integration required contributions from all branches of science, he implied, and many of those contributions would come from experiments conducted in phytotrons. Frankel forecast the very opposite of the reductionist agenda, envisioning instead a synthesis between molecular biology and disciplines that studied organisms at higher levels of organization.

The Canberra phytotron turned out to be one of the longest-operating phytotrons, but its continued operation required it to adapt to the evolution of biological science. Its history exemplifies the kind of integration Frankel was imagining in 1962. Although CERES was originally meant for plant physiology, biochemistry, and agronomy, the emergence of molecular biology and its applications to genetics opened a new era of genetically modified organisms, so the phytotron was upgraded to allow for experiments with transgenic plants. By the twenty-first century, as agricultural science strove to meet the challenges of climate and ecosystem change, the Canberra phytotron, refurbished in time for its fiftieth anniversary in 2012, would evolve to serve the emerging science of "high-resolution plant phenomics." By applying state-of-the-art imaging techniques borrowed from medicine, along with advances in robotics and computing, high-resolution plant phenomics delves deeply into the developing plant's phenotype and its relationship to the genotype and to the environment.[79]

Phytotrons and the Green Revolution

CERES was also the "god-parent," in Frankel's words, of the phytotron that was dedicated in September 1974 at the headquarters of the International Rice Research Institute (IRRI) in Los Baños in the Philippines. IRRI was founded in 1960 by the Rockefeller and Ford foundations as an international research center for the development of new varieties of rice, which it started to introduce in the mid-1960s as part of the agricultural initiative that became known as the "Green Revolution." The phytotron, a gift of the Australian government, used the overall design concept of CERES except that the growth cabinets were a Japanese rather than Australian design. At the opening conference Frankel explained the purpose of the phytotron much as Went had done nearly three decades earlier: to figure out how to increase plant productivity. He also lauded the benefits of being part of a multidisciplinary group and breaking down the barriers between biological sciences.[80] One particular advantage that he saw in IRRI's phytotron was the ability to build on IRRI's extensive rice bank, which included traditional rice varieties from across Asia and Africa as well as wild rice species. With the phytotron IRRI was in a unique position to expand research on one of the world's most important food crops. Research at IRRI also raises questions about how phytotrons served as instruments of the green revolution, or how we should place such instruments within the history of the green revolution.

In 1968 William S. Gaud, administrator at the U.S. Agency for International Development, coined the term "green revolution" to refer to the record high yields of wheat and rice in developing countries (India, Pakistan, Turkey, and the Philippines) in the 1960s.[81] These countries had planted what were touted as "miracle seeds," high-yielding semi-dwarf varieties of grains produced by agricultural programs funded by the Rockefeller and Ford foundations, starting in the 1940s. The programs were supposed to make developing nations more secure in food supply and also enable them to export grains. Norman Borlaug, who headed the Rockefeller wheat-breeding program in Mexico, received the Nobel Peace Prize in 1970 for his work championing the methods that were instrumental in producing these grain surpluses.

The benefit of dwarf varieties arises with the application of chemical fertilizers. In traditional varieties applying fertilizer leads to more vegetative growth and causes the stalks to bend ("lodging"), which makes harvesting difficult. Dwarf varieties, with stiffer stalks, were able to accept fertilizers without lodging and therefore were more suited to modern agricultural practices. A second element of the green revolution strategy was to eliminate the varieties' sensitivity to photoperiod (day length), which meant they could be grown over a much larger area because they possessed what was known as "wide adaptability." Jonathan Harwood has referred to this emphasis on creating near-universal varieties as a "cosmopolitan" strategy, as opposed to a "local" strategy aimed at adapting varieties to specific local conditions of climate or soil.[82] In addition to being able to accept fertilizers to maximize yields, these varieties required application of pesticides and often irrigation. Such varieties were less suited to rain-fed areas, where both drought and flooding could occur.

The green revolution programs adopted the cosmopolitan strategy for both wheat and rice, and IRRI's high-yielding rice varieties in the 1960s were of this type, performing well as long as there was adequate rainfall or irrigation. But this form of agriculture, with its need for fertilizers and pesticides, came at a higher cost, and often the widely adapted varieties were viewed as inferior to local varieties. By 1970 the claims and methods of green revolution programs were drawing increasing criticism, which intensified after poor harvests in 1972.[83] One criticism was that the superiority of so-called miracle seeds compared with traditional varieties had been exaggerated. Many viewed the programs as failures because they did not make a significant impact on hunger or poverty. They benefited larger-scale commercial farmers but did little for subsistence farmers.

Historical scholarship has focused closely on these programs and their antecedents, the critical reactions to them, and the continuing legacies of these development strategies.[84] Both Harwood and Marci Baranski (who examined wheat agriculture in India) have observed that the policies of the green revolution, particularly the emphasis on near universal varieties with wide adaptability, persisted even in the face of criticisms and viable alternatives.[85] However, Baranski observed that although wide adaptability remained codified in India's wheat-breeding program into the 1990s, it was in decline at IRRI and at other centers.

How might we fit the IRRI phytotron into these debates? At the conference to celebrate the opening of IRRI's phytotron, speakers did not address the economic, political, or cultural problems of transferring these Western technologies to diverse nations; instead, they focused directly on the scientific problem of creating varieties that had to be grown in a very wide set of climatic conditions. The overall message, made plain in Frankel's opening address, was that phytotron-based research, along with field research, was needed to create varieties that were more precisely adjusted to their environments. That adjustment required "*detailed understanding of the developmental phases* of the plant which determine its productivity, as of those phases which are susceptible to injury or breakdown." Indirectly, Frankel was making an argument against the idea of trying to create near-universal varieties and in favor of concentrating on varieties adapted to specific conditions. He returned to this point more directly later in his talk. While complimenting both rice- and wheat-breeding programs for their successes in producing crops grown over wide areas, he also commented that skilled breeders "might produce *locally adapted* cultivars with even higher productivity." The phytotron was intended to serve this purpose, to gain better understanding of "the problems facing the rice plant in diverse environments."[86]

Three papers discussed the need for a more finely tuned local strategy to achieve a better match between variety and environment. One pointed out that climate change was likely to become severe in the future and therefore more research was needed in regions where rice was not irrigated and was subject to drought.[87] Widely adapted varieties were not suited to such regions. Two other papers examined the trade-offs between the perceived advantages of wide adaptability versus specific adaptation to local conditions. In the discussion Frankel posed the question: "In the short term, wide adaptability is desirable, but is it in the long term?"[88] In asking this question, he was not challenging the overall goals of the green revolution. The phytotron was meant precisely for the type of high-cost, high-production agriculture that had resulted from the green revolution and from large-scale commercial farming, as Frankel made clear in his opening address. Increasing plant productivity and efficiency remained the goal, and he agreed that the breeders of widely adapted varieties had achieved success. But future advances, he thought, might come from locally adapted cultivars. While he was not mounting any major challenge to the principles underlying the green revolution, he was urging some course corrections.

Lloyd Evans's address reinforced the same idea. He pointed out that relatively little was known about the physiological basis of adaptability of many successful crops, that more research was needed using facilities like phytotrons, and that one contribution phytotrons could make was to clarify the goals of plant breeding and agronomy.[89] Although both rice- and wheat-breeding programs had focused on insensitivity to photoperiod, he explained, there were many situations where close adaptation required sensitivity to photoperiod—for instance, in wheats grown at high latitudes, where flowering time needed to be responsive to day length to avoid frost injury. These comments were gentle nudges in the direction of greater attention to local adaptation and away from adherence to the principle of wide adaptability. Evans was indicating that the gains of any future green revolutions would be won only by more insight into plant physiology as well as any ecological problems that affected productivity, such as pests and diseases.

Kaori Iida, in her historical work on Japanese biology, drew attention to the involvement of Japanese scientists with IRRI, which raises questions about how Japanese approaches to breeding might have influenced IRRI's programs.[90] Hitoshi Kihara, who became the second director of Japan's National Institute of Genetics (NIG) in 1955, took a keen interest in rice research and in IRRI from its beginnings. Although Kihara had made his reputation largely on the basis of wheat genetics, he turned his attention to rice in the 1950s. On a visit to the United States in 1953 he met with the Rockefeller Foundation's deputy director of agriculture, J. George Harrar, to discuss research on rice improvement, an area in which the Japanese excelled. In the late 1950s and early 1960s Kihara also received Rockefeller funding for rice research at NIG and used this funding to build environmentally controlled greenhouses for rice research. Iida considers Kihara's networking and institution-building activities to be part of a strategy to reconstruct his scientific program after the war and to assert the authority of Japanese scientists in fields such as rice research.

When IRRI was founded in 1960, Kihara was made a member of its board and also chaired its program committee, which gave him a say in determining the nature and scope of the institute's scientific program. He was also in a position to recommend Japanese scientists for IRRI's staff. Kihara was especially interested in collecting and preserving cultivated and wild varieties of rice to serve as genetic resources for plant research, as he had done earlier for wheat, and in this respect, as Iida suggests, he seems to have had an impact on IRRI's work. He proposed a multi-institutional effort to collect strains of seeds and herbarium specimens from around the world, with the idea that a duplicate set of these strains would be held at IRRI. Preserving biodiversity by collecting seeds of cultivated and wild varieties is one starting point for developing improved local varieties. Kihara further ensured that NIG and IRRI maintained a close collaborative relationship in the 1960s. Although Kihara was getting his funding from Rockefeller, his research strategy was not focused on creating near universal varieties of rice but on understanding all the climatic conditions that affected rice, including drought and other unfavorable conditions, as well as studying disease and pest resistance.

By the time of the IRRI conference Kihara had retired from NIG, but one of the talks about breeding for local environmental conditions and soil types was given by T. T. Chang from IRRI and Hiko-Ichi Oka, an applied geneticist from NIG. If that talk reflects the way the research program at NIG was developing in the 1960s (and more research would be needed to determine whether this is true), then it would appear that scientists at NIG, such as Oka, understood that the emphasis on wide adaptability had to change. Chang and Oka pointed out that high-yielding varieties were largely confined to favored areas, but there was a need to assist farmers in less favored areas, where sensitivity to photoperiod or tall stature would be beneficial traits. As they said, this kind of breeding, which meant putting back traits that had been deliberately removed, "may appear to be a reverse step in crop evolution, but [it is] a necessary one."[91] The phytotron therefore could be seen as a vehicle to address and correct some of the flaws of the green revolution's cosmopolitan approach. As Iida points out, we do not have a clear picture of what was or was not passed on to IRRI as a result of the close connection to NIG. She suggests that more research is needed on the Japanese side of this story, and Harwood agrees that a fuller account of these continuing debates needs further scholarship.

Construction of phytotrons also prompted new ideas about how such facilities could speed up the production of food. This kind of production usually involved fruits and vegetables (that is, more expensive crops) and does not fall strictly within the definition of the green revolution, which focuses on grains. But if we expand our view of the green revolution to include other kinds of crops, then phytotrons and related facilities were precursors to what is now known as controlled-environment agriculture. In the 1960s, it was feasible to use climate-controlled greenhouses or artificially lit growing facilities to raise plants to the stage of transplanting to the field, but was it also possible to raise plants to harvest? The Phyto-Engineering Laboratory constructed in 1966 at the USDA's research station in Beltsville, Maryland, stimulated interest in this possibility.

Predictions were that controlled-environment facilities would be widely used commercially by the end of the century and that future farming would be guided by computers and enhanced with space-age technologies. The USDA's yearbook for 1968 projected a futuristic vision of fully mechanized farming and the creation of nuclear-powered "food factories" in which every aspect of a plant's life was controlled in the interests of maximum productivity.[92] Dana Dalrymple, an agricultural economist, reviewed the prospects for controlled-environment crop production in 1973 and noted that there were experiments being made in different locations, for instance, the construction of greenhouse complexes in northern Mexico and Abu Dhabi by the Environment Research Laboratory of the University of Arizona.[93] Completely controlled environments, although technologically possible, were not economically feasible at that time. Given the interest in controlled-environment agriculture today, it would be valuable to explore this subject as a different aspect of the green revolution.

The Israeli Phytotrons: The Future of Arid Lands

The conference in Jerusalem that Went attended in 1952 was planned in cooperation with UNESCO and reflected widespread interest in the development of agriculture in semi-arid and arid regions. Arid and semi-arid lands seemed the most promising places for development, the "last frontiers of opportunity in land use" worldwide.[94] Study groups convened by UNESCO began to discuss ways to promote research on arid zones in 1949, with the aim of creating an international body to deal with arid zone problems.[95] This interest in arid regions, which accelerated during the 1950s, meant that Went's research in southern California, where the environment resembled Israel's, would inevitably be seen as relevant to these global concerns. Some of the earliest visitors to Went's new laboratory came from Israel and Australia.

Walter Clay Lowdermilk, who was a consultant to the Food and Agriculture Organization of the United Nations, gave the opening address as president of the conference in Jerusalem. Lowdermilk had long been interested in agricultural development in Palestine. Before the war, when he was assistant chief of the U.S. Soil Conservation Service, he had been sent at the request of a congressional committee to survey agricultural practices in Europe, Africa, and the Middle East in 1938–1939. Although the outbreak of war interrupted his survey, he did manage to visit several countries and was particularly enthusiastic about the environmental stewardship of the Jewish settlers in the Palestine Mandate. In a book about Palestine published in 1944, Lowdermilk discussed the possibility for agricultural development in the region after the war, anticipating the establishment of a Jewish homeland and a rapid increase of population.[96] In his conference address in 1952, he looked forward to a time when science and technology would help people "reconquer" the desert and put an end to what he viewed as past misuses of arid land that had caused deserts to expand.[97]

In the closing address, Sir Ben Lockspeiser, secretary of the Department of Scientific and Industrial Research in the United Kingdom, developed the military metaphor implied in the phrase "conquest of the desert." He argued that, in addition to the "generals" or elite scientists whom he was addressing at the conference, the operation called for an army of trained troops, by which he meant technologists, engineers, and other technicians who would undertake the hard work of conquering the desert.[98] UNESCO had an important role to play, he said, in helping to guide such a campaign, along with the FAO, which took an active interest in arid zone problems. The FAO was also drawing attention to problems of irrigated agriculture throughout the Middle East.[99]

In the 1950s, cultivated lands covered about one-tenth of the global land surface, whereas desert, arid, and semi-arid lands represented over one-third of the land surface. With the world population at more than 2.6 billion by the mid-1950s and with the expectation that it would double within fifty years, there was a growing sense of urgency about the need to develop agriculture in under-used lands, including the humid tropics, frigid regions, and arid and semi-arid regions. Because a large number of countries contained arid and semi-arid regions, and because such areas made up a large proportion of the world's land mass, the focus sharpened on the future of these arid lands.

In 1950, UNESCO took the lead in promoting international cooperation in research on natural resources with particular attention to arid and semi-arid zones. It established an advisory committee on arid zone research, which held its first meeting in April 1951 in Algiers.[100] UNESCO had initially thought of creating an international institute for the arid zone, but that idea was quickly shelved, which in Lockspeiser's view was a good decision because the greater need at that time was to support existing *national* institutes or build new ones. UNESCO's arid zone program continued to expand through support of international conferences during the 1950s, concentrating on different aspects of research in each year: hydrology and underground water in 1952, plant ecology in 1953, energy sources in 1954, human and animal ecology in 1955, and climatology and microclimatology in 1956. This focus on the fragile ecosystems of dry lands remained at the center of UNESCO's work for the next half century.

As the military metaphors suggest, there was great optimism in the 1950s that nature could be "conquered" by the use of technology. However, relatively little was known about dryland ecology at that time. The dominant ecological theories adopted an equilibrium viewpoint, which assumed that natural systems evolved in a linear way toward stable final states unless they were disturbed. Drylands were assumed to be highly disturbed and degraded systems, and the cause of this supposed degradation was often assumed to be nomadic pastoralists and their herds of cattle. Despite the fact that these pastoralists had occupied the same lands for centuries, traditional uses of drylands were seen as an obstacle to improvement. Policies therefore favored extending agriculture into marginal lands, which often required irrigation, as well as planting trees, managing rangelands, and settling nomadic populations. Expansion of deserts, or desertification, came to be seen as a newly emerging crisis, and UNESCO's programs helped to disseminate and entrench these ideas about the presumed crisis.

A half century later, the policies based on these assumptions, which often failed and led to even greater problems, were understood to be flawed. New non-equilibrium perspectives showed that fluctuating conditions and cyclical periods of rain and drought were normal for such landscapes, not evidence of ruin. With this new understanding came new respect for the knowledge and traditional practices of Indigenous people. Scholars have closely examined this shift in viewpoint over several decades, and two books in particular provide good guides to these changes. Charles F. Hutchinson and Stefanie M. Herrmann's study, *The Future of Arid Lands— Revisited,* published under the auspices of UNESCO, offers a point-by-point comparison of the prevailing wisdom of 1956 and 2006, while Diana K. Davis's recent book, *The Arid Lands,* covers some of the same ground but adopts a more critical perspective, arguing that the idea of drylands as wastelands has been politically motivated.[101] My interest here is not to review this ongoing discussion but rather to examine how these postwar debates generated opportunities to press for greater resources for plant sciences, including construction of phytotrons.

With UNESCO's sponsorship, as well as support from the U.S. National Science Foundation and the Rockefeller Foundation, a conference on the future of arid lands was held in 1955 in New Mexico. Among the international group representing seventeen countries was Michael Evenari (1904–1989), a plant physiologist and ecologist from the Hebrew University in Jerusalem. In the published conference proceedings, his paper was co-authored with his former doctoral student Dov Koller (1925–2007), who did not attend the conference but was about to come to the United States to spend two years working in Caltech's phytotron.[102] Evenari and Koller built a case for more scientific research involving all branches of natural science to address the problems of developing the desert. Evenari focused on the need for international centers and laboratories—funded through agencies such as UNESCO—whereas Koller argued more specifically for improved scientific facilities at the Hebrew University, including a phytotron. The Israeli discussions illustrate the interplay of internationalism and nationalism in the context of growing interest in the future of arid lands.

Evenari was a desert botany specialist who had been one of the first botanists to work in the Caltech phytotron during a seven-month sabbatical in 1949.[103] His career is recounted in his autobiography, *The Awakening Desert*, from which most of my information is taken. Evenari was born Walter Schwarz in Elsass-Lothringen (Alsace-Lorraine), then part of Germany, in 1904. He received a Ph.D. in botany in 1926 from Johann Wolfgang von Goethe University in Frankfurt. He then held positions as assistant in plant physiology at the German University in Prague and the Technical University in Darmstadt, Germany. At Darmstadt, under the influence of Bruno Huber, he also developed an interest in ecophysiology. After Hitler's rise to power he was fired from his job in 1933 and immediately left for Palestine, where he joined the plant physiology section of the Hebrew University. In 1935, when he took on Palestinian citizenship, he also changed his name to Michael Evenari, the surname being the Hebrew equivalent to Loewenstein, his mother's family name. In 1940 he volunteered for the British army, returning to Palestine after demobilization in August 1945.

After the war he divided his time between science and work for Haganah, the Jewish self-defense organization.[104] Evenari was sent to the United States in 1947 to investigate laboratories and equipment in preparation for plans to build a biology building at Hebrew University's campus on Mount Scopus. Accompanied by zoologist George Haas, he traveled across the country giving lectures at universities, and his lecture on germination-inhibiting substances at Caltech resulted in an invitation to return there for a year. Back in Jerusalem in May 1947, he found Palestine in a state of anarchy as conflicts among the British, Jews, and Arabs intensified. In February 1948 Hebrew University sent him on a fund-raising mission to the United States and to South American cities that had large Jewish populations. After his fund-raising trip ended in October 1948, he was advised not to return right away to Jerusalem.[105] The British Mandate had ended in May 1948, leading to the new state of Israel, but immediately the neighboring Arab countries invaded it. With the road to Mount Scopus under Arab control, travel along the access road was restricted. "In this way," he later wrote, "we were effectively cut off from the University premises, housing all our books, instruments, and collections."[106] A makeshift university was established, distributed among fifty-five buildings all over Jerusalem. Plant physiology was in a former British police building, with the rest of the department in a hut three kilometers away. Evenari decided to accept Caltech's offer of a visiting professorship, and in January 1949 he and his wife, Lieselotte, drove from New York to Pasadena, where they stayed until the end of September. He was one of the first foreign scientists to work in the phytotron.

Evenari was already interested in germination inhibitors, and at Caltech he took up the problem of how different wavelengths of light affected germination of lettuce leaves, a subject that remained the theme of his research for the next two decades. He remembered his time at Caltech as one of the happiest and most productive of his career. Bonner introduced him to plant biochemistry, and he had productive interactions with other members of the Caltech faculty, including trips to the surrounding desert with Went. George Beadle, the division's head, offered him a continuing position at Caltech, but he chose to return to Israel. On his return to Jerusalem conditions were not much improved. As he later wrote, "We had only the most primitive means available for teaching and research. Breaking a Petri-dish, pipette, or measuring cylinders was a disaster, because we could usually not replace even these small items. . . . There were no lecture halls, so that we taught students in private homes. Despite all these difficulties, we continued research and teaching to the best of our abilities. Under these conditions, there was no way I could continue the work I had started at Cal. Tech., which I deeply regretted."[107]

Although Evenari was hardly in a position to build anything like a phytotron in Jerusalem, he did help to make it possible to bring a phytotron to Israel in other ways. He became vice president of Hebrew University in 1953 and during the 1950s raised funds to build a new and more centrally located campus at Givat Ram, which opened in 1958. It was during this period as vice president that he participated in the conference in New Mexico on the future of arid lands, and with Dov Koller contributed a chapter to the published proceedings that assessed Israel's experience in the previous five years with experimental desert agriculture.[108] Developing desert agriculture was one of the highest priorities of the new state, and a large team of researchers across many disciplines was engaged in this problem, aided by a grant from the Ford Foundation. Israel consisted of three phytogeographical and climatic areas, the Mediterranean, Irano-Turanian, and Saharo-Arabian. The Mediterranean region was semi-arid, and the other two were arid or very arid, with annual rainfall ranging between 25 millimeters and 300 millimeters. In the Negev desert, in the southern portion of Israel, irrigated farming was deemed practical only in the northwestern part of the desert. Elsewhere agriculture had to rely mainly on local rainwater. Because this dry environment was very uncertain, the scientists found it difficult to anticipate the kinds of problems they would encounter.

Evenari and Koller's first point therefore was that desert agriculture required planned, thorough scientific fact-finding involving all the branches of natural science. The information-gathering had to entail scientific teamwork and could not depend on ad hoc methods. A second key point was to have experimental settlements for practical work located in desert areas but also to use physiological laboratories to complement field work. A third key point was the need for a central international organization to collect and correlate information from different local projects. In this paper they made no mention of building phytotrons, which might have seemed prohibitively expensive.

While serving as vice president of Hebrew University, Evenari kept up his research in plant physiology, especially the physiology of germination, but by the mid-1950s he was also exploring the fascinating question of how ancient cultures flourished in the desert. Prompted by Koller, he had become interested in ancient farming techniques in the Negev desert, and in 1954 he started to do research on the Nabataeans, who had developed systems for channeling and storing runoff water.[109] The Nabataean kingdom was at its height between the first century BCE and the first century CE. The Nabataeans built five large towns in the Negev at Avdat, Shivta, Khalutza, Kurnub, and Nitzana, but their capital city was Petra, further east. The cities survived into the seventh century under Roman and then Byzantine rule.

Originally it was believed that the Nabataeans imported their food, but evidence that they practiced agriculture led to the surmise that the past climate had been much wetter. The noted biblical scholar and archeologist Nelson Glueck challenged this idea in the mid-twentieth century when he showed that they had built elaborate systems for catching rain and irrigating, suggesting that they were truly farming in a desert climate.[110] In the mid-1950s this knowledge became suddenly relevant to Israel's goals. As Evenari and Koller explained, "When the new state of Israel began its attempts to push agriculture into the Negev desert, the Nabataean methods of farming became a matter of highly practical interest. We felt that an intense study of their agricultural system could teach us a great deal to help our own endeavor to recreate fertility in those areas of the desert where the Nabataeans had created it once before."[111]

Evenari teamed up with Naphtali Tadmor, a student at Hebrew University's School of Agriculture, and Leslie Shanan, a hydrologist and water engineer, and together they investigated more than a hundred farm systems to learn how this ancient civilization managed to farm in the desert. In 1956 they decided to put their theories to the test by reconstructing an ancient farm. The first experiment was at a site near the ancient city of Shivta, an isolated location where it was unsafe to live at that time. In 1959 they started a second project at Avdat, an ancient city in the Negev highlands that was already being reconstructed and was in a safer location. With funding from Hebrew University, the Rockefeller Foundation, and private donors, these projects flourished, and by the 1970s the farms were boasting abundant yields of a large variety of fruits, vegetables, and cereal crops.[112]

Although engaged in this ambitious field research, Evenari did not by any means dismiss the idea of building phytotrons in the early years of this reconstruction work. In 1960 he wrote a report for UNESCO's Arid Zone Project, and presented it at a symposium on arid zone problems held in Paris in May.[113] This meeting was meant to assess the past effectiveness of UNESCO's Arid Zone Project and make recommendations about its future course. At the symposium two films were shown highlighting arid zone developments in Pakistan, Israel, Morocco, and the Soviet Union. Pierre Chouard also invited a group of delegates to visit the Gif phytotron, then under construction. About 250 scientists from 33 countries attended.[114]

In his report, Evenari made a strong case for the value of phytotrons as important tools to solve problems of arid zones. Two supporters of phytotrons, Chouard and Paul Kramer, had read and critiqued the report, and Evenari acknowledged their contributions. He also quoted from a letter from Harvard physiologist Kenneth Thimann, who suggested that UNESCO "should set up an international laboratory for this kind of work which would compare with what the physicists do for nuclear studies (e.g. CERN at Geneva)," referring to particle accelerators operated by the European Organization for Nuclear Research.[115] "Again and again," Evenari observed, "we have come to the conclusion that the most important 'break through' we can expect will come mainly through phytotronic research. . . . A phytotron would not only enable the plant physiologist to find some answers to the basic problems of arid zone physiology, it would at the same time permit the precise testing of all methods to improve arid zone agriculture (e.g. increased drought resistance, etc.)."[116]

His report also concluded that the ideal scenario was to create an international research institute with a phytotron for arid zone research, and he clearly had in mind that UNESCO should support such an enterprise. The institute he envisioned would combine both a phytotron and a laboratory with experimental fields and normal glasshouses. If that ideal could not be obtained, another option was to enable one of the existing phytotrons to serve the function of the "ideal" research institute for arid zone problems. Chouard contributed a report also encouraging international phytotrons and ecotrons, while proposing the creation of a special advisory committee on physiology and genetics to help guide UNESCO's activities related to arid zone research.[117] UNESCO did not build an international phytotron, but it did continue to support the work of scientists within the phytotron community through its conferences. Notably, its conference on "Plant Response to Climatic Factors," held in Uppsala, Sweden, in 1970, concluded that phytotrons were becoming increasingly important research tools.[118]

While Evenari looked to UNESCO to help fulfill the promise of phytotrons to further arid zone research, his doctoral student, Dov Koller, was responsible for bringing phytotrons to Israel. Koller was born in Tel Aviv, obtained his Ph.D. in plant physiology from Hebrew University in 1954, and then joined the faculty and became a full professor in 1969. From 1957 to 1959 he was a research fellow at Caltech's phytotron, with funding from the Rockefeller Foundation, and there, as his wife, Ditza, recalled, he caught the phytotron bug.[119] Like Evenari, he was interested in germination and more broadly in photobiology and morphological responses of plants to light signals.

The problem of germination, which Koller studied at Caltech and which was the focus of the Hebrew University's research in plant physiology, had both practical significance and general ecological importance for understanding how plants succeeded in colonizing their environments. As Koller explained, there is a big difference between domesticated and wild species when it comes to germination of seeds.[120] In species used for agriculture or horticulture, easy germination is highly desired and these species have been specifically bred for reliable germination. Domesticated plants, in other words, have only minimal control over germination. But wild plants are the opposite: seeds do not germinate easily, and they do not germinate all at the same time. It would not be adaptive for all seeds to germinate too quickly, for if germination were followed by drought or disease, all would die. It is better for the survival of the species if the seeds were inhibited from germinating, so that a reserve of seeds could build up, extending the likelihood that some of the plant's offspring would survive. Control over germination is one way that a plant adapts to a variable environment.

The mechanism behind the inhibition of germination was important to understand, not only to explain the abundance and distribution of species in the wild, but also to understand how regions might be reseeded and made more abundant and hospitable to human occupation. This was a particularly important problem in dry regions. In Israel, for example, reseeding experiments were being done with a perennial grass, known in the United States as "Smilo" (*Oryzopsis miliacea*), which was considered a promising pasture plant. These experiments were often completely unsuccessful. The experiments Koller undertook in Caltech's phytotron were in part designed to figure out what environmental conditions (especially what combination of light and temperature) inhibited or promoted germination in this and other species.[121] His work in the phytotron was directly connected to the experimental efforts that he and Evenari had described at the New Mexico conference (published in 1956), which related to the broad problem of desert agriculture and the future of arid lands.

The active campaign for an Israeli phytotron appears to have started soon after Koller's return to Israel in 1959, when he became director of the department of botany at Hebrew University. On the basis of a proposal from Koller and Eliezer Tal (from Israel's National Council for Research and Development), the National Council took up the question of building a phytotron in Israel in 1961. It formed a study group of Israeli scientists and brought in Frits Went for a month to advise the council. The report, submitted in May 1961, combined both Went's recommendations and that of the study group and, unsurprisingly, made a strong case for building a national phytotron, identifying Jerusalem, Beersheba (where Koller was director of the Desert Plant Research Laboratory), and Rehovot (at the Hebrew University's Faculty of Agriculture) as possible locations.[122] A total of seventy-four interviews were conducted, with Went interviewing fourteen people, including the minister of agriculture, Moshe Dayan.

Went's arguments, in addition to rehearsing the usual reasons for supporting better research facilities, also drew attention to the much faster rates of progress possible in branches of biology that did not require work on mature plants. Lines of research that focused on molds, yeasts, and bacteria, and work that required only seedlings in experiments, had made tremendous advances in the study of metabolic pathways, physiological processes, and hormones. Work that required mature plants, on the other hand, lagged behind, largely due to the need for controlled-environment laboratories. The study group emphasized the importance of being able to work year-round, instead of discontinuing experiments in the summer because of excessive heat. One department had to send its scientists overseas every year to test locally produced flower bulbs and tubers under European climatic conditions. In addition, Israel was getting more orders from abroad, especially from Africa and Asia, and needed to study the adaptations of plants to their destinations.

At first Koller succeeded in building a relatively small phytotron in Jerusalem at Givat Ram, which was more for his personal use and was not the national laboratory intended in the report's recommendations. In 1969, however, he became a professor in the department of agricultural botany at Rehovot, and there he was able to build a full-size phytotron that served a broader community of scientists. After twenty-five years of operation, it was replaced in the mid-1990s by a new, larger and more fuel-efficient phytotron that is still operating.[123]

The Israeli case study illustrates three themes. One is the link between experimental laboratory science and field biology, which embraces both ecological and agricultural research. While the type of historical reconstruction in which Evenari was engaged at Avdat might appear to be a move in the opposite direction from phytotronics, the two types of scientific work were actually complementary and served the same goals. The reconstructions and experiments on the desert farms were equivalent to ecological field studies, but with an archeological twist and a focus on the human ecology of a vanished civilization. The reconstruction work was part of a broader study of the desert ecosystem and the problem of how plants (and animals) adapted to the stress of the desert environment. Experimental work that was done in phytotrons also focused on how plants coped with stresses such as drought, heat, or cold, and how certain mechanisms, such as control over germination, gave plants an adaptive edge in an uncertain environment. Experimental work in laboratories (both phytotrons and conventional laboratories), experimental field studies, and experimental reconstructions of past agricultural systems all served the same goals of understanding how organisms were adapted to their environments and ultimately how humans could use this knowledge to obtain more from apparently barren lands.

The ecophysiological work that Evenari and his colleagues undertook in the desert did involve instrumentation, as Evenari described in his autobiography. Otto Lange, an ecologist from the University of Würzburg, had done ecophysiological studies of desert plants at Avdat in 1967. In 1971, Lange returned to Avdat with his assistants, Ludger Kappen, Uwe Buschbohm, and Ernst-Detlef Schulze (who later was a founding director of the Max Planck Institute for Biogeochemistry in Jena). They outfitted a special truck "with the latest instruments for measuring water loss (transpiration), photosynthesis and water potential (the force of the plant for absorbing water from the soil)."[124] The plants were connected to the instruments by long cables. "The instruments," Evenari recalled, "worked day and night. For eight months we stayed together in Avdat without outside interference and felt like [we were] living in a scientific kibbutz."[125] The farm experiment was not an attempt to return to ancient customs but rather an effort to recover ancient knowledge in the light of modern science and with all the help that modern instruments could provide. Their research between 1967 and 1971 resulted in fifty-three scientific publications on the ecophysiology of desert plants.

The second theme illustrated by this case study is the role of agencies such as UNESCO in furthering work of this kind by encouraging international exchanges and collaborations, sponsoring conferences, and publishing reports and conference papers. The focus in Israel was on development of the desert, which had direct relevance to the economic security of the nation, but the problem of the future of arid lands was of much broader concern. Through its project on arid lands UNESCO provided support for experimental botany and for the kind of work done in phytotrons, even if the idea of having an international collaborative enterprise equivalent to CERN did not materialize. In their evaluation in 2008 of the assumptions that guided scientific work in the 1950s, Hutchinson and Herrmann observed that the outlook in 1956, with some exceptions, was exceptionally optimistic about what might be achieved through "science and technology, if pursued in the spirit of international cooperation."[126] They viewed that international spirit as marking a major difference from what they observed fifty years later: "Now, there is less general optimism. There is also less commitment to international cooperation, and more reliance on market forces to effect change in drylands—and everywhere else—through globalization. The degree to which this will affect the drylands is unclear."[127]

The third theme, illustrated especially in Koller's work, is the direct connection between these modern areas of research and very old Darwinian questions that focused on plant adaptations—their mechanisms and their meaning. Koller's posthumously published book *The Restless Plant* (2011), for instance, explored the problem of plant movement. Plants are stationary but they move constantly, evincing all kinds of adaptations designed to exploit their environments and disseminate their seeds. Koller took up this question with a thoroughly modern spirit, giving detailed descriptions of the cellular motors that drive plant behaviors. What his book illustrates particularly well is the great difficulty these questions pose, and how answering them requires meticulous experimental work, the sort that phytotrons were meant to support as well as field observations. People who caught the phytotronic "bug," including Frits Went himself, remained driven by curiosity to understand plant behavior in all its aspects, to appreciate what kind of creature the plant was, and to learn how it came by its remarkable adaptive abilities.

FIVE

Phytotrons Under Communism

THE LAST TWO CASE STUDIES—Russia and Hungary—are of particular interest because their stories suggest new perspectives on the history of "Lysenkoism," a movement in Soviet science associated with the controversial ideas of Trofim Lysenko, a Soviet agronomist. Lysenkoism is generally viewed as a tragic episode of political interference in science that did long-lasting damage to Soviet biology, thanks to its dismissal of modern Mendelian genetics and the promotion of the idea that acquired characters are inherited. Russia's first phytotron, built in Moscow in the 1950s, was an effective way to counteract the destructive influence that Lysenkoism was having on Soviet biology after 1948.

In the case of the Hungarian phytotron, which opened in 1972, the goal was to advance an unorthodox biological argument that actually had its origins in Lysenkoism and belief in the inheritance of acquired characters but also responded to new biological developments of the 1960s. These two case studies offer fascinating examples of phytotronics in the service of opposing positions, one anti-Lysenkoist and the other pro-Lysenkoist. Both episodes shed new light on the phenomenon of Lysenkoism, revealing aspects of the history of life sciences that have either been ignored or given scant attention.

The Soviet Phytotron: An Answer to Lysenko?

The phytotron built in the 1950s by the Soviet Academy of Sciences in Moscow is the most intriguing of all the phytotron projects because it was conceived and built at a time when an approach to biology known as "Lysenkoism" or "Michurinism" was favored by the Soviet state, resulting in the suppression of scientific freedom. Trofim Denisovich Lysenko (1898–1976) was an agronomist who advocated an approach to biology that was strongly opposed to Mendelian genetics, which he saw as a foreign import to the Soviet Union from the West. His rise to power in the 1940s, culminating in Stalin's formal approval of his theories in 1948, had a devastating impact on the pursuit of genetics in the Soviet Union for many years, and most analyses of Lysenkoism focus on this impact. However, Lysenko was also critical of some of the leading Russian plant physiologists, and we should ask what effect these criticisms had on the field of plant physiology. One historian has argued that physiologists suffered the humiliation of pretending to accept Lysenko's ideas and survived by hiding behind this façade. It is this idea that I wish to probe and challenge, for there is evidence that plant physiologists developed an effective strategy for gaining resources and doing the research they wanted to do. They achieved success, quite remarkably, by building a phytotron in the suburbs of Moscow.

For those unfamiliar with Lysenkoism, a bit of background: Lysenko was the son of peasants and received a poor early education, graduating as an agronomist from the Kiev Agricultural Institute in 1925. In the late 1920s he claimed to discover a technique known as vernalization (although others had known of the technique earlier). "Vernalization" refers to the requirement that a plant be exposed to a prolonged period of cold during the winter months, in order to induce growth and flowering in the spring and summer. The term also referred to an artificial method of accelerating a plant's growth and development by exposing seeds to moisture, germinating them, and then chilling them for a period of time before planting them in the spring. This treatment caused the plant to develop more quickly and therefore had potential applications in cold climates with short growing seasons. Lysenko's original work was done on peas and then was extended to wheat and other crops. Winter wheat was normally planted in the fall, but it was vulnerable to late spring frosts that killed the young plants. Farmers could adapt by vernalizing winter grains and planting them in the spring after the threat of frost was past. The cereals would still develop quickly enough to be harvested in the summer. An untreated winter cereal would not ripen if planted in the spring, yet if treated it ripened just like spring grains.

The question was whether the technique could be used on a large scale to increase agricultural productivity. Lysenko's method was referred to as "yarovization," from the Russian word for "vernal." Claiming to have invented the technique, he promoted the method as a way to increase yields of cereal crops in the Soviet Union. Soviet "yarovizers" were touting their accomplishments in the early 1930s, boasting that they were able to harvest grain crops in less time, push cultivation of cotton and corn farther north, and speed up plant breeding experiments in greenhouses.[1] What intrigued scientists was the idea that a winter grain could be made to acquire the properties of a spring grain, suggesting that there might be some form of permanent transformation occurring in the plant.[2] Lysenko thought characters acquired during an organism's lifetime could be inherited, and therefore he believed a permanent change was occurring. He is notorious in the history of science, not because of his interest in vernalization but because of the theoretical ideas that accompanied this work and the dogmatism with which he promoted his ideas.

Lysenko's base in the mid-1930s was in Odessa at the All-Union Institute for Genetics and Breeding, where his theories about vernalization, although vague and not well grounded in science, started to gain followers. His experiments, along with other Soviet work on vernalization, were becoming known in the West and around the world through translations of Soviet discussions of the subject. The Imperial Bureau of Plant Genetics in Aberystwyth, Wales, sparked interest in Soviet biology by publishing a series of essays in the 1930s, starting with a review of vernalization and Lysenko's work in 1933.[3] That review garnered attention in Europe, Canada, Australia, Ceylon, Brazil, and China. As a result, the Imperial Bureau published further updates reporting on research in the Soviet Union and elsewhere. It also translated and published a review of the subject by Nikolai Alexandrovich Maximov (1880–1952), a leading Russian plant physiologist, in 1934.[4] Zhores Medvedev, a Russian dissident biologist who wrote about the Lysenko controversy, noted that Maximov "was about the only scientist in 1929–1931 who criticized Lysenko's work on plant development, since he was working in the same area and saw more clearly Lysenko's erroneous tendencies and methodological mistakes."[5] When the eminent agricultural breeder Nikolai Vavilov took an interest in Lysenko's work in the 1920s, even proposing that Lysenko be given a post at Vavilov's All-Union Institute of Plant Industry in Leningrad, Maximov, who was head of plant physiology at the institute, objected and blocked the appointment.[6]

In his review in 1934, Maximov pointed out that a considerable body of scientific research had preceded Lysenko, including work by Maximov himself. Since Lysenko had not yet published a treatise on the theoretical basis of vernalization, a detailed critique of his views was not then possible. Maximov gave Lysenko credit for his contributions to the subject but also pointed out that numerous people in the Soviet Union had challenged Lysenko's views, and he noted that the idea that vernalization actually accelerated plant development might not be correct. Maximov's discussion in the 1930s made Lysenko seem relatively unoriginal. Lysenko's work was considered part of the larger subject of developmental plant physiology, as Nils Roll-Hansen has observed. Although Russian scientific leaders were willing to give him some attention at first, over time biologists became more critical as Lysenko's scientific weaknesses became evident.[7]

Lysenko's theory, known as the "stage theory" or "phasic theory," focused on the environmental conditions (such as cold or heat) that caused a plant to develop through successive stages or phases. Lysenko tried to derive a mathematical law that would explain this developmental process. But he was not knowledgeable about plant physiology or biochemistry and was therefore resistant to alternative explanations that were based on ideas about plant hormones or biochemical processes. Most controversial, at least in Western opinion, was his acceptance of the inheritance of acquired characteristics, or Lamarckian heredity, which Western observers saw as a throwback to nineteenth-century views of heredity that by the 1930s had been rejected by the majority of geneticists.

However, Lysenko presented his theories as being in keeping with Russian and Soviet traditions, especially as represented by the revered Soviet plant breeder Ivan Vladimirovich Michurin, who died in 1935, and who also accepted the inheritance of acquired characters. These Soviet traditions were rooted in nineteenth-century Russian Darwinism, which placed emphasis on the inheritance of acquired characters as one of the causes of evolutionary change.[8] It was often pointed out that Darwin himself accepted the idea that the environment could have a direct inherited effect on organisms. In fact, Darwin's own hypothesis about heredity, called "pangenesis" and proposed in 1868, implied clearly that influences external to an organism could affect what Darwin called the "gemmules" within the organism, or the material substance that carried hereditary traits. Thus Lysenko could present his ideas as part of a long Russian tradition that combined Darwinism, Lamarckism, and Michurinism. He argued that the Soviet approach was more successful in improving crop productivity than methods that followed Western ideas of genetics, which were based on Mendelism. In particular he objected to the Mendelian interpretation of heredity developed by Thomas Hunt Morgan and others, which was based on the idea that hereditary traits were carried by genes located on the chromosomes. Genetic theory, as advanced by Morgan and his collaborators, did not accept the validity of the inheritance of acquired characters. Lysenko and his followers also attacked Morgan's school of genetics because it used the fruit fly *Drosophila* as the model organism and therefore could be dismissed as irrelevant to agricultural productivity.

By the mid-1930s, Lysenko had grown strongly opposed to "Mendelian-Morganist" genetics, insisting that Soviet science, based on Michurinism, was fundamentally different from and superior to Western science. He strove to create a genetic theory in line with Marxian concepts, while his cohort of "yarovization geneticists" dismissed classical Mendelian genetics as nothing more than a game "like chess or football."[9] His theories purported to explain how environmental treatments like vernalization created permanent hereditary effects in cereals, transforming one species into another. Fundamental to his ideas by the late 1930s was belief in the inheritance of acquired characters and denial of the concept of the gene that was so central to Western genetics. The growing dogmatism of the Lysenkoists began to turn the early sympathizers, Vavilov among them, against Lysenko.

Nils Roll-Hansen, in a step-by-step account of the clash between Michurinism and Western genetics, has charted the tightening political control over science in the Soviet Union as Lysenko rose to power in the late 1930s.[10] His analysis emphasizes the interdependence of science and politics to explain how poor science gradually drove out good science. In 1939 Lysenko was elected to the USSR Academy of Sciences. Vavilov was in control of the Academy's Institute of Genetics, as well as the All-Union Institute of Plant Industry, but he was by this time leading the opposition to Lysenko. In 1940 Vavilov was arrested on espionage charges and he died in prison in 1943. A Lysenkoist plant cultivator became head of the Institute of Plant Industry, while Lysenko himself became the new director of the Academy's Institute of Genetics in Moscow.

Until this time, it was still possible to criticize Lysenko, but that changed soon after the war. At a conference held at the Lenin Academy of Agricultural Sciences (of which Lysenko was president) in August 1948, Lysenko presented a long report denouncing Western ideas in genetics. This time Stalin, who was sympathetic to Lamarckian views, gave official state sanction to Lysenko's theories and brooked no opposition. In this way Lysenko was transformed, as Medvedev has described, into a kind of dictator of biology and agricultural sciences.[11] In a detailed analysis, Nikolai Krementsov points out that the mounting tensions of the early Cold War in 1948 (the time of the Berlin blockade) are crucial to understanding Stalin's decision to back Lysenko.[12] Politicians, not scientists, would thenceforth determine what was permissible in scientific thought and practice.

After the 1948 conference, there began a period of repression of Western genetics in the Soviet Union and satellite states in Eastern Europe. Nikita Khrushchev, who quickly achieved preeminence in the Soviet leadership after Stalin's death in 1953, also supported Lysenko from 1956 on. It was not until after Khrushchev was ousted in 1964 that Lysenko was forced to resign from the Institute of Genetics and other Lysenkoists were dismissed from their posts, but it still took many years to erase the damaging effects of Lysenkoism, especially since it had become the basis for biological education. As Roll-Hansen points out, Lysenko did not lose his membership in the Academy of Sciences, and he kept a research institute in Moscow until his death in 1976.

While Lysenko is notorious for his clash with Western genetics, plant physiology also suffered under the hegemony of Lysenkoism. David Joravsky noted its effect on plant physiologists, who were among the first to feel the weight of Lysenkoism.[13] One victim was Maximov. Medvedev reports that in the 1930s Maximov was briefly arrested, banished to Saratov (a closed city), and forced to "confess" his errors in criticizing Lysenko.[14] In Saratov he headed the plant physiology section of the All-Union Institute of Grain Economy and was also in the department of plant physiology at Saratov University.[15] By 1939 he had moved to Moscow to work in the K. A. Timiryazev Institute of Plant Physiology, part of the Academy of Sciences, and was director of the institute at the height of Lysenko's power. (Timiryazev was a highly esteemed Russian biologist of the nineteenth century and a Darwinian.) The scientists at this institute included some of the leading plant physiologists in Russia, and the institute was a hotbed of anti-Lysenkoism. Medvedev recalled that while other institutions, such as the K. A. Timiryazev Moscow Agricultural Academy, were completely transformed by Lysenkoism, the Institute of Plant Physiology's faculty remained mostly opposed to Lysenko: only one of the thirteen voting members of the science council was a follower of Lysenko.[16]

Joravsky reported that when Maximov took exception in 1948 to Lysenko's criticism of the institute's work on plant hormones, he was quickly forced to apologize and recant, promising that the institute would be loyal to the Lysenkoist school. In discussing these events, Joravsky characterized Maximov as an accommodating hypocrite, someone willing to adopt Lysenkoist camouflage in order to keep his position, which Maximov did keep until his death in 1952. Joravsky explained that such behavior was necessary to protect colleagues and that Maximov had little choice. Joravsky also characterized his successor, Andrei Lvovich Kursanov (1902–1999) as a younger version of Maximov, referring to him as a "pliable man of principle."[17] He used these two cases to illustrate the pathological condition of the scientific community. Scientists like Maximov and Kursanov were able to accommodate without succumbing to Lysenkoism, Joravsky argued, while in other cases people surrendered completely to pseudoscience.

Joravsky contrasted physiologists and geneticists on the basis of their different responses to Lysenko. He saw the geneticists as being intransigent in their opposition, with the result that they were "overwhelmed by complete disaster."[18] The physiologists, on the other hand, led by "pliable men of principle," managed to appear accommodating. Joravsky argued that the nature of plant physiology made it possible just to hunker down, do one's empirical research quietly without raising "theoretical red flags at the Lysenkoite bulls," and thereby survive the period of repression in a way that was impossible in genetics.[19] The price was to humiliate oneself by seeming to accept Lysenkoist nonsense.

Krementsov's more recent analysis significantly revises this picture of the demise of genetics. Some geneticists who lost jobs were able to find positions in other departments, and scientists adopted various subterfuges in order to continue their research. In addition, genetics research continued in the context of Soviet atomic bomb development. As Krementsov concludes, "the party monopoly over decision making in science policy, then, did not reduce Soviet scientists to passivity. They retained the ability to influence decision makers at the highest level of the party apparatus, and some managed to so do with great ingenuity."[20] Mark Adams has similarly noted that especially during the time in the Khrushchev era known as the "thaw," lasting from late 1955 to late 1958, many initiatives were taken to restart Soviet genetics and oust Lysenkoism, including some led by physical scientists and mathematicians. Working through informal networks, these scientists maneuvered cleverly to gain resources from the state. Speaking of the leaders of the Soviet Academy of Sciences, Adams writes that by the 1960s they had achieved their agenda for Soviet science: "So long as they stayed clear of politics, science was theirs: they could manage their ever-growing scientific empire, exercise freedom of scientific thought, increasingly travel abroad, and generally enjoy the secure perks of a highly-privileged Soviet caste."[21] Adams was not referring to plant physiologists, but he could have been. In the light of these later analyses, I believe that Joravsky's interpretation of how plant physiologists responded to Lysenko also requires revision. Although they may well have appeared accommodating, as Joravsky suggests, we should also ask what was at stake for them and what they did subsequently. Were they as submissive and self-effacing as Joravsky suggests? Or were they engaged in the kind of empire-building that Adams describes in reference to other scientists?

This is where the Soviet phytotron comes in, for it is the missing piece in the stories of Lysenkoism told so far. It was not just any phytotron but one of the largest in the world. It was built by the very anti-Lysenkoists who had come under fire after the 1948 conference. During the turbulent time after 1948 there was a crackdown at the Timiryazev Institute of Plant Physiology: some lines of research were terminated, others were curbed, and the closing of the institute was discussed.[22] But was the response to lie low and stay quiet? It seems not. By one account Maximov was campaigning for the construction of a phytotron for his institute by 1950, and after his death Kursanov continued this campaign, assisted by Pavel Aleksandrovich Henkel, deputy director of the institute.[23] Although we do not know exactly when the plan to build a phytotron was first proposed, the *Great Soviet Encyclopedia* entry for "Fitotron" says that it was built between 1949 and 1956, placing its beginning soon after the Lysenkoist crackdown and close to the opening of the first phytotron at Caltech.[24] If we accept this date, and if Soviet biologists were hoping to construct a phytotron during Maximov's tenure as the institute's director, then both Maximov and Kursanov would have had good reason to behave in an accommodating manner to protect their mission and gain resources for their research programs. But it was not apparently a quiet accommodation.

The phytotron, called the Artificial Climate Station, opened in 1957 in a suburb of Moscow. It consisted of two buildings, a main building with experimental chambers, laboratories, and offices, and another building for work with radioisotopes. The exterior design was neoclassical, including a grand entryway with imposing columns, very different from the subdued California vernacular of the Earhart laboratory or the modern look of other phytotrons but in keeping with Stalinist architectural styles. (A photograph of the front of the institute is found on Wikimedia; interior photographs appear in an article in *Priroda,* the popular journal of the Academy of Sciences.)[25] The architectural design was probably the work of the Academy of Sciences Design and Research Institute, which dates back to a prewar design organization meant to develop the resources for the Academy of Sciences.[26] In 1953 the Academy of Sciences established a State Institute for Designing Research Institutes, Laboratories, and Research Centers. Its projects included particle accelerators, nuclear reactors, computer centers, telescopes, laser facilities—and phytotrons. Its divisions also designed science cities in various locations, including Novosibirsk's Akademgorodok.

The Moscow phytotron, the model for other phytotrons in the Soviet Union, was highly ambitious. Its director, Ivan Ivanovich Tumanov, described it as wider-reaching than other phytotrons because it was intended to serve experiments in many areas of physiology, including vernalization, photoperiodism, and many other problems relating to plant growth in diverse Soviet environments. Hence its organization was more complex than other laboratories. When Frits Went visited the Soviet phytotron for the first time in 1965, when the laboratory was undergoing renovation, he discovered that none of the copies of his 1957 book on the Earhart laboratory had made it through to the intended recipients. Despite this block in communication, the Soviet achievement impressed Went greatly, and he thought he might benefit from some of the technical innovations that he found in Moscow.[27]

In an article titled "The Soviet Phytotron," published in 1959 in *Priroda,* Tumanov described the advantages of the laboratory. *Priroda* had been taken over by Lysenkoists after 1948, and in the early 1950s it became the main mouthpiece for Lysenko.[28] But Tumanov seems to have used *Priroda* to deliver an oblique anti-Lysenkoist message. His description of the equipment and buildings of the phytotron made two telling points. One was that this laboratory would provide an important corrective to conclusions deriving from field research, which required many years and could be "impeded by many failures." Second, he contended that the phytotron made it possible to "avoid one-sided and simple-minded conclusions."[29] These remarks could be seen as comparable to any phytotronist's boosterism. But might we also see them as a dig at Lysenko's simple-minded conclusions, which were derived from field experience and not careful laboratory study?

Quite apart from the fact that scientists everywhere were becoming excited about phytotrons, the Soviet scientists' early response and ambitious plans suggest that the phytotron was in effect their way of responding to Lysenko through research infrastructure rather than through direct debate, which would have been futile. Lysenko's theories were vague, they lacked scientific backing from disciplines such as physiology, biochemistry, and biophysics, and they were being promoted in a dogmatic manner. The phytotron was about precision experimentation and uniting classical whole-organism science with the newer fields of biochemistry and biophysics. It is possible, therefore, to see the physiologists' strategy in a different, less humiliating light and to propose an alternative hypothesis: every piece of precision equipment in the Soviet phytotron was the physical embodiment of its researchers' determination to bury Lysenko's vague ideas through exact science. The plant physiologists did not react by hunkering down; they built an empire.

Kursanov's mission was to modernize plant physiology by introducing the methods of biochemistry, biophysics, and cytology.[30] The phytotron was a means to this end. The physiologists may not have engaged in theoretical debates with Lysenkoists and were restrained in that sense, but otherwise they were going on the offensive. They could succeed because they too, like Lysenko, were interested in the relations between organisms and environment, and they too were interested in improving Soviet agriculture. As was the case with all other groups working in phytotrons, the organism's development in relation to its environment was the central problem. These goals could be pursued entirely independently of Lysenko and his followers, and they also drew on deep Russian traditions.

An anonymous historical review of Russian and Soviet plant physiology, published in the Soviet journal *Plant Physiology* at the time that the phytotron opened in 1957, illustrates how biologists at the Institute of Plant Physiology linked the need to modernize plant science to the historical strengths of Russian biology as well as the goals of the Soviet state.[31] The article praised the strong Russian traditions in plant and soil science going back to the nineteenth century and to the work of Timiryazev himself, but it also pointed out that botanical research had not been closely related to agricultural needs under the Czarist government. After the Bolshevik Revolution, new emphasis was placed on agriculture. In keeping with this trend, the article noted, plant physiologists started to do more work in the field and to take "an ecological approach to plant physiology."[32] The shift toward ecological plant physiology, the article continued, in turn led to closer ties between plant physiology and agronomy. During the 1920s and 1930s many Soviet physiologists, Maximov and Kursanov among them, contributed simultaneously to ecology and plant physiology with studies on both wild and cultivated plants. The article credited the Michurin doctrine with influencing plant physiology, parenthetically referring to Lysenko, but at the same time linked Michurin's practical goals in plant breeding to the teachings of his predecessor, Timiryazev.

However, the article observed that this shift toward an ecological and agronomical approach had some negative consequences, one being decreased "use of more exact methods of investigation, thus shifting the load to more gross and rapid comparative methods."[33] Under Kursanov's leadership, the article continued, the Institute of Plant Physiology had taken on the task of developing and disseminating new methods of research, including greater attention to biochemistry, a discipline that had been neglected for two decades. The contributions of Soviet physiologists, not only in expanding research efforts but also integrating findings from the study of different physiological processes, were emphasized and again linked back to the traditions established by Timiryazev, who sought to connect research findings into "one great, integrated whole."[34] The article affirmed that Soviet biologists were doing work of great importance in plant physiology, including developing lines of research on topics such as photoperiodism, which, although discovered in the United States, had found "its second home in the Soviet Union."[35] Moreover, it was Soviet research on "physiologically active substances" such as plant hormones that showed the greatest promise for being able to control, retard, or stimulate physiological processes, and this research had also influenced biology in other countries.

Certainly, the writer affirmed, plant physiology should be "closely linked with agronomy." But for exactly this reason, it was also essential "to make a deeper theoretical study of the causative relations and to discover those multiple links of a biochemical and physiological order which lie at the base of living processes."[36] The modernizing goals of Kursanov's institute, which included its new phytotron and greater attention to the study of fundamental processes, were presented as the direct path forward for science but also for meeting the practical goals of the Soviet state and demonstrating leadership in the development of plant physiology in the world. The article's passing reference to Michurinism and Lysenko implied that Lysenko's contributions were relatively minor compared to the cumulative results of other scientists over forty years. And it left no doubt who the leaders of Soviet botanical science really were.

In 1959 the Institute of Plant Physiology hosted a conference on the physiology of plant hardiness in Moscow, bringing together scientists from the Soviet Union and the satellite republics.[37] Maximov had been interested in the resistance of plants to cold and drought, and after his death the Moscow phytotron became known for its "remarkable work" on the physiology of plant resistance to very low temperatures, which was one of Tumanov's interests.[38] This meeting showcased the work being done in the phytotron and it generated demand for similar facilities elsewhere. The conference noted the progress in plant physiology achieved because of the Artificial Climate Station and recommended increased funding of fully equipped physiological laboratories in other locations. Because phytotrons were meant to address agricultural problems, it was easy to link the need for such facilities with the larger goals of the state. Studying plant hardiness and resistance to cold, drought, and salt—subjects central to plant physiology—were all part of the larger goal of increasing crop productivity.

This conference was well timed, coming just a few months before Khrushchev, the Soviet premier, toured the United States in September 1959. Khrushchev was impressed by the agricultural research facilities in the United States, notably those of the USDA in Beltsville, Maryland, where he made a two-hour tour and listened to lectures on chemical control of weeds, photoperiodism, and plant growth regulators. As reported in the *Baltimore Sun,* the scientists were "impressed by the close attention Khrushchev gave them."[39] Superintendent Charles Logan said, "He was quite interested in what we had to present. I think he was impressed. But he doesn't miss a trick to try to impress you back about what he and his country are doing." The Soviet phytotron must have appeared fully competitive with what the Americans were offering.

In addition to using the phytotron to build a modern research program in the Soviet Union, Henkel and Kursanov founded a journal, *Plant Physiology,* in 1954. Starting in 1957, with the help of the American Institute of Biological Sciences, it was translated into English. It published research papers, articles on the latest equipment, and summaries of conferences, as

well as occasional historical articles or biographical sketches celebrating
the progress of Soviet physiology. These publications gave the work of the
Timiryazev Institute prominence, and much of that work would have been
done in the phytotron. Apart from the historical overview in 1957 that mentioned Lysenko in passing, Lysenkoism was absent from the journal. Kursanov published an English-language review of Soviet plant science in 1956
in which he devoted a paragraph to Lysenko's work, making a comment in
passing that Lysenko's stage theory of development had been confirmed by
several authors, but offering no details.[40] Tellingly, he cited just one article
by Lysenko among four hundred citations in total. His review showed that
there was a vast amount of other research being done, much of it at the
Timiryazev Institute. In 1974 a historical article on plant physiology at the
Soviet Academy of Sciences credited Maximov with initiating the first
phytotron in the USSR and Kursanov for modernizing the laboratory and
increasing interest in biochemistry, biophysics, and molecular biology.[41]
The construction of the phytotron was seen as an important event ushering
in a new era of modern science.

By joining the growing worldwide phytotron movement but also
by exerting leadership with a more ambitious laboratory than existed
anywhere else at that time—resulting in a distinctly Soviet enterprise—
physiologists devised an effective strategy for gaining resources for their
research. Other phytotrons followed. The Siberian branch of the Academy
of Sciences built a phytotron at the Institute of Plant Physiology and Biochemistry in Irkutsk (near Lake Baikal) in 1970. In 1977 a large phytotron
was built at the All-Union Institute for Breeding and Genetics in Odessa
(where Lysenko had been), one year after Lysenko's death in 1976. Robert
Downs's survey of phytotrons worldwide listed six in the USSR by 1980.[42]
Over time the laboratories adjusted to fit emerging trends in biology. In
1991 an Artificial Climate Station Biotron, built in Pushchino (about 96
kilometers south of Moscow) as a branch of the Shemyakin and Ovchin-
nikov Institute of Bioorganic Chemistry of the Russian Academy of Sciences, opened to support research on transgenic plants.[43] In this case,
modern genetic engineering was being applied to achieve the long-desired
transformations and improvement of crops.

Perhaps most indicative of the cultural importance of the phytotron in the Soviet Union, compared with elsewhere, is the way it became a household word. In the early 1980s a mass simulation game was played via Soviet television; it was a social science experiment to test how people made economic and management decisions. Participants were asked to simulate the management of an agricultural business and were required to make various decisions and send in their responses by mail in order to win the game. The game was called "Cybernetic Phytotron." Its designers must have assumed that everyone would understand the reference.[44]

I offer these observations about how the Moscow plant physiologists were affected by Lysenkoism to encourage more research on plant physiology and related sciences in the Soviet Union after 1950. One might argue that the phytotronists were successful because there was no incompatibility between their goals and those of Lysenko; improving agricultural productivity was the common objective, and plant physiologists worked on plants of economic importance. From the vantage point of the Soviet state, it would make sense to support all these enterprises and wait to see which paid off. Soviet plant physiology was not a foreign import, like Mendelian genetics; on the contrary, it had a long history extending back into the nineteenth century. There was nothing anti-Soviet about supporting physiological research that had clear agricultural applications.

However, these physiologists had been under attack, their theories had drawn criticism from Lysenko just as they had criticized Lysenko, and they had experienced some of the repressive effects of Lysenko's rise to power. We do not know what was said to sell the idea of the Soviet phytotron to political leaders, or whether hypocritical statements were needed to justify the arguments. But clearly Joravsky's view that the physiologists were "accommodating hypocrites" who hunkered down and kept quiet needs revision. They were, as Krementsov says about Soviet geneticists, quite creative and successful in drawing new resources into their enterprises. Retrospective views of Kursanov by a former student reinforce this judgment by arguing that he was keen to bring Russian science up to par with worldwide plant physiological standards and to counteract the damaging effects of Lysenkoism.[45]

The Hungarian Phytotron: Lysenkoism Evolving

I do not mean to suggest that none of the work being done in the phytotrons of the Soviet Union or satellite states was influenced by Lysenkoist ideas and belief in the inheritance of acquired characters. The Hungarian case is an interesting example of a phytotron project that was promoted by a biologist, Sándor Rajki (1921–2007), who had been a firm follower of Lysenko. He subsequently developed his own theory of metabolism and heredity, which he saw as different from Lysenkoism but also different from standard genetic views of the time. Lysenkoism in Hungary has barely been studied, but Miklós Müller, professor emeritus at Rockefeller University, witnessed many of the aftereffects of Lysenkoism in Hungary and has investigated its history.[46] He points out that Michurinist ideas about plant breeding were already current in Hungary before the Soviet occupation during World War II, so that Lysenko's ideas would not have appeared especially novel to Hungarians. In 1948–1949, a period of political upheaval in Hungary, the Hungarian Academy of Sciences was reorganized and became an organ of the Communist party. One result was the extremely rapid takeover of biology by Lysenkoists under the guise of Michurinism.

Müller notes that there was a range of responses to this takeover. Some geneticists simply adopted Lysenkoist camouflage but continued to use standard breeding methods. However, Rajki, who became a prominent agronomist specializing in wheat, was deeply influenced by Lysenko's ideas, an influence that stayed with him for his entire career. He was also the person who, after many years of effort, brought the first phytotron to Hungary in 1972, partly in the hope of settling some of the controversies that stemmed from his unorthodox views.

Rajki was nearly twenty years younger than the other phytotron leaders I have mentioned. Educated in Hungary at the Agricultural College in Kolozsvár, he received training in classical genetics from his teacher Alajos Mudra, professor of genetics. He was assistant lecturer at the University of Magyaróvár after the war and in 1948 worked briefly in the education department of the Ministry of Agriculture.[47] He then went to the Soviet Union in October 1949 for further studies, during which time he traveled extensively to learn about the Soviet agricultural program. He received his Ph.D. in Moscow from Lysenko's Institute of Genetics, although he did not work under Lysenko himself. Returning to Hungary in 1955, he became the director of the Agricultural Research Institute of the Hungarian Academy of Sciences at Martonvásár, near Budapest, a post which he held until 1980. As Müller notes, he was a firm believer in Lysenkoism, and he remained sympathetic even after Lysenko's downfall, although by then he was proposing his own theory of heredity based on his reaction to new developments in genetics and molecular biology.[48]

Rajki's focus was the development of new wheat varieties, with the goal of doubling Hungarian production of this important crop. He concentrated especially on what he called "autumnization," or the conversion of spring wheat into winter wheat, the converse of Lysenko's vernalization, but with the same emphasis on the inheritance of acquired characters. In his view, environmental effects that were inherited had caused the permanent conversion of spring wheat into winter wheat. In this case, the environmental effect arose from repeated sowing of spring grains in the autumn, but exactly what caused the purported hereditary change was not known. Unlike Lysenko, Rajki was not interested in squelching scientific freedom. He considered his perspective a legitimate scientific stance and invited discussion and debate from scientists who opposed his views, seemingly in the belief that open debate would help to resolve these questions. In 1962 his institute hosted a symposium on wheat breeding and genetics. The opposing views of genetics were divided into two camps, with Rajki supporting a Lysenkoist-Michurinist theory against the "geneticists-formalists," terms that echoed Lysenkoist characterizations of the two sides. Although most participants were Hungarian, there were scientists attending from Yugoslavia, Bulgaria, Czechoslovakia, Romania, Poland, Italy, Great Britain, the USSR, the German Democratic Republic (East Germany), the German Federal Republic (West Germany), and Sweden.

Rajki presented a long paper on his autumnization experiments, in which he concluded that the "cardinal thesis of materialistic [Lysenkoist] biology on the inheritance of acquired properties has been proved."[49] J. Mac Key, from the Swedish Seed Association in Svalöf, found the experimental results themselves of interest but thought the Lysenkoist explanation was difficult to understand or accept.[50] Opposition to Rajki's ideas was not confined to foreign scientists. Rajki pointed out in the discussion that even at Martonvásár "both the representatives of classical and Micurin [*sic*] genetics work beside one another in full creative freedom," and he hoped that the atmosphere of the symposium would reflect a similar freedom of thought.[51] But he agreed that the processes underlying the autumnization process were not perfectly understood, in part because temperature and light were not fully controlled in field studies. Therefore he argued at this symposium for the incorporation of phytotron experiments to obtain a "more complete picture."[52] Rajki would have to wait another decade before his phytotron finally opened, and during that period his position evolved in response to new developments in biology. Since his belief in the inheritance of acquired characters was not in step with new ideas coming from molecular biology, his strategy was to challenge those ideas by branding them as a new type of dogmatism. By attacking some of the central claims of molecular biology, he seemed to think he might create room for alternative theories, including his own.

In 1953, James Watson and Francis Crick had published their discovery of the double helical structure of DNA, and in the following years their model for DNA replication was shown to be correct. At the same time Crick theorized about how the information contained in the DNA molecule could result in the synthesis of proteins through an intermediary molecule, RNA. In a lecture in 1957 and then in publications in 1958 and later, Crick argued that there was a one-way flow of information from DNA to proteins via a carrier molecule (RNA). The term "information" referred to the sequence of the units of these molecules: the sequence of nucleotides in the case of DNA and RNA, and the sequence of amino acids in the case of proteins.[53] Crick asserted that once information had gone from DNA into the protein, it could not get out again; the protein could not pass its information to another protein, or to DNA or RNA. This theory about the one-way flow of information also implied that the inheritance of acquired characters was impossible, although Crick was not concerned with that question. He intended the idea about one-way flow of information to be treated as a hypothesis or an assumption, but he referred to it as a "central dogma," a term that became a fixture in the biological literature. Crick had simply not understood what the word "dogma" meant, he later confessed.

In Rajki's view, Crick was in fact being dogmatic. He had left the impression that biologists were being encouraged to accept his theory, regardless of whether there was evidence to support the hypothesis. To Rajki in the 1960s it seemed that indeed molecular biologists were imposing their views in a high-handed way. In an essay titled "On the Situation in Genetics" in 1966, followed by a short book in English on his autumnization experiments, Rajki challenged this so-called dogma, which was standing in the way of any sympathetic hearing of his own views.[54] If he could defeat the "central dogma," that would open a path to the possibility of the inheritance of acquired characters. He therefore looked for any evidence that seemed to contradict the central dogma.

Although he continued to refer to Lysenko's ideas and the teaching of Michurin, he also drew on other contemporary work that challenged the central dogma, notably the work of the American plant physiologist Barry Commoner. In the early 1960s, Commoner critiqued the accepted idea that the nucleotide sequence of DNA explained biological specificity, or the uniqueness of each organism and species, and instead offered a more complex explanation.[55] Critiques like this provided Rajki with ammunition to support his own unorthodox views. Drawing on Commoner's critique and the scientists he cited, Rajki claimed that his autumnization experiments proved that there was a genuine genetic transformation occurring, which transformed spring wheat into winter wheat through the action of the environment, after the spring wheat was repeatedly sown in the autumn. The organism could not be separated from its conditions of life, he reasoned, and therefore change in the conditions of life resulted in change in the organism.

Rajki acknowledged that the Lysenkoist stance had itself become hardened and dogmatic by 1948, resulting in the repression of classical genetics in the Soviet Union and the abuse of power by Lysenkoists. He viewed this dogmatism as also harmful to the Michurinist genetic viewpoint because it discouraged research. Since 1948, he observed, Lysenko and some of his leading collaborators had grown "more and more detached from direct experimentation which is the vital element of all sciences," with the result that they had become "more or less mere office holders."[56] But he viewed the strong reaction after the fall of Lysenko, when Michurinism was in turn condemned, as an equivalent power play that made it difficult to have free scientific discussion in genetics. The lessons of 1948 had not been learned, he charged, as shown by the reorganization of institutes and dismissals and transfers of scientists in order to reinstate classical and molecular genetics after 1964. His preference was to encourage and develop both genetic points of view equally.

As long as Rajki attached his views to Lysenko's, it was difficult for Western reviewers to take them seriously. G. Ledyard Stebbins, a noted American botanist, dismissed the autumnization studies as a "relic of the discredited Lysenko theory of the inheritance of acquired adaptive modifications."[57] Stebbins complained that although Rajki believed that the DNA code could be altered by "feedback" from the "assimilatory systems" via proteins and RNA, he did not have any concrete proposal about how this feedback would operate. Ralph Riley, a British wheat geneticist, regarded Rajki's writing as a polemic in the Lysenkoist vein and countered that his results with wheat could be explained as a result of genetic mutation, hence falling easily into conventional genetic frameworks.[58] He admitted, however, that experiments on other plant species (flax and tobacco) were puzzling in that they suggested a genetic change was being induced by the environment, but without evidence of the mutations that he thought were occurring in wheat. Riley's recommendation was to repeat the work "outside the sphere of Lysenkoist influence."[59]

Rajki was by this time moving toward an alternative theory, which he called the "metabolism-biochemical concept of heredity" and which was meant to be distinct both from Michurinism and from the molecular gene concept. He still asserted that proteins could be modified by the environment and that this information could be transferred to RNA and DNA. In 1972 he wrote another polemical essay on the subject, "Metabolism and Heredity, or Autumnization as a Microevolution," published in English, Hungarian, and Russian. This time he cited the work of Howard Temin, an American studying the Rous sarcoma virus, an RNA virus. Temin's experiments seemed to contradict the central dogma because the results suggested that RNA could synthesize DNA. Although Temin's ideas were at first greeted with much skepticism, in 1970 he and a postdoctoral fellow, Satoshi Mizutani, identified an enzyme in the virus (called reverse transcriptase) that provided a way to transcribe viral RNA into DNA.[60] These results were independently confirmed by David Baltimore the same year. In Rajki's view, Temin's results meant that it was no longer possible to ignore evidence that the central dogma might be wrong. Crick continued to defend his "dogma" because Temin's work on viruses was seen as exceptional.[61]

Precisely because his views were controversial, Rajki saw the benefits of having a phytotron to test them in the hope of confirming them.[62] Although the idea of building a phytotron was first broached in 1959, in the 1960s it was only possible to build smaller homemade growth chambers. By 1969 the Hungarian Academy of Sciences decided to build a full phytotron, in this case using equipment supplied by the Canadian company Conviron. Rajki had contacted the company through an intermediary in 1967, and negotiations began to decide on the design and the equipment.[63] After representatives from Conviron visited Hungary to explain what they could offer, Rajki went to Canada to visit the phytotrons at the University of Guelph (Ontario), Agriculture Canada Ottawa, Agriculture Canada Winnipeg, and the Conviron factory in Winnipeg. He decided he wanted the Hungarian phytotron to be like the one in Guelph; the Canadian company would supply packaged growth rooms and chambers that would be assembled on site, while the Hungarians would be responsible for the building itself.

Construction began in April 1971 and the phytotron opened in November 1972, representing an investment of one million dollars. For Conviron it was seen as a "benchmark installation which set the standards for controlled environment facilities."[64] The phytotron, although originally meant for researchers at the Martonvásár Institute, was open to scientists from any country. In his 1972 essay Rajki hoped that the new phytotron, in the planning stages for over a decade, would help him to make the exact tests that would prove his theory. His enthusiasm for the phytotron reflected the same idea expressed elsewhere, that greater precision could help to resolve controversial questions. But for him, resolution did not mean giving up his belief in the inheritance of acquired characters. In 1973 he was publicizing the phytotron's early research program in a way that revealed the lingering imprint of Lysenkoism, writing that his own research was "built essentially on our metabolism-biochemical concept of heredity, being developed on the basis of the Lamarckian-Darwinian-Michurinian theory."[65] He pointed out that the work included evaluation of experiments by researchers holding "genetic views opposed to ours," but clearly he did not anticipate having to give up his views to achieve the much-sought resolution. The phytotron also served the practical needs of plant breeding and production.

Temin went on to receive a Nobel Prize for his work in 1975, but Rajki was not successful in persuading his opponents, who continued to view his theories as examples of an outmoded line of thought. The phytotron did not lead to the scientific breakthrough Rajki hoped for, but it still found its place in experimental biology just as it did in other countries. In this case, the phytotron helped especially with studies of how cereal crops adapted to the stress of low temperatures. Rajki eventually became disillusioned by Communist rule in Hungary, after the state made trumped-up charges against a colleague who ran an experimental farm. In 1979 he resigned from the Communist party, knowing that this would destroy his scientific career, and after his retirement in 1982 he spent most of his time abroad, especially in Italy and the United States. After his death in 2007, Rajki was remembered by his student Zoltan Bedő for his work on improved wheat varieties, for the phytotron, and for the research community he established at Martonvásár, where he "constantly encouraged us to ignore fashionable research topics, where it was easy to make headway, and to give our attention to problems that appeared to be insoluble and represented a real challenge."[66]

Whereas the Soviet phytotron in Moscow can be seen as a way to overwhelm and render irrelevant Lysenkoist dogmas that threatened the progress of Soviet biology, the Hungarian phytotron was a way to clarify and advance a line of research on metabolism and heredity that had originated in Michurinist-Lysenkoist theories but evolved in the wake of advances in molecular biology. It is impossible in retrospect to figure out exactly what might have been the basis for Rajki's conviction that he was right, given the lack of knowledge at the time of gene expression and gene regulation. But it is also important to bear in mind that phytotron-based experiments sometimes yielded results that did point toward the inheritance of acquired characters, even though interpretations of such results did not challenge standard genetic knowledge. Rajki's claims did have intriguing parallels in other studies.

This was the case even in experiments done in Caltech's phytotron. One example came from studies of plants in maladaptive environments, where they were kept in constant conditions in order to chart the deleterious effects of non-varying conditions. Many plants required variation in their environments to thrive. Harry Highkin at Caltech investigated the opposite condition, the damaging effect of constant temperature, which inhibited plant growth.[67] These experiments, performed over several years on peas, turned up two surprising results. First, the inhibitory effect of constant temperature appeared to accumulate over five generations. Second, when the plants were returned to a fluctuating environment, the inhibitory effect carried forward to a plant's offspring through three or four generations.

This result suggested that a plant's growth was affected not only by its own environment but also by the environment of the parental generations. Highkin had no explanation for this effect, but he compared it to what the German-American zoologist Victor Jollos in 1921 had called "lasting modifications" in studies of protozoa.[68] Jollos had also experimented on plants, showing that treatment with chloral hydrate caused changes in the plant that persisted over several generations, after which the plants eventually returned to normal. Highkin suggested that a constant environment was abnormal and comparable to the toxic chemical that Jollos had used. He related the toxic effect tentatively to the plant's "biological clock," arguing that a constant environment was incompatible with the circadian clock operating within the plant.

Alan Durrant, working on the effect of mineral nutrition on flax at the University College of Wales, Aberystwyth, had found a similar effect extending across several generations. He hypothesized that changes in the cytoplasm had caused changes in the nucleus, or an altered balance between cytoplasm and nucleus. Durrant's experiments had started in 1953 on several varieties of flax and linseed, one goal being to examine gene-environment interactions. He found that applying fertilizer to a single variety of flax produced inherited changes. He subsequently spent a year at the Caltech phytotron to test these results, which were exciting attention because of their possible relationship to Lysenko's controversial claims about transforming species of wheat. By 1962 Durrant was confident that the environment was producing heritable changes lasting several generations, but he demurred on the question of whether his results supported the findings of "eastern geneticists."[69] He argued that in flax there were three types, termed genotrophs. One of the genotrophs was plastic and could be changed into either of the other forms, depending on fertilizer treatment. Anton Lang, who replaced Went as the Caltech phytotron's director, believed that the extended effects probably came from changes in gene expression, but not in the gene itself.[70] At any rate, as he felt obliged to point out, these experiments gave no support to Lysenko's theories. Durrant's later interpretations supported Lang's: in 1981 he argued that these were unstable genotypes, by then well known in cultivated plants, and that the changes occurring were changes in gene regulation, which could be reversible or maintained for indefinite periods.[71]

Although Rajki's work was clearly aligned with the Michurinist-Lysenkoist camp, it is not necessarily the case that his experimental results were fraudulent or his claims completely bogus, but they were dismissed at the time because of his habit of invoking Lysenko. Other phytotron experiments were also turning up puzzling results, but in those cases the scientists were careful to distance themselves from Lysenko, even while admitting they did not have a full explanation of what was occurring. Anton Lang's suggestion that these results were probably the result of changes in gene expression is likely to come closest to explaining what was occurring in Rajki's experiments.

Today the study of such mechanisms falls within the field of epigenetics, which refers to changes in gene expression that do not involve a change in the sequence of the bases of DNA, and it is also well recognized that such changes can be responses to environmental stress. Such modifications can be inherited through mitotic cell division, and in plants they are also inherited through meiotic cell division, providing a mechanism for transmitting an environmentally induced change to the next generation. Barbara McClintock's discovery of transposable "controlling elements" in maize, for which she shared the Nobel Prize in 1983, is an example of an epigenetic mechanism, and these mechanisms have been found also in cereal crops such as rice and wheat. It might be the case that Rajki was observing an inherited epigenetic effect, although we cannot be sure. It is interesting to see that the growth of epigenetics has also stimulated historical revisions of Lamarckian theory and Lamarck's legacy. These revisions have provided opportunities to reframe the history of Lamarckism: Lamarckian heredity can be viewed not so much as a failed theory that fell into disrepute, but as a theory that has undergone various transformations as scientific knowledge has improved.[72]

Conclusion

By considering the history of biology through the lens of phytotronics, we are encouraged to view the development of modern plant science from an international and comparative perspective. One common theme is the way botanists fought for greater resources for the pursuit of what they considered to be "basic" science. While support for the study of economically important plants could be taken for granted, for most first-generation phytotronists the support of basic research was less certain, especially within plant sciences. Therefore the campaigns to build phytotrons emphasized the reasons to foster basic science as the surest route to solving practical problems. The phytotron campaigns were also a means of generating support for biological disciplines that had been neglected due to overemphasis on agricultural improvement. This motive was especially important for the champions of the Soviet phytotron, but we can see these arguments being made in other countries too. The phytotron was a vehicle for linking basic and applied science and for demonstrating that basic research should not be understood as disinterested science but as essential for progress on any applied problem.

The individual case studies reveal how the science that involved phytotrons also reflected different national goals. Each of the case studies reveals something different about what phytotrons meant for the communities and nations that undertook these projects. The most striking feature of this history is that phytotron construction quickly expanded into an international movement. Scientists worked to form cooperative social networks and used organizations like UNESCO to further their common goals. Around the world there were successful efforts to bring resources into whole-organism biology through promotion of laboratories like phytotrons, which dared to embrace the idea of Big Science. For such enterprises to succeed, it was important to engage in community-building activities that emphasized cooperative relationships between different groups and disciplines. It was also important to link communities that had different roles in these enterprises: engineers, technicians, and all those who designed, built, and operated the instruments, and the scientists from various disciplines who created the research programs. These efforts to communicate, share knowledge, and build an international network meant that very soon after the opening of the first phytotron in 1949, a full-scale international movement arose. As a result, plant sciences were transformed.

Phytotronics and related laboratory innovations were also meant to improve ecology as an experimental science, and, as we saw in chapter 4, the movement stimulated discussions about the need for special laboratories designed for ecological research. Part 3 examines how ecological science expanded its scope, starting in the 1950s, and looks at the many innovations that supported this expansion. The field of physiological plant ecology, which seeks to explain the adaptive significance of traits, developed in the postwar decades into a large and vibrant branch of ecology. Part 3 charts this progress through selected case studies; all involve exploration of the relationship between field and laboratory research, as well as the fundamental problem of synthesis.

PART THREE

Synthesis

Physiological Ecology Comes of Age

DURING THE 1950S AND 1960s physiological plant ecology developed into a major field in ecology. Although plant physiology and ecology were closely linked in the late nineteenth century, the botanists who first took physiological studies into the field to explore problems of adaptation initially viewed this approach as a way of broadening the scope of physiology rather than a way to define a new discipline of ecology. Physiological plant ecology emerged only later as a recognized field in the discipline of ecology, starting in the 1950s. Its emergence coincided with the development of population ecology and ecosystem ecology, and one problem that ecologists confronted was how interpretations at these different levels could be integrated.

Frode Eckardt coined the term "ecophysiology" in 1962 at a conference held in Montpellier, France, and in America the term "physiological ecology" came into broad use in the 1960s as well. William Dwight Billings, an American ecologist, thought the two terms meant the same thing; both referred to the study of organisms in nature and their relations to the environment and other organisms.[1] Billings perceived physiological plant ecology as a convergence of disciplines that had been building for several decades. Part 3 examines several connected case studies that illustrate key challenges that faced physiological ecologists, especially during the decades from the 1950s to the 1970s that saw the emergence and maturation of this field. Biologists met these challenges in part through innovations in experimental technique, through cross-disciplinary collaborations, and through efforts to reconcile laboratory and field studies. These activities also entailed the problem of synthesis; I use the term "synthesis" to embrace all activities that seek linkages across disciplines and hierarchical levels, as well as efforts to reconcile different interpretations.

In exploring the multiple ways in which botanists tackled the problem of synthesis, Part 3 considers not only major laboratories like the Earhart laboratory at Caltech, the world's first phytotron, but also smaller-scale controlled-environment growth chambers, laboratories designed specifically for ecological research, and mobile laboratories that took scientists into the field. Chapters 6 to 8 take up different ecological themes, carrying the story past Frits Went's tenure at Caltech and into the next stages of his career at the Missouri Botanical Garden in St. Louis (1958–1963), Washington University in St. Louis (1963–1965), and the Desert Research Institute in Reno, Nevada (1965–1985). Although Went's interests are woven into these stories, I also discuss new lines of research that were engaging ecologists and evolutionary biologists in the second half of the twentieth century. I take a close look at particular ecological problems that required not only a blend of laboratory and field research but also long-term commitments to the pursuit of those research questions.

Each case study looks at the science of adaptation from a different perspective. My goal is to show how the practice of ecology was changing in this postwar period, when the discipline of ecology was maturing at the dawn of the modern environmental age. As new techniques and new laboratories entered the scene, ecology increasingly required a sophisticated balancing act that incorporated observations and experiments in the field and the laboratory. Physiological ecologists moved continually from one to the other and back again. Even when scientific controversies were not quickly resolved, new ideas and discoveries energized new fields of research and expanded ecology's scope.

This chapter examines the widening reach of physiological plant ecology through case studies that gave rise to controversial interpretations. These involve plant distribution and the idea that some plants may produce toxic substances to ward off other plants, the adaptive significance of plant emissions of volatile organic compounds, and discussion about whether such emissions can affect the atmosphere and possibly the climate. These fields of research illustrate how physiological plant ecology expanded in the postwar period and how it became entwined with broader studies of eco-system function and global climate change, or what is known as "earth sys-tem science." The turn toward earth system science was also a leap into Big Science. I consider two projects that exemplify different visions of larger-scale ecology, both involving controlled experiments at the ecosystem level. One was the Biosphere 2 Laboratory during the time it was operated by Columbia University (1996–2003), and the second is a project in Colorado that studies biological, hydrological, and atmosphere interactions.

My case studies bring to light two general themes. The first is the dif-ficulty of translating or extrapolating experimental results from greenhouse and laboratory experiments to the natural environment, or, in short, figuring out the *ecological significance* of laboratory studies. The second is the chal-lenge of linking physiological ecology to larger-scale problems involving the structure and function of ecosystems, and changes in those systems over time. Physiological ecology is the foundation for all of ecology, and therefore its discoveries and conclusions must be connected to the ecological under-standing of groups, communities, and ecosystems. Making this connection is not easy. Each chapter will explore different ways that ecologists, along with colleagues from many other disciplines, have tried to forge those links.

Defining a New Field

William Dwight Billings (1910–1997), one of the fathers of physiological plant ecology in the United States, published reviews of the field at intervals during his long career. His perceptions, which align with those of Went, form the starting point for this chapter. He was a graduate of Duke Univer-sity, an early center for botanical research, and obtained his Ph.D. there in 1936, working under Henry J. Oosting. From 1938 to 1952 he was on the biology faculty of the University of Nevada in Reno, where his explorations of the surrounding region strengthened his interest in the ecology of deserts and mountains. In fact, Went's suggestion that the desert taxa east of the Sierra Nevada possibly gave rise to certain alpine taxa in that mountain range stimulated Billings's interest in exploring the physiological require-ments that would be needed for such an evolutionary connection. These were long-term studies that occupied most of his career; much later (in 1980) he remarked that "the work is far from finished."[2] In 1952 he returned to Duke and over the next three decades helped to build the botany department into one of the premier centers of physiological plant ecology.[3] Duke acquired a phytotron in 1968, which along with a twin phytotron run by North Carolina State University formed the Southeastern Plant Environmental Laboratories. Duke botanist Paul Jackson Kramer led the campaign to build these phytotrons, having been inspired by a sabbatical semester at Caltech in the 1950s.[4]

In 1957 Billings published an essay that highlighted some of the research areas of interest in the emerging field of physiological ecology. James Bonner, who was on the editorial board of the *Annual Review of Plant Physiology*, had taken notice of Billings's work and invited him to write an essay defining this new field.[5] Billings obliged but observed that physiological ecology was "as broad and diverse a subject matter in biology as it is possible to imagine."[6] It was not a narrow or highly specialized field centered on specific problems or questions. Billings perceived the field as emerging from the convergence of disciplines and identified three main sources as plant geography, plant physiology, and ecology. He recognized that there were also contributions from geology, climatology, soil science, and agronomy. As my case studies will show, other disciplines contributed as well. The point is that this discipline arose from a convergence or synthesis of disciplines that occurred along several fronts.

In 1957, Billings did not analyze what was promoting this convergence, but he returned to this topic in 1985 in a historical sketch of the field.[7] First, he identified the importance of new sources of patronage that had appeared early in the twentieth century to support scientific programs combining plant physiology and ecology. Of particular significance for the later development of physiological ecology was the Carnegie Institution of Washington's Desert Botanical Laboratory, founded in 1903 outside Tucson, Arizona. (The Carnegie Institution of Washington was renamed the Carnegie Institution for Science in 2007, but I will use its historical name.) Carnegie's early decision to make botany one of its priorities for research was crucial in lending support to the new science of ecology. Second, Billings noted the importance of instrumentation and laboratory innovations, especially in the decades following World War II, and he gave credit to Went and his phytotron, among other innovators and innovations, for helping to advance the field. Third, he observed that from the 1920s to the 1940s there were many plant physiologists who focused on whole-plant research and plant-environment relationships and who had strong ecological interests. Among these he included Frits Went and James Bonner, and he credited this group with having the "breadth of training and knowledge to bring physiology and ecology together."[8] Paul Kramer would also fall within this group, as would most of the scientists who led the campaigns to build the first phytotrons around the world.

Billings's observations support the thesis that I advance in this book, namely that Went can be considered one of the immediate ancestors of the modern field of physiological ecology. I will explore several examples of ecological research that combined laboratory and field investigations and that involved Went in some way. Although Went's stamp is on the problems to be discussed here, many lines of research that had different intellectual and institutional origins contributed to physiological plant ecology. There remains much for future scholars to explore. To gain a sense of that broader history, good starting places, in addition to Billings's two essays in 1957 and 1985, are periodic assessments of the state of the field, biographical essays, and literature reviews by some of the leaders in this field.[9] As these sources emphasize, the development of this field benefited greatly from improvements in laboratory design and instrumentation from the mid-twentieth century onward. The link between the emergence of physiological plant ecology and improvements in laboratory designs and instruments will be central to the case studies of these three chapters, and this link helps to explain why this field gained prominence in the postwar decades. Although one could consider this convergence of disciplines as a gradual development occurring over the course of the twentieth century, I will focus on how changes in the postwar landscape of science gave particular impetus to the growth of this field.

By the 1960s, Billings had come to realize that understanding adaptation and how organisms interacted with their environments and with other

organisms—the central problems of physiological ecology—were also foun-
dational problems for all of ecology. As he put it, "Physiological ecology is
not a science in isolation but an integral part of the whole of ecology. If I did
not realize it earlier, it was brought home to me by F. H. Bormann in 1964
during a midnight walk around the countryside at Brookhaven while we
were both spending some time in George Woodwell's laboratory. His arguments were logical and persuasive; one should realize that physiological
ecology is just one of the bonds that hold ecology together in its attempt to
understand the biosphere."[10] In 1982 Billings elaborated on this insight and
identified two major roles for physiological ecology in general ecology.[11]
One was understanding "what makes an ecosystem tick" or how the function of the entire ecosystem depends on the activities of its component organisms. The second involved trying to predict, as far as was possible, what
would happen if the biosphere changed, causing migrations, extinctions, or
shifts in distribution of species that would also affect how ecosystems functioned. Physiological ecology was important, therefore, not just for understanding the world as it was, but what it might become.

Laboratories on the Move

Frits Went was always thinking of ways to improve laboratory facilities for
biologists, and when it came to ecology, these innovations extended beyond
phytotrons to mobile laboratories that could travel to remote locations and
support experimental work in the field. When Went started his ecological
studies of the desert, about 1941, the United States had just lost its longest-
lived research center for desert biology, the Carnegie Institution's Desert
Botanical Laboratory. This laboratory, founded in 1903 and located a couple of miles outside Tucson, Arizona, had grown into a vibrant research
community that supported scientific work extending from plant physiology
to ecology. The high quality of research done there helped to bring credibility to the new science of ecology that was emerging in the early twentieth
century. Some of the most prominent ecologists in the country, such as
Frederic Clements and Forrest Shreve, were associated with the Desert Botanical Laboratory.[12]

However, the Carnegie Institution had shifted its priorities in the late 1920s and built a larger laboratory at Stanford University, which became its center for experimental botanical research under the direction of Herman A. Spoehr, a plant physiologist and specialist in the study of photosynthesis. The Tucson laboratory under Forrest Shreve's direction continued to be a home base for the study of desert ecology for another decade, but a fire in one of the buildings in 1939 provided an excuse to close the lab in 1940, and it was handed over to the U.S. Forest Service, under the Department of Agriculture. Shreve wrapped up his long-term desert studies and retired in 1946.

Went believed the loss of the desert lab had left a void in American studies of desert ecology, which by the 1950s was a subject of growing importance, given worldwide interest in the problems of arid regions. But rather than recreating a similar facility in the desert, he had a different idea: to build a mobile laboratory that could travel widely and take advantage of sudden changes in vegetation, as might occur after a heavy rainfall. The scientists at the desert laboratory had roamed the desert regions of the United States and Mexico, conducting research over a large area. Went proposed the idea of making the laboratory itself mobile, so that experimental studies could be done in the field. His idea was realized with the creation of Caltech's Mobile Desert Laboratory in the summer of 1956 (figure 9).

The mobile laboratory was funded by a donation from Pearl McCallum McManus of Palm Springs, California. She and her husband, Austin McManus, had helped to develop Palm Springs into a fashionable resort town, and after her husband's death in 1955 she provided the funds for the mobile laboratory that Went had been trying to raise for a long time. The laboratory, which was based near Palm Springs, consisted of a combination truck and house-trailer. The house-trailer furnished creature comforts for the scientists with a kitchen, stove, refrigerator, shower, and air conditioning but also served as a laboratory with benches and instruments. The truck, with an electric generator, large water tank, and air compressor, supplied power for the laboratory. Lloyd Tevis, a specialist in ant ecology from Caltech, was in charge of the lab. As he explained, the lab's goals were to complement the work of the Earhart laboratory but also to attract workers

Figure 9. The first mobile laboratory that Went built for ecological research at Caltech in the 1950s. (From *Engineering and Science* 20, no. 5 [1957]: 23; resolver. caltech.edu/CaltechES:20.5.0)

from all over the world to study desert ecology in a region of exceptional diversity. Because it encompassed large areas of protected parks and national monuments, the region offered many opportunities for ecological studies, both zoological and botanical, in relatively pristine environments.[13]

    After Went left Caltech in 1958 and became director of the Missouri Botanical Garden in St. Louis, his ambitions took new directions aimed at public education, but his research interests in ecology continued. He had an expansive vision for the botanical garden, wanting to develop it both as a cultural center for St. Louis and as a research center. As a public attraction he built the Climatron, a dramatic exhibition greenhouse that opened in 1960 and was designed to exhibit plants from different climatic regions.[14] He was clearly thinking about educational reforms along several lines, for in the same year the National Science Foundation (NSF) awarded him a grant for the design of a simple classroom plant growth chamber.[15] To help with the research side of the garden's work, he also created another mobile laboratory, constructed in 1963 along the lines of the Caltech laboratory. By this time he had grown frustrated by the slow pace of the garden's development and did not see eye to eye with its trustees. He resigned his directorship in 1963. He remained in St. Louis on the faculty of Washington University for the next two years, and the mobile laboratory, with financial support from NSF, was taken over by the university.[16]

In 1965, Went moved to the Desert Research Institute, founded in 1960 at the University of Nevada in Reno, to head up its new Laboratory of Desert Biology, which he proceeded to fashion into an updated version of the Carnegie Institution's former desert laboratory. The University of Nevada had at best a middling reputation in 1960, and the new Desert Research Institute was intended to bolster its prestige. Its first director, meteorologist Wendell A. Mordy, was determined to create a top-notch research center. Fields of research initially encompassed atmospheric physics, desert biology, water resources, and anthropology but also extended to medical and industrial research. The connecting link was research related to the Nevada environment. Assisted by generous funding from the Nevada-based foundation created after the death of "Yeast King" Max C. Fleischmann, the Desert Research Institute prospered under Mordy's leadership. In 1968 the institute became a separate division of the state university system and could operate independently of the campuses in Reno and Las Vegas. Despite this promising start, Mordy's tenure as director did not last much longer. As funding sources dried up, he clashed with the university's new chancellor and was forced to resign in 1969.[17]

To build his institute, Mordy had sought out scientific stars to head each of the labs. He was good at sensing when people were ready to leave their home institutions and pouncing with an attractive offer; that is how he snagged Frits Went in 1965. Went in turn brought substantial grants to Nevada from NSF. He built air-conditioned greenhouses, a controlled-environment laboratory on a smaller scale than the phytotron, and not just one mobile lab but a complex of mobile laboratories and living trailers designed to study a variety of problems in biology across several disciplines.[18] The mobile laboratories could accommodate groups of five to ten researchers and could travel widely. Among them was the same trailer laboratory that he had commissioned while at the Missouri Botanical Garden. In 1966 another trailer laboratory, also funded by NSF, began operation. The mobile laboratory complex included a living trailer, trucks serving as traction units for the trailers, and a gasoline tank trailer. Finally, a set of air-conditioned greenhouses with a central head-house was built on the Reno campus next to the Desert Research Institute's building. The two laboratory trailers could be stationed on campus, connected to the head-house, and available to the scientists working at Went's desert laboratory. Clearly Went had lost none of his entrepreneurial energy in the move from St. Louis, and yet in addition to his administrative work he maintained his research interests. The following sections consider ecological problems that started with Went's research at Caltech and continued through his years at Reno.

How Do Plants Interact?

One of the reasons Went gave for building the world's first phytotron was to support ecological research.[19] Many of the physiological problems studied were linked to ecology because they involved responses of plants to environmental conditions. Several of the practical problems had ecological aspects—for example, projects that focused on how to protect landscapes threatened by fire and erosion or studies of the impact of photochemical smog on vegetation. Ecological studies remained a core interest throughout Went's long career.

Starting in 1945, Went and his collaborators made several trips to Joshua Tree National Monument, Death Valley, and the Coachella Valley to study plant distribution and abundance, one of the central problems of ecology. One series of investigations focused on germination of seeds, or the conditions of rainfall and temperature that caused sudden blooms of desert annuals, carpeting the ground with millions of fragrant flowers. His collaborators included Mogens Westergaard, a Danish plant geneticist, Marcella Juhren of Caltech, and Edwin Phillips, botanist at Pomona College in Claremont, California. Their inquiries combined field observations with experimental studies at the Clark greenhouses and later the Earhart laboratory. The phytotron studies were meant to pin down what triggered germination, because control over germination appeared to be a key factor determining the distribution range of annual plants. In research that continued into the

1950s, Went pursued the idea that it was important to know what caused or inhibited germination, rather than to look to other causes such as competition or differential dying of seedlings and older plants. He would use these studies to develop an unorthodox view of the Darwinian struggle for existence that greatly diminished the role of competition in plant ecology and evolution.[20] The flowers blooming together in the desert, he thought, did not appear to compete with each other, for they all flourished, although they remained small. What he saw in the desert was not a war to eliminate the less fit but a general sharing of resources by millions of tiny flowers.

Another way to tackle the question of distribution and abundance is to ask how species form associations or communities, or what kind of influence one species has on another. Went had studied the mutual influence between plants in the tropics during the five years he spent at the Dutch botanical garden in Java before coming to Caltech.[21] He was impressed by the way that epiphytes (plants, such as orchids, that grow on other plants) were found in association with particular host trees. Some grew exclusively on certain trees. He concluded that this close relationship was not caused by the physical properties of the host or by microclimatic differences in the habitat but was instead the result of another cause, which he thought might be chemical. This idea led him to ask whether other kinds of associations between plant species might be the result of some type of mutual influence, perhaps also chemical in nature.

This investigation required looking at the relationship between annual plants and shrubs in the Colorado and Mojave deserts, which Went started studying in 1941.[22] Some annuals seemed to have close relationships to shrubs, whereas others grew in open spaces between shrubs. There were several possible reasons for the observed associations. The shrub might offer a favorable microclimate, such as shade. Environmental conditions, such as wind, might cause organic matter to accumulate beside shrubs, favoring seed germination. Local disturbances, such as floods, might produce distributions different from those in undisturbed environments. Finally, the shrub might be conferring some benefit to the annual plant that was not physical in nature. Went selected an undisturbed location for deeper study and surveyed other regions less extensively for comparison. He concluded that there was a strong pattern of association between annuals and certain shrubs. For example, *Rafinesquia neomexicana* (desert chicory) occurred often in connection with *Krameria canescens* (rhatany), and *Delphinium parishii* (desert larkspur) occurred near shrubs that were unfavorable for *Phacelia distans* (distant scorpionweed).

Went zeroed in on two main causes explaining plant associations. First, detritus from shrubs supported many annuals, which explained why some shrubs became better media after their death. Second, he concluded that the living shrub could be giving off materials that had a specific effect on other plants. At this stage he did not guess what that material might be, for this would require more study "both in natural surroundings and in the laboratory."[23] Went was well aware that animals also affected plant distributions and did not claim that the relationship between annuals and shrubs was the only, or even the most important, explanation for the distribution of desert annuals.

The answer proved elusive, however. In 1945, Went and his co-author Harlan Lewis admitted that they could not correlate plant responses under controlled conditions with plant responses in nature.[24] The translation of results from the highly controlled environment of the laboratory to the field would continue to be a problem after the Earhart laboratory opened in 1949. Asking too broad a question about why plants were found together stymied their efforts to find conclusive answers in the laboratory, because there were too many variables operating. A more focused approach was needed. One observation suggested a more precise question that appeared amenable to laboratory analysis: one desert shrub, *Encelia farinosa* (brittlebush), did not harbor annuals that were commonly found in association with other shrubs. Was it producing a toxic substance that inhibited growth?

James Bonner, Went's colleague at Caltech, took up this problem. The idea that plants might inhibit surrounding plants was not new, for botanists had proposed in the nineteenth century that plants gave off toxic substances, and in the early twentieth century some were trying to identify the toxic compounds. In 1937 the Austrian plant physiologist Hans Molisch coined the term "allelopathy" to describe chemical interactions among plants. Although his interests mainly focused on the effects of ethylene (a hormone), his term came to define the study of plant toxins and their effects on other plants. At that time, the notion that such toxic effects were significant in limiting plant growth was not widely accepted. Since the 1850s, the dominant view was that propounded by the chemist Justus von Liebig, who had argued that mineral content of the soil was much more important for the growth of plants. Although interest in toxic secretions revived in the early twentieth century, even into the mid-twentieth century scientists focused on mineral nutrition or else competition for water rather than on toxic secretions.[25]

Bonner's interest in reviving the question of toxic secretions came about as a result of a wartime project on the growth of guayule plants, which were being studied as a source of rubber. He found that they produced organic compounds (such as transcinnamic acid) that inhibited the growth of guayule seedlings, but he concluded that this toxic effect was not important in agricultural settings, where guayule was growing successfully. The effect seemed to be limited to pot culture.[26] Working with Reed Gray, a biochemist, he investigated whether a similar toxic compound was being produced in *Encelia*. Bonner and Gray found that the leaves had a strong and quickly acting inhibitory effect on plants such as tomato, pepper, and corn, but the substance did not inhibit certain other plants.[27] Bonner suggested that such inhibition might explain why *Encelia* was not associated with desert annuals in the field. Other studies suggested that *Artemisia absinthium* (wormwood) also contained an inhibitor in its leaves that affected the growth of other species. Bonner proposed in 1950 that the production of toxic substances that remained in the soil might have ecological significance, might be more common than was thought, and therefore was worth further investigation.[28]

Connecting Lab and Field

Despite Bonner's confident assertions, there was a possibility that these laboratory studies had no relevance to field conditions and therefore lacked ecological significance. Cornelius H. Muller (1909–1997), a plant ecologist at the University of California, Santa Barbara, was skeptical of Bonner and Gray's hypotheses and decided to take a closer look. Muller had also done research on guayule for the U.S. Bureau of Plant Industry during the war. In 1949 he investigated the same regions and species that Went, Bonner, and Gray had studied. In the mid-1950s he confirmed that Went's pattern of associations was correct and that Bonner and Gray's identification of a toxic chemical produced by the leaves of *Encelia* was also correct. In fact he found that another shrub, *Franseria dumosa* (white bursage, synonym *Ambrosia dumosa*), which was associated with annual plants, produced an even more toxic substance. But there remained the question of showing *ecological* significance: did these toxins have a role in determining species distribution or the succession of species?

Muller looked carefully not just at the general patterns of plant associations but at the occasional instance where an association deviated from the common pattern. In his initial studies, some conducted with his colleague Walter H. Muller (no relation), he concluded that chemical toxicity was *not* ecologically important in explaining these plant associations.[29] He thought that the growth habits and longevity of the two shrubs were more important. *Encelia* grew higher off the ground than *Franseria* and also had a shorter lifespan. As a result, it did not accumulate organic debris in the soil around its base. It appeared that the accumulation of organic debris, rather than presence or absence of toxic chemicals, determined whether annuals could grow around these shrubs. He postulated that the toxic substances were either broken down by microbes or became attached to the surface of soil particles and were not affecting other plants.

Harold A. (Hal) Mooney, who later became a leading physiological ecologist at Stanford University, was a sophomore at the University of California, Santa Barbara, at the time that Muller was doing this research. He recalled that Muller liked to make his students feel part of the "grand battles" in which he was always engaged. Muller used this battle to demonstrate "that there was no substitute for good field observations and that the sophisticated chemical analytical techniques used by Gray & Bonner were inappropriate and had led to an erroneous conclusion."[30] Mooney's studies were interrupted for two years when he was drafted into the army, and on his return to Santa Barbara he found a surprising change: "C. H. Muller et al. were now engaged in research that purported to demonstrate that allelopathy did indeed exist, but in this case, in the coastal sage community near Santa Barbara."[31] The coastal sage community referred to the soft chapparal of coastal southern California, which had vast areas of *Salvia leucophylla* (purple sage) and *Artemesia californica* (California sagebrush) growing close to natural grasslands. A striking aspect of the landscape was the appearance of bare zones separating the *Salvia* and *Artemesia* shrubs from the adjacent grasslands. But it took a student to draw these zones to Muller's attention with a pointed question: could these circular bare areas indicate instances of chemical inhibition?[32]

Muller, still dubious, promised to investigate it. He recalled that efforts to get a grant for the project from the National Science Foundation elicited the same objections that he had raised earlier, but he did finally get his funds.[33] The first results were published in 1964 in *Science*, long after the departure of the student whose question had prompted the research.[34] With his collaborators Walter Muller and Bruce L. Haines, Muller now argued that volatile organic compounds given off by certain plants, and possibly deposited on the ground in dew, inhibited the growth of other plants. Muller considered these substances to be waste products of the plant's normal metabolic processes, which had to be eliminated because of their toxicity. Although the production of growth inhibitors was not a new idea, they believed their study to be the first to show that a volatile inhibitor might be effective in the field.

A more thorough discussion published in 1966 concluded that the inhibition effect helped to explain how these shrubs invaded grasslands. Completely reversing his views from a decade earlier, Muller now argued that plant chemical inhibition could have profound ecological significance and therefore should be incorporated into models of community dynamics. The presence of inhibitory toxins, he suggested, could "seriously skew the pyramid of numbers, upset the biomass proportions of trophic levels, and thus determine the cycling of energy and materials."[35] In short, chemical inhibitors could have a major effect on ecosystem structure and function. The Ecological Society of America, in giving Muller its Eminent Ecologist Award in 1975, recognized his role in developing chemical ecology and making "allelopathy a part of ecological understanding."[36]

In accounting for his change of mind, Muller pointed to the central role that new chemical laboratory techniques had played in making it possible to isolate and identify organic compounds given off by plants. Techniques such as gas-liquid chromatography, paper and thin-layer chromatography, electrophoresis, and spectrophotometry all revealed "evidence of subtle chemical interactions" for the first time.[37] But his later recollections also emphasized the importance of field experience. At one point, he wrote, it had dawned on him that he had "permitted himself to be too thoroughly seduced by the joys of controlled experimentation" and had short-changed field research, which had become too organized and hurried.[38] The remedy, he wrote, was to revert to the habits of his "cowboy youth" and put himself into the field, "contemplating the scene" before him. The result was the synthetic paper of 1966 that pulled together and reconciled the laboratory and field studies, culminating in the conclusion that allelopathy had a place in the discipline of ecology. This paper gained considerable notice, becoming a "citation classic" in 1982. Muller noted, however, that the controversy over allelopathy had continued and surmised that the paper's popularity had more to do with the photographs included than with the evidence cited. He also thought that in the earlier publication in *Science* in 1964, the cover photograph showing the bare zones around shrubs had been more convincing than the data.

Muller was careful not to imply that allelopathy was a panacea for all ecological problems, and he warned against overly simplified ecological explanations, or what he called "single-factor ecology."[39] Chemical inhibition of annual herbs was, in his view, "strongly influenced by other factors, both physical and biotic." Those other factors included drought and weather patterns, soil type, shading, and animal activities. He also accepted the idea that toxic chemicals could have arisen for other reasons, such as protection against herbivores. Despite such caveats, in the 1960s and 1970s Muller became embroiled in controversies with other biologists who offered interpretations that played down the role of allelopathy. As recounted by Rachel N. M. Dentinger, Paul Ehrlich and Daniel Janzen proposed convincing alternative explanations that focused more on coevolutionary relationships between plants and insects. Bruce Bartholomew, Mooney's graduate student at Stanford,

interpreted the bare zones as the result of increased herbivory in the protected regions near shrubs. Bartholomew's studies convinced even Muller's
own students after he and Mooney visited Santa Barbara, but Muller
(not present for those discussions) upheld the view that allelopathy was of
ecological importance.[40]

Muller's students encountered strong resistance from their adviser
whenever they proposed alternative explanations. Richard Halsey, in a recent review of these controversies, raised the question of the seeming con-
tradiction between Muller's professed belief in open-mindedness and his
"frequent dismissal of evidence contrary to the allelopathic model."[41]
Halsey thought Muller's stubbornness arose from the determination that it
took to champion an unpopular idea. Muller's comments and his reactions
to critics suggest that his strong responses stemmed from his "distaste for
invasion of ecology by reductionism," which was exactly why he had earlier
rejected Bonner and Gray's chemical hypothesis in the early 1950s.[42] However, once he had decided that allelopathy was occurring, he was not willing
to dismiss it just because alternative explanations were available; dismissing
it would presumably put ecologists on the slippery slope to reductionism.
Allelopathy had to remain in play, and by sticking to his guns Muller also
helped to build the field of chemical ecology.

In the 1950s there was still considerable distance between ecologists
who prioritized field observations and experiments and physiologists like
Bonner, who were more accustomed to working in laboratories. How could
that boundary be crossed? How could laboratory results be reconciled with
field observations? Muller accepted the challenge of connecting these
worlds, and in the process completely reversed his original opinion. It was
important to take advantage of new instruments and laboratory techniques
but also important not to be seduced by them and to know when to head
back out to the field. The availability of new techniques encouraged more
experimental research, but this research had to be synthesized or reconciled with field studies. Muller had confidence in his reconciliation and in
the importance of chemical ecology as a new field. But not all agreed.

Controversies continued to swirl around Muller's theories of
chemical inhibition, for the larger problem of determining the ecological

significance of allelopathy was not easily settled. Robert Whittaker, a plant ecologist at Cornell University and one of the preeminent community ecologists of the mid-twentieth century, suggested why this lack of resolution might be expected. He reviewed the literature dealing with allelopathy for a book on chemical ecology published in 1970. He argued that allelopathy was likely to be very important, given the large number of secondary chemical compounds that plants produced—that is, compounds not directly needed for the plant's growth and development. But he also speculated that apart from conspicuous cases, such as the production of the inhibitory substance juglone by walnut trees, many of these effects would be subtle, lying "below the surface of the community relationships we are able to observe."[43] He compared the position of ecologists to that of geographers trying to understand the typography of the ocean bottom from data on islands alone.

Despite the difficulty of getting beneath the surface, Whittaker believed that allelopathic substances were probably important for explaining the function of plant communities and that they were part of an "extensive traffic in chemical influences relating organisms of all the major groups to one another in natural communities."[44] Whittaker saw the study of all these chemical interactions as the basis of a new field that he called "allelochemics." He and his co-author P. P. Feeney followed up with a lengthy discussion of allelochemics in *Science* in 1971, which included a sympathetic review of Muller's work along with other examples of chemical interaction.[45]

However, the study of the ecological role of such chemicals was still in its infancy, which left room for other authorities to throw the weight of their opinion against the thesis that allelopathy had ecological significance. John L. Harper, an eminent and authoritative British plant ecologist, penned a sharply critical review of a monograph on allelopathy that appeared in 1974.[46] He found fault with the book's lack of balanced reporting. It had failed to mention Muller's own early criticism of allelopathy and had not cited Bartholomew's persuasive alternative to Muller's later interpretation of bare zones surrounding shrubs. Alastair Fitter, an ecologist at the University of York, commented that Harper's review was "unusually influential" and "convinced most ecologists that chemical warfare between plants was not a profitable field of study, despite its undeniable importance in microbiology."[47] Harper published a book on the population biology of plants in 1977, which again cast doubt on the significance of allelopathy. He noted that the hypothesis that plants released toxic materials that harmed or killed other plants "might seem easy to test but in fact there are very great difficulties in demonstrating that toxicity effects are other than laboratory artefacts."[48]

While Harper's authority may have put a damper on ecologists' interest in allelopathy, it did not squelch their interest entirely. Innovative research continued at the University of California, Santa Barbara, and received Muller's encouragement. A good example was the dissertation research of Ragan (Ray) Callaway, who received his Ph.D. in 1990, working with his adviser Bruce Mahall and with Nalini Nadkarni, assistant professor at Santa Barbara from 1984 to 1989 (and now professor of biology at the University of Utah). That research explored both the positive (or facilitative) and negative (or interfering) effects between a tree species, *Quercus douglasii* (blue oak), and the grasses and forbs (herbaceous flowering plants that are not grass-like) in the understory, and it included allelopathy as well as competition for resources among the negative effects. In an article based on this research, Callaway, Nadkarni, and Mahall acknowledged Muller for contributing to the interpretation of their data.[49]

Callaway and Mahall also published further studies in the early 1990s exploring the relationship between *Larrea tridentata* (creosote bush) and *Ambrosia dumosa* (white bursage), two desert species that Mahall had been studying since the mid-1970s. They focused on interactions between the roots of the two species, a subject not well studied at that time. The experiments, conducted in a greenhouse, involved constructing plexiglass chambers into which the plant roots could grow and then connecting the chambers so that roots of the "test" plant could grow into the rhizosphere (the region surrounding the roots) of the "target" plant. Viewing windows enabled the researchers to see how the roots grew. They found evidence suggesting that the *Larrea* roots released a toxic substance that acted over a short distance and inhibited other plants. This work, they concluded, "may represent the strongest evidence to date for root-mediated allelopathy."[50] The *Ambrosia* roots, in contrast, also inhibited other roots, but only after direct contact between roots and not by release of a toxic substance. *Ambrosia* roots were able to detect and avoid other roots, which helped to avoid competition for water.

As Mahall and Callaway concluded in 1992, "simple competition for limiting resources among neighboring plants has been recognized and investigated for years, although there are still very few instances in which it has been unequivocally demonstrated in the field under natural conditions." Nonetheless, such competition had been "adopted as the basis of much theory and many models of plant-plant interactions," while "the possible existence of other forms of interaction, such as allelopathy" had been met "with impressive resistance." Yet, they continued, "the ecological advantage and evolutionary plausibility of allelopathy and even more subtle forms of interactions, referred to here and elsewhere as 'communications,' are inescapable, and new and stronger evidence for such interactions is mounting."[51] The field that Whittaker had envisioned in 1970 was coming into existence by the 1990s.

In 1993, Callaway joined the faculty at the University of Montana, Missoula, where his research group has continued to explore the direct and indirect interactions among plants, including positive as well as negative interactions. In the late 1990s and early 2000s that research took on a new focus with the study of species that were considered "invasive," meaning those that were new to a region and proliferated quickly, competing with species native to the region. The common explanation for the success of invasive species was that in their new locations they lacked the natural enemies that had kept them in check in their native lands. Callaway's group, studying invasive knapweed species (*Centaurea diffusa,* diffuse knapweed, and *Centaurea maculosa,* spotted knapweed) argued that allelopathic effects were involved in these invasions.[52] In a study with ecologists from Colorado State University and Pennsylvania State University, Callaway and his collaborators integrated ecological, physiological, biochemical, cellular, and genomic approaches, identified the phytotoxin involved, and provided strong evidence that, in the case of the spotted knapweed, the toxin exuded by the invader's roots accounted "at least in part for the displacement of native plant communities."[53] Commenting on this study, which was published in 2003, Alastair Fitter hoped that it might "give renewed respectability to the study of allelopathy, and encourage others to use new techniques to assess its ecological significance."[54]

Clearly, for ecologists at the University of California, Santa Barbara, the subject was always respectable. Muller's own persistence and the continued research interest in allelopathy at Santa Barbara kept the subject alive, leading to innovative experiments on root systems that provided clearer evidence of long-suspected allelopathic effects. This area of research gained energy when focus on new problems (such as species deemed invasive) was combined with new methods that yielded stronger evidence in favor of the ecological significance of allelopathy. But this case study also reveals that it can take a very long time, decades in fact, to persuade a broader community of ecologists that an idea has merit.

What does it mean to prove a hypothesis? What does it mean to demonstrate ecological significance? What does it take to convince skeptical colleagues that an entirely new field, in this case chemical ecology, deserves recognition and support? These questions are fundamental to the pursuit of ecology, a discipline that is constantly shifting, forming new links to other disciplines and expanding into new kinds of problems. One of the ironies, or perhaps paradoxes, of ecology as a discipline stems from its hybrid or synthetic nature. By seeking different forms of synthesis, ecology also continually reinvents itself, forming new connections and adding new branches. To answer the questions it sets, ecology must constantly seek to move across disciplinary borders yet somehow retain its integrity as a discipline, so that the ecological point of view will be recognized as distinct and valuable. It must, to some degree, hold firm yet also be prepared to shift its perspective and directions as new knowledge emerges. The need for reinvention inevitably generates controversy.

Do Trees Cause Pollution?

The history of allelopathy illustrates how complicated it can be to try to pin down the ecological significance of a given trait, in this case the emission of toxic compounds that could plausibly be seen to serve different functions. Understanding plant adaptations can take many decades and sometimes the results may demand complete rejection of old ideas, but such changes can be slow. As Jack Schultz remarked in 2002, the long habit of classifying entire groups of plant chemicals as "secondary" metabolites deserved to be re-evaluated in the light of improved knowledge about primary plant functions by that time.[55] Metabolites once thought to have no function might later be found to have important adaptive functions. Controversies can therefore persist for a long time. But sometimes these disagreements, even when they last for many years, can lead to a satisfactory resolution through synthesis, as opposed to adjudicating between alternative hypotheses. What is important is to recognize when there is an opportunity and need for synthesis.[56] I will use the example of plant emissions of volatile organic compounds to illustrate how progress toward a synthetic interpretation was achieved in this field.

Debate about the adaptive function of secondary metabolic compounds was further stimulated by the observation that plant emissions were not just small-scale and localized in impact but possibly occurred on a large enough scale to affect the atmosphere and regional climate. This startling idea prompted a closer look at the emissions of volatile organic compounds by plants and led to focused efforts to pin down the mechanisms underlying these emissions. How exactly were these compounds synthesized in the plant? That "how" question was quickly followed by a "why" question: why were plants secreting these substances? If they were not merely waste products, they might have an adaptive function, helping plants to cope with stressful conditions.

These questions about the adaptive significance of traits were exceedingly complicated, requiring decades before clear answers emerged. Finding answers prompted collaborations across disciplines, and to encourage such collaborations efforts had to be made to organize the relevant research communities. The invention of new instruments and techniques also helped. Thinking synthetically—putting the puzzle pieces together and drawing on ideas and findings from different fields of research—was also crucial in moving the discussion forward. Here and in the next two sections, I follow in broad outline some of the turns and twists of this discussion within selected research groups. I focus mainly on two biologists who sought an answer to the "why" question: Thomas D. Sharkey (at the University of Wisconsin and Michigan State University) and Russell K. Monson (at the University of Colorado), both of whom worked with many collaborators for more than four decades. The full story has international scope, but that must be left for future scholars.

The story begins in the 1950s with Went's research on photochemical smog and the realization that this type of pollution might have a natural counterpart in the volatile organic compounds that many plants emit. These compounds include terpenes, aromatic oils that impart a characteristic scent, for instance, the scent of pine, lavender, or orange peel. Cornelius Muller had considered these emissions in his work on allelopathy, but he was interested only in the inhibitory effects on other plants. Went agreed with Muller's findings about plants inhibiting other plants, but he also began to see a similarity between these natural emissions of terpenes and the larger-scale smog episodes that were plaguing the citizens of Los Angeles. Were volatile emissions from plants also sources of air pollution?

One of the earliest lines of research in Caltech's phytotron concentrated on photochemical smog, as discussed in chapter 3, especially the impact of smog on plants. Caltech chemist Arie Haagen-Smit had advanced the idea that "Los Angeles smog" was the product of chemical reactions occurring on sunny days, when hydrocarbons in the presence of nitrogen oxides were broken down, producing ozone and haze and creating the typical sharp smell and irritation of photochemical smog. Did the volatile compounds emitted by plants undergo similar reactions and produce comparable hazes? While investigating the damaging effect of anthropogenic smog on plants in the 1950s, Went became interested in the blue hazes produced by forests. In February 1954, while flying over Central and South America, Went viewed the tropics from the air for the first time and noted an unusual haze hanging in layers over the jungle. The captain of the plane told him the hazes appeared when there were no fires within hundreds of miles, and Went concluded that they were coming from the jungle itself. After seeing similar hazes over Panama and Brazil he decided that there was not the slightest doubt "that vegetation produces a natural smog of moderate intensity."[57]

His first discussion of the subject was in a *Scientific American* article in 1955. Titled "Air Pollution," it moved from a discussion of industrial causes of photochemical smog to the idea that there was a similar type of smog produced by plants, which he proposed to call "natural smog."[58] He argued that organic emanations from plants underwent the same chemical reactions as man-made smog and produced a similar haze, but without the irritating effects of man-made smog. Haagen-Smit had estimated that sage-brush in the southwest released terpenes that, like hydrocarbons in gaso-line, could form peroxides and ozonides in the air. These products might be toxic, Went surmised, but did not reach harmful concentrations in the air. This natural smog, he posited, explained the haziness of the Amazon basin, the fiery red sunsets in the tropics, and the atmosphere's absorption of more of the sun's light in summer than in winter.

While visiting Australia in 1955, Went found a typical natural haze in the Blue Mountains of eastern Australia and thought it was a result of eucalyptus oils.[59] He speculated that the haze particles might act as precipitation nuclei around which cloud droplets were formed. It might therefore be the case that clouds formed in areas with certain kinds of vegetation. Thinking of how to use this knowledge, he proposed that finding good nuclei producers would provide a way to alter the local climate. (In recent years scientists have made similar observations and are now actively studying the problem of how forests affect the water cycle, a topic to which I return at the end of this chapter.)

Went also started wondering about the ultimate fate of these mole-cules, this "natural pollution," in the air. He did not think that they would be completely oxidized to carbon dioxide and water but instead might eventually just settle out of the atmosphere when they got heavy enough or be brought down by rain and snow. Then, making a bold leap, he specu-lated that these tiny drops falling back to earth could be the "parent material for petroleum."[60] That idea, which Went maintained for the rest of his life, never caught on, and Went had to admit that it did not account for many facts about petroleum deposits, but the notion tempted him nonetheless. He continued to ponder the possibility.

Went knew about John Tyndall's experiments in the mid-nineteenth century that involved creating a "blue cloud" or a blue haze by passing strong light through a tube filled with air containing organic vapors. Tyndall discovered that the blue light was polarized, and he suggested it was reflected by submicroscopic particles. Went argued that the blue hazes in the countryside were similar to what Tyndall had described. In 1960 he published two articles on the meaning of these natural blue hazes.[61] Current ideas that they were produced by smoke, dust, water vapor, or fog could, he thought, be ruled out easily, leaving the hypothesis that the hazes were related to the "blue cloud" of Tyndall and the smog that Haagen-Smit would regularly produce experimentally in his laboratory, to demonstrate that his theory of photochemical smog was correct.

Those earlier experiments gave Went the idea of emulating Haagen-Smit's approach by creating "natural smog" in the laboratory. He filled jars with dilute ozone, placed them against a black background, and shone a strong light through the jars. As soon as he dropped crushed pine or fir needles into the jar, blue smoke formed around the needles and filled the jar. Plants with high levels of essential oils could react dramatically: "A spectacular display of smoke streamers in the jar was caused by twisting an orange peel inside; every ejected oil droplet left a trace of blue smoke, like high-energy particles in a Wilson cloud chamber."[62] Went argued that plants gave off significant amounts of these compounds, and he again hypothesized that they did not fully decompose but eventually came back to earth to become the source materials for petroleum formation. He argued that the hazes and clouds that they produced were important in the heat balance of the earth and perhaps provided part of the energy for thunderstorms.

These ideas about blue hazes attracted the attention of Reinhold Rasmussen (1936–2019), then working as a geologist in a potash mine in Utah.[63] He decided to go to Missouri to work with Went on the problem of terpene emissions and blue hazes, which became the subject of his Ph.D. dissertation in 1964.[64] Rasmussen's field work was in part conducted near the biological station at Highlands, North Carolina, near the corner where North Carolina, South Carolina, and Georgia meet. He used the mobile laboratory to get to a secluded stand of oak, pine, and hemlock forest, far removed from any sources of industrial or urban pollution. The laboratory was equipped with a three-thousand-dollar gas chromatograph that could detect and analyze chemical compounds in the air. The local mountaineers, who referred to the machine as "the sniffer," suspected its real purpose was to sniff out stills, and Rasmussen quickly realized he was not entirely safe moving from one isolated area to another with his sniffing machine.[65]

Rasmussen published his results jointly with Went in *Proceedings of the National Academy of Sciences* in 1965.[66] They analyzed air from Virginia, Missouri, and Colorado, in addition to North Carolina, as well as samples from Holland that had been collected by other scientists. They found that emissions of volatile compounds increased in the morning, plateaued from late morning to early evening, and then decreased. They also found that oak forests could produce almost as many aromatics as a pine forest, and with the same absorption characteristics as isoprene (an unsaturated five-carbon molecule that is the building block of terpenes), although the human nose could not detect the odor. By providing more specific information about these plant emissions, they hoped this study would furnish concrete support for the idea that plants emitted volatile organic compounds that underwent chemical reactions in the air, producing the blue hazes commonly seen over forests in the summer, in the same way that photochemical smog was generated in industrial and urban areas.

Time magazine picked up on Went's claims in a report on a talk he gave at a joint meeting of the American Meteorological Association and the International Biometeorological Congress in August 1966. He discussed the emission of toxic substances by plants such as western sage, which inhibited the growth of other plants. But the report also cited Went as stating "flatly that trees foul the air with ten times more pollutants than all of man's fires, factories, and vehicles," a more startling claim.[67] Went continued to stress the significance of plant emissions of volatile chemicals into the 1970s.[68]

Rasmussen's early work had only tentatively identified isoprene among plant emissions. In 1970 he confirmed that isoprene was being emitted by certain plants (such as oaks) under natural conditions, but only in the presence of sunlight, which suggested a connection to photosynthesis.[69] He was by then aware that earlier research in the Soviet Union had pointed toward similar conclusions. In the mid-1950s, Guivi A. Sanadze was leading a small group at the Institute of Botany of the Academy of Sciences of the

Georgian Soviet Socialist Republic in Tbilisi. In their experiments on allelopathy, these scientists found that isoprene was being emitted in relatively
large amounts compared to other hydrocarbons. Sanadze credited Andrei
Kursanov at the Timiryazev Institute of Plant Physiology in Moscow with
first appreciating the significance of isoprene emission and providing support for research on this subject. Kursanov was a leader in modernizing plant
physiology in the 1950s, and he was the force behind the building of the Soviet phytotron in Moscow. Sanadze's research on this subject over the following decade earned him a doctoral degree from the Timiryazev Institute
in the 1960s. His Laboratory of Photosynthesis at Tbilisi State University
became a center for the study of isoprene emission and its connection to
photosynthesis.[70]

Rasmussen's analysis in 1970 confirmed what Sanadze's group had
found earlier. Rasmussen also looked into the biological fate of these emissions, finding that they could be used by wild populations of fungi as sole
carbon sources, and he hypothesized that microbial life growing on vegetation under forest canopies might also use these volatile compounds. He became a leading specialist in the atmospheric chemistry of trace gases and the
emission of terpenes by plants, although it took a few more years for these
ideas to be fully accepted.

In the 1960s and 1970s Rasmussen was busy with a number of advisory and consulting positions and in the 1970s he was working on air pollution at Washington State University in Pullman (until 1977, when he
accepted a faculty position at the Oregon Graduate Institute's School of
Science and Engineering). While at Washington State he received funding
from the Environmental Protection Agency (EPA) for studies of terpene
emissions; he agreed that these were real and significant, but the fate of
these gases in the atmosphere was still unknown.[71] With EPA funding he
initiated a project to measure organic emissions from vegetation, leaf litter,
and water surfaces in order to get a sense of the magnitude of such emissions within a given region. He handed this project to a graduate student,
Patrick R. Zimmerman, who compiled the data and submitted the final report to the EPA in March 1979.[72] The study reported on measurements
made on selected plant species in California, Washington, North Carolina,

and Florida. Based on his limited survey, Zimmerman also calculated what worldwide emissions of volatile organic compounds might be. His estimate was about ten times what Went and Rasmussen had estimated based only on the emission of terpenes.

These findings gained notoriety during Ronald Reagan's campaign for the presidency in 1979–1980. In 1980, Reagan outraged environmentalists by defending a remark he had made in a radio broadcast that trees and other vegetation caused more pollution than man-made sources.[73] Reagan was opposed to setting or enforcing tough emissions standards to control air pollution, a stance he had also adopted as governor of California.[74] In the ensuing uproar, Arthur Galston, who had worked with Went at Caltech and was then professor of botany at Yale University, wrote a letter to the *New York Times* defending the idea that mother nature was indeed a polluter. Galston pointed out that Went had documented massive release of hydrocarbons by vegetation, which could react with ozone to produce blue hazes. These naturally occurring emanations, he affirmed, could certainly be considered "a form of atmospheric pollution."[75]

The mid-1980s was a turning point in the study of these emissions, as scientific opinion started to shift, following recognition that strategies to reduce ozone pollution in cities were not working. Since the 1970s the EPA as well as state agencies had been focusing on hydrocarbons released by human activities in efforts to control ozone levels in urban areas. Although these emissions had been reduced, often at great cost, ozone levels did not decrease. Dozens of cities were in violation of the National Ambient Air Quality Standard, with about 40 percent of those cities in the South. Scientists at the Georgia Institute of Technology took up this problem, analyzing sources of ozone pollution in Atlanta. They also brought together an interdisciplinary group of research scientists, engineers, and air quality managers for discussion of the problem, which led to a long-term monitoring and assessment program called the Southern Oxidants Study or SOS. The Georgia Tech group concluded that one flaw in ozone abatement strategies was the neglect of the impact of biogenic or natural hydrocarbon emissions. That study, published in September 1988, cited Went's and Rasmussen's work and argued that in Atlanta natural emissions of isoprene had a large effect in producing ozone and therefore contributed to urban photochemical smog. The role of what Went had called "natural smog" could no longer be neglected in developing effective strategies for controlling ozone.[76]

Bringing Communities Together

Research on the emission of volatile organic compounds intensified in the 1980s as the extent of these emissions became more apparent. The underlying biochemical mechanisms of such emissions were not well understood in the 1970s, and isoprene in particular would pose a puzzle for a long time. How prevalent was this trait, which was found only in some plant families? Why were plants emitting such large amounts? Forming these compounds required energy, and in giving off these compounds the plant was losing carbon. How could a plant afford such losses? And what was happening in the atmosphere when these reactive compounds were released? Answering these questions required interaction among communities that had not normally worked together in the past. During the 1980s and 1990s these groups started to form closer collaborations.

Pat Zimmerman received his master's degree from Washington State University in 1979 and was then recruited as a staff scientist by Paul Crutzen, a Dutch atmospheric chemist (and later Nobel laureate) who was the director of research at the National Center for Atmospheric Research (NCAR) in Boulder, Colorado. Zimmerman worked at NCAR until 1997 (during which time he completed a Ph.D. in 1996 at Colorado State University). At NCAR he organized a section called the Biosphere/Atmosphere Interaction Project and continued research on measurement of biogenic hydrocarbon emissions. His research program involved designing better computer models to predict the emissions of volatile organic compounds from plants. Zimmerman's group made important contributions to the study of trace gas emissions by devising instruments to make continuous observations of isoprene emissions and detect rapid changes in the emissions in response to changes in the leaf's environment.[77]

Zimmerman had realized that in order to improve atmospheric trace gas inventories and modeling approaches, he needed to understand the processes underlying isoprene synthesis.[78] In 1987 he approached two members of the University of Colorado faculty, Russell K. Monson, who was studying photosynthesis, and R. Ray Fall, who was a biochemist. Zimmerman proposed that they work with him to design a physiological and biochemical model to predict the fluxes of isoprene in relation to temperature and light, and they agreed to join forces. In 1988, Monson and Fall began a study of isoprene emission rates in quaking aspen. These emissions were known to occur in the light and were therefore thought to have some relationship to photosynthesis, but the mechanisms linking these two processes were not understood. To study this problem Monson and Fall not only put their heads together but also their equipment. Fall moved his gas chromatograph into Monson's laboratory, and they combined it with the advanced instruments that Monson was using to measure the carbon dioxide and water fluxes from isolated plant leaves. The two spent the summer of 1988 measuring the isoprene fluxes from isolated aspen leaves. They related those fluxes to photosynthesis rates and movement of water vapor and carbon dioxide through the stomata of the leaves, as functions of temperature, light, humidity, and atmospheric carbon dioxide concentration. That study was published in 1989.[79]

While they were preparing their manuscript in the fall of 1988, they received a letter from Thomas D. Sharkey, a plant biochemist. He had recently joined the faculty at the University of Wisconsin, Madison, after working at the Desert Research Institute in Reno from 1982 to 1987. Sharkey had been interested in isoprene emissions from his graduate student days at Michigan State University in the mid-1970s, when he had done his first isoprene experiments, but he had not yet published anything on the subject.[80] Sharkey and one of his postdocs, Francesco Loreto, had in mind a project similar to Monson and Fall's, but on oak leaves. Monson shared the main results with Sharkey, and the paper by Loreto and Sharkey appeared a year later in 1990.[81] The two studies were closely related; both were trying to understand how isoprene emissions were linked to other metabolic processes in the plant, especially photosynthesis.

Attention was now increasingly directed toward the biological emission of trace gases, not just isoprene but also such gases as methane, carbon

dioxide, and nitrous oxide. The role of isoprene emissions in the tropo-
sphere, the lowest level of the earth's atmosphere, was coming under scrutiny as scientific evidence pointed to the impact of these emissions on air
quality. Isoprene, for instance, did not remain long in the atmosphere but
reacted with the hydroxyl radical (OH), known as the "detergent of the atmosphere," to generate carbon monoxide, ozone, and organic peroxides.
These same reactions also reduced the effectiveness of the removal of methane (a greenhouse gas) from the atmosphere. To understand the exchanges
between the biosphere and the atmosphere, the research communities
studying these two levels had to come together.

Sharkey was instrumental in helping to link those communities. In
January 1990 he organized a workshop at Asilomar, California, that brought
together scientists from several disciplines to discuss the biological generation of trace gases. Elizabeth Holland, who was working with Zimmerman's
research group at NCAR, organized the atmospheric side of the conference.
The subsequent volume of papers, edited by Sharkey, Holland, and Harold
Mooney, focused on how emissions from the biosphere affected atmospheric chemistry and how these atmospheric chemical reactions in turn
affected the habitability of the biosphere.[82] Monson viewed this conference
as a "primary event in bringing together the atmospheric and physiological
perspectives on this issue."[83]

Ray Fall summarized the workshop discussions about isoprene, noting that although three decades had passed since Sanadze made his discovery of isoprene emissions, "we still do not know how or why plants
synthesize and emit this volatile hydrocarbon."[84] Isoprene appeared to be
different from other secondary metabolic products. Other secondary products were known to have a variety of roles: they repelled animals and microorganisms that attacked plants, attracted pollinators, or repelled other
plants. But isoprene's role was hard to pin down. So little was understood
that Fall envisioned that a major multidisciplinary effort would be needed,
which he thought should include "genetics, biochemistry, and physiology,
as well as related work in plant ecology, evolution, pathology, and microbiology." Moreover, he argued, "this work should be carried out in collaboration with scientists concerned with measuring and modeling impacts of

isoprene emissions on larger systems, including forest canopies, geographic regions, continents, and even on a global scale. This is especially important considering isoprene's potential for modifying the chemistry of Earth's atmosphere."[85] Finding ways to link scientific communities would be crucial to putting these puzzle pieces into place, and this workshop was an effective first step.

At the workshop Sharkey and his collaborators discussed the biochemistry of isoprene synthesis and emission, while Monson and his collaborators, Ray Fall and Alex B. Guenther, also discussed the problem of scaling up, or how to estimate isoprene fluxes at scales ranging from a single leaf to the entire globe. Guenther was a postdoctoral fellow who had just finished his Ph.D. in Civil and Environmental Engineering at Washington State University in 1989.[86] He worked in Monson's lab with the same instrument setup that Fall and Monson had used for the aspen work, but instead he worked on eucalyptus. Emission rates from plants changed throughout the day and throughout the season, and they were sensitive to environmental conditions such as light and temperature. Models of regional and global tropospheric processes could not be developed without more accurate assessments of emission rates. The group's research tackled this problem. Two articles that came from this research, published in 1991 and 1993, were highly influential: Guenther's algorithms laid the foundation for later regional-to-global models of biogenic volatile organic compound emissions.[87] Guenther became an international leader in atmospheric and terrestrial ecosystem research. After his postdoctoral year he joined NCAR as a senior scientist in November 1990; he later worked at the Pacific-Northwest National Laboratory from 2013 to 2015; and since 2015 he has been a professor at the University of California, Irvine.

Partly in response to that workshop, Ellis B. Cowling, a forestry specialist from North Carolina State University, invited Sharkey, Monson, and Fall to a meeting in Atlanta that included most of the contributors to the Southern Oxidants Study. At the meeting, Sharkey, Fall, and Monson formed a collaboration and proposed to the SOS leadership a project to study the physiological processes and associated mathematical models underlying isoprene emissions from trees. In 1990 the SOS was just starting a multiyear study to monitor the formation of ozone in ten southern states and evaluate alternative strategies for its decrease. Cowling was involved in that study and he became its director in 1993. As Sharkey and Monson recalled, after the initiation of the SOS project "a fruitful collaboration between plant physiologists and atmospheric chemists resulted in rapid progress in both fields. Many papers describing new insights into the factors that regulate isoprene emission were published, often with authors from the EPA and universities."[88]

Sharkey used SOS funding to support the work of his graduate student Eric Singsaas, who was tackling a problem having to do with the adaptive function of these emissions. New questions were prompting a shift in the field of isoprene research, as scientists started to ask why these emissions occurred. Were plants that emitted isoprene fitter? What was the benefit to the plant? Sharkey and Singsaas proposed in 1995 that isoprene protected plants from thermal damage.[89] Monson wanted to follow up on these ideas and hired a postdoc, Barry Logan, to repeat some of that work, but Logan used slightly different experimental methods and did not get the same results.[90] Logan and Monson emphasized that their work did not disprove Sharkey and Singsaas's hypothesis about heat tolerance, but it weakened the experimental evidence in its favor. More research was needed.

By this time in the late 1990s, there was increasing interest in hydrocarbon emissions on several fronts and the field was expanding dramatically. The University of Virginia hosted an international workshop in 1997 on hydrocarbon emissions of plants in atmospheric chemistry, and a special feature of *Ecological Applications* in 1999 (for which Sharkey was guest editor) focused on two of the dominant emissions, isoprenes and monoterpenes (ten-carbon molecules consisting of two isoprene units).[91] Sharkey then took an important step in receiving funding from the Gordon Conferences to start a Gordon Conference on Biogenic Hydrocarbons and the Atmosphere. The first, which Sharkey chaired, was held in 2000 in Ventura, California, and conferences have since then been held every two years.[92] As Monson recalled, "Of particular importance to the isoprene storyline is that Tom and I had a very open debate [at the first Gordon Conference] about whether the thermotolerance hypothesis that he and Eric Singsaas had published was correct."[93] Over the next fifteen years the adaptive role of isoprene remained a topic of debate. The thermotolerance hypothesis made sense, but it never quite accounted for all the research findings. The "why" question remained in play and prompted more research using newer methods. As that research uncovered new puzzles, biologists also took on the challenge of synthesis. Could theories and experimental findings from different disciplines be pulled together and reconciled?

Toward a Synthetic Understanding

As experimental research expanded, the understanding of isoprene's function broadened. During the first decade of the 2000s, the techniques of genetic engineering were being applied, creating possibilities for new kinds of experiments. Plants that were not emitters could be engineered to emit isoprene in order to study the effects.[94] At the University of Queensland in Australia, Claudia Vickers and her group engineered tobacco plants (normally non-emitters) that were able to emit isoprene in the same quantities as natural emitting species. Isoprene emissions could also be inhibited to see what happens. A group in Germany led by Jörg-Peter Schnitzler studied isoprene metabolism using a gene-silencing technique called RNA interference (or RNAi), which is based on the natural mechanisms that cells use to silence genes. Understanding evolutionary history was important, for only some plants emitted isoprene. Monson and his group explored this problem, finding that the trait had been gained and lost several times in independent evolutionary events.[95] It became clear that isoprene affected metabolism in many ways. Christopher M. Harvey and Sharkey proposed in 2016 that it acted as a signaling molecule, modulating gene expression. Sharkey's group found that isoprene led to changes in the expression of several gene networks that were involved in stress responses as well as in plant growth.[96]

With findings coming from multiple studies of genetically altered plants ranging from poplar trees to tobacco plants to the humble *Arabidopsis thaliana* (mouse-ear cress, the first plant to have its genome sequenced), and with new insights about the role of isoprene in altering gene expression, the stage was set for a synthetic approach that would add a new dimension to the "why" question. Monson, Sharkey, and Schnitzler, along with Sarathi Weraduwage (a member of Sharkey's group) and Maaria Rosenkranz (a member of Schnitzler's group) proposed a novel synthetic interpretation in 2021.[97] Monson had just retired from the department of ecology and evolutionary biology at the University of Colorado (where he is now professor emeritus of distinction). Sharkey is distinguished professor in the department of biochemistry and molecular biology at Michigan State University. Schnitzler is director of the Research Unit Environmental Simulation at the Institute of Biochemical Plant Pathology, Helmholtz Zentrum, in Munich. I will refer to their joint article as the isoprene synthesis paper.

Their proposals built on some of the findings that had emerged from experiments with genetically engineered plants. Scientists agreed that isoprene had adaptive value in some plants and that it enhanced photosynthesis in the face of stress. But genetic engineering experiments had revealed inconsistences that did not fit conventional theory. For instance, in some species growth had increased and in others it had decreased, even when conditions were not stressful. The authors of the isoprene synthesis paper suggested that earlier studies had taken too narrow a view in thinking of isoprene as acting alone, rather than interacting with several other metabolites. The paper therefore posed a larger question: what is the broader adaptive scope of the trait and how does stress tolerance fit into the broader scope? To find an answer it was necessary to link research findings coming out of plant physiology, genetics, and biochemistry with theoretical approaches coming from evolutionary biology and ecology.

The authors placed the problem of adaptation within the context of theories that had developed in ecology and evolutionary biology over the previous decades. This body of theory examined adaptive strategies in terms of costs and benefits. In the words of one such analysis, plants faced a "dilemma": they had to grow enough to compete, but they also had to put resources into various defenses needed for survival.[98] There was a trade-off between these different needs, and how this would work depended on a host of conditions. The isoprene synthesis paper built on this idea and argued that isoprene's role was not limited to stress tolerance alone but to "stress tolerance within the context of the growth-defense tradeoff." "In essence," the authors contended, "we make the case that isoprene has evolved in certain plant lineages as a means to stage an effective form of chemical defense, with minimal costs to growth, in the face of climate stress."[99] The plant, in other words, had evolved a very effective trade-off, where it gained an important benefit without a high cost. The isoprene synthesis paper presented no new data but drew solely on the existing literature, pulling together the findings of thirty years of research with multiple collaborators. It represented, the authors argued, a major shift in thinking from direct effects of isoprene to indirect effects, where the indirect effects came about through changes in gene expression and protein abundance.

The synthetic activities that I have so far described in this case study involved bringing scientists from different disciplines together in workshops and conferences to try to work out the extent of biological emissions of isoprenes as well as their adaptive significance. New techniques, including mathematical modeling and genetic engineering, made it possible to explore these problems in new ways and helped greatly to advance understanding. Finally, the participants in one of the central debates concerning the adaptive function of these emissions combined forces to develop an explanation that placed isoprene emissions within the broader context of an evolutionary trade-off between costs and benefits. The key point is that the resolution of a long-standing controversy came about not by adjudicating between alternative hypotheses but rather by placing the original debate within a different conceptual context, adopting a broader view of the question to be answered, and seeking a synthetic interpretation. In general, I wish to suggest that these kinds of synthetic activities deserve closer analysis from historians, philosophers, and sociologists of science.

Scaling Up: From Leaf to Ecosystem

The problem of scaling up presents another important opening for synthesis in ecological sciences. Knowing what is occurring at the level of the leaf is one thing, but what is occurring in the canopy as a whole? By the early twenty-first century, scientific study of biogenic volatile organic emissions had established that these compounds underwent chemical reactions in the atmosphere and produced a variety of compounds, including ozone, nitrogen oxides, organic acids, and organic peroxy radicals (which are involved in the production of ozone). To understand these processes, biologists had formed alliances with atmospheric chemists, but there was still a lot to be learned about what was occurring at the tree canopy level. As Monson commented in a review of the subject in 2002, "Quite a bit is known about the controls over isoprene emission at the level of individual leaves, but much less is known about the scaling of isoprene emission to the canopy."[100] The magnitude of these emissions was striking. As Sharkey and his co-authors explained in 2008, isoprene emission was the main biogenic source of hydrocarbon in the atmosphere: "roughly equal to global emissions of methane [CH_4] from all sources. This surprising finding of such a large flux of isoprene from plants to the atmosphere raises a number of questions, including what happens to the isoprene in the atmosphere and why plants emit isoprene."[101] An interdisciplinary field was emerging devoted to the study of the "breathing of the biosphere" and the physical, chemical, biological, and ecological processes that controlled it.[102] A subject this large demanded larger-scale experiments. This section will look at a couple of such projects in which Monson participated, projects designed to scale up from leaf to forest.

The first example is a collaborative study involving Monson and Fall, Todd N. Rosenstiel (Monson's graduate student at Colorado State University), and Mark J. Potosnak and Kevin L. Griffin, both from the Lamont-Doherty Earth Observatory at Columbia University.[103] This study considered the impact of large-scale agroforestry on the environment. To meet the world's demand for lumber, large plantations were being created of fast-growing trees, but those species, such as poplar, acacia, and eucalyptus, were also known to emit large quantities of isoprene. Those trees might offer a benefit in removing carbon dioxide from the atmosphere, thereby helping to ameliorate a problem that was quickly becoming a crisis as carbon dioxide levels continued to rise. But if carbon dioxide promoted photosynthesis and tree growth, would its removal at the same time lead to much higher isoprene emissions from these plantations? Would the costs outweigh the benefits? The answers could only be determined through experiment.

The experiments, which started in 1998, were run at the Biosphere 2 facility in Oracle, Arizona, north of Tucson. Biosphere 2 had begun as an experiment in the 1980s to see whether humans could survive in a totally enclosed system. An environmental group led by John Allen came up with the idea, funding came from Texas billionaire Ed Bass, and building began in 1987. An ambitious project costing about $250 million (well above the original budget of $30 million), Biosphere 2 tried to create an earth in miniature, an enclosed multi-ecosystem environment (the earth being Biosphere 1).[104] It was a complex of greenhouses that contained agricultural fields as well as different ecosystem types such as rain forest, savanna, and even a miniature ocean with coral reefs. It was almost totally closed to the outside but was not self-sufficient because it needed climate control systems, which consumed a lot of energy.

The original plan was for a group of eight people to seal themselves inside for two years, without any direct physical contact with the outside, and live only from what they could grow themselves. The experiment, which started in 1991, quickly ran into problems. Oxygen levels dropped because they had not counted on the consumption of oxygen by soil microbes. Oxygen had to be pumped in. The occupants lost weight; they were not able to grow enough food. The group survived the two years but the scientific merit of the experiment was being sharply questioned. A second experiment in 1994 ended early, and by 1995 the project was in jeopardy, plagued by high costs. In December 1995, Columbia University took over the lease under the aegis of its Earth Institute, and in 1996 it became possible to use the facility for scientific studies on whole ecosystems, after some re-engineering to fix some of the flaws in the earlier designs. During Columbia's tenure the Biosphere 2 Laboratory had a coastal marine mesocosm, tropical forest mesocosm, and plantation (temperate) forest mesocosm, and it attracted a wide variety of researchers. Columbia originally planned to operate the facility until 2010, but it withdrew support prematurely in December 2003. The University of Arizona eventually took over the complex to use as a research facility.

The plantation experiment that began in 1998 was not run in the main Biosphere 2 complex but in a separate forestry section that had independent climate and carbon dioxide control. Three groups of cottonwood trees,

about fifty per plantation, were planted in three controlled-environment mesocosms of 12,000 cubic meters each. Carbon dioxide levels were set at
three concentrations: one the same as current atmospheric levels, another at
double that level, and a third at three times that level. The semi-enclosed
nature of the structure allowed for the continuous measurement of isoprene
production. The study showed that although increased carbon dioxide did
result in more growth, isoprene emissions were reduced, an unexpected result. The authors proposed metabolic explanations for what might be causing the decreased isoprene emission. Monson, Rosenstiel, and a group of
collaborators would later develop this line of research using gene-silencing
techniques, which reduced isoprene emissions to negligible levels. This research, partly done at the plantations of the Biosphere 2 campus, by then
operated by the University of Arizona, confirmed that removing the trait did
not interfere with plant growth. The productivity of their agroforestry plantations was the same in isoprene-emitting and non-emitting forests.[105]

At the time that Columbia University was operating Biosphere 2, biologists were optimistic about the potential of doing whole-ecosystem experiments of this kind. The facility appeared to open a new era for ecological
science. C. Barry Osmond, who accepted an appointment as research leader
at the Biosphere 2 Laboratory starting April 1, 2001, and remained for two
years, was excited by what this unique tool offered.[106] Osmond was an Australian scientist; as a graduate student he had attended the conference marking the opening of Canberra's phytotron in 1962. Later he formed part of the
"Carnegie-Canberra pipeline," an ongoing series of collaborations between
botanists at Canberra and at the Carnegie Institution of Washington at Stanford. Another member of that pipeline, Joseph Berry from the Carnegie
group, also led a research team at the Biosphere 2 Laboratory.

Before coming to Biosphere 2, Osmond had held positions in the
United States in the 1980s and early 1990s. He succeeded Frits Went at the
Desert Research Institute, where he worked from 1982 to 1986 on a half-
time appointment, and then accepted a research appointment at Duke University's botany department from 1987 to 1991, after which he returned to
the Australian National University in Canberra as director of its Research
School of Biological Sciences. Returning to the United States in 2001 to

work in the Biosphere 2 Laboratory, he was enthusiastic about the kinds of projects that could be run from such facilities. As he said, "It seemed that at last we could test model simulations of canopy/ecosystem carbon fluxes in a real simulator."[107] He also believed Biosphere 2 would encourage multi-institutional projects, and indeed scientists from around the world were coming to work in this prototype apparatus. It had the potential to transform ecological research in the era of earth system science. Osmond, Berry, Monson, and other like-minded biologists joined up to make the case for the importance not just of scaling up but of doing *controlled experiments* at larger scales. Their message, expressed in an opinion piece published in 2004, was that in view of the current global climate crisis, business as usual was not an option: "Experimental ecosystem science has to expand the size and duration of controlled experiments with complex natural systems, such as coral reefs and forests, to test hypotheses leading to mechanistic understanding of large-scale processes in the biosphere."[108]

Ecologists did not at that time have access to the apparatus or funding needed for manipulations of ecosystems. Except for NASA's Earth Observing System, they noted, "we have yet to see significant investments on Earth in the facilities needed to scale up observations from organisms to ecosystems." The Biosphere 2 Laboratory was a good start, but more investment was needed. Comparing the climate crisis to an older problem, the HIV-AIDS plague, the authors asked: "Where would we be in controlling the epidemic today if we had left the response to epidemiologists alone? Just as the full arsenal of experimental biomedical research has been mobilized to address the pandemic, so we need now to mobilize the whole arsenal of experimental capabilities in natural sciences and engineering in support of climate change science, from the molecule to the biosphere." Since Columbia had ended support of the Biosphere 2 Laboratory by the time of this appeal, the authors noted ruefully that "peer endorsement has been slow to mobilize, the promised private institutional support has been terminated prematurely, and the enterprise has foundered." But if peer support lagged within the American community, they pointed out, it was not lagging elsewhere: the need for large-scale closed systems was "already recognized in Japan, in Europe and elsewhere as more optimal iterations of B2L are being explored."[109]

Getting peer support within the United States was a major obstacle. Osmond recalled that at an international workshop in December 2001, although there was optimism about the potential of the Biosphere 2 Laboratory, "one was left with the impression of strong resistance from some ecologists, especially from the US national laboratories, where the concept of controlled ecosystem-scale experiments replicated in time seemed to threaten traditional long-term, spatially replicated observation of natural ecosystems."[110] Joseph Berry made a similar observation: even though Osmond's vision for Biosphere 2 had met with some success, "it failed to garner sufficient support from the peer-review process to receive continued funding."[111] Berry remained convinced of the need for science at this scale.

But Biosphere 2 was not the only experiment in earth system science. Another interdisciplinary research project that began in 2008 in Colorado epitomized the kind of large-scale effort that the authors of the opinion piece had been advocating. This project was initiated by the National Center for Atmospheric Research in Boulder, along with scientists from the university community. Its purpose was to investigate the exchange of trace gases and aerosols between the terrestrial ecosystem and the atmosphere and to look for potential feedbacks between biogeochemical cycles and water cycles. The project was called the "Bio-hydro-atmosphere interactions of Energy, Aerosols, Carbon, H_2O, Organics, and Nitrogen" or BEACHON for short. To serve this project's goals, a fixed field site was established in 2008, the Manitou Experimental Forest Observatory, located in the ponderosa pine forests of the Manitou Experimental Forest, a few miles northwest of Colorado Springs. The observatory was maintained through a cooperative agreement between NCAR and the U.S. Forest Service, and it was available to the scientific community for training, scientific research, and model development and evaluation. An overview of its scope and scientific work was published in 2014 by an international group of forty-four scientists, including Monson, who also produced a short film describing part of the project. This is truly a Big Science enterprise on a grand scale.[112]

BEACHON comprises an international team of about a hundred biologists, chemists, and atmospheric physicists. Among the hypotheses being tested is that terpenes emitted by the pine forest have an important role in nucleating or seeding cloud formation and that these emissions are important for understanding the regional water cycle. While the main adaptive function of terpenes is to repel or kill insects that feed on pine needles, they are produced in such large quantities that they leak into the atmosphere and then bind to other molecules, which in turn combine to form clusters, creating an aerosol. The aerosols collect water molecules and eventually become clouds. Scientists are studying the process of cloud seeding and how variation in environmental conditions (such as rainfall) affects the production of organic molecules. Scientists now know that terpene emissions by forests have to be factored into studies about how forests affect climate.

A project like this requires a highly sophisticated level of instrumentation to analyze these processes. There is a field-based scientific camp that consists of a village of field laboratories housed in trailers, placed around a metal tower rigged with instruments. Several types of specialized instruments collect data, which the scientists must piece together like a puzzle. In tandem with field studies, research also includes analysis of processes in a more controlled environment, in this case a large laboratory in Boulder at NCAR, which allows for control of light, temperature, and mixture of gases. As Monson explained, "this combination of very highly controlled laboratory analyses and then a number of observations and analyses that we can do at the site in real time" enables scientists to figure out what chemical reactions are occurring in the atmosphere and what the results of those reactions are.[113] The goal is to link these molecular processes to the water cycle, since the forests could be triggering the formation of some of the clouds that produce rain. It is particularly important to understand the role of these forests in the water cycle at a time when climate change might lead to shortages of water.

Conclusion

Physiological plant ecology has come a long way since the world's first phytotron opened at Caltech in 1949. I have used case studies focused on plant emissions of volatile organic substances to chart some of the challenges of this discipline. Continual innovation in laboratory designs and instruments has been central to the field's development, while balancing field experiments and observations with experiments conducted in controlled environments has been a constant feature of this enterprise. The study of adaptation, which is central to physiological ecology, has posed problems of great complexity and led to decades of research by interdisciplinary groups. Organizing the scientific community around a set of focused questions through regular conferences and workshops has been important in creating cross-disciplinary links between communities that were previously separate, especially between atmospheric scientists and biologists. New techniques such as genetic engineering have steered that research along new paths. Creating a satisfactory explanation of isoprene's adaptive significance pushed biologists to adopt a broader perspective, resulting in the synthesis of many lines of research within an evolutionary framework. The search for synthesis has been a central preoccupation in this science.

Another challenge that advanced the movement toward synthesis is that of "scaling up" or moving from the organism to the ecosystem and even to the entire earth. Scaling up has spurred new cross-disciplinary connections and recognition that controlled experiments must also be done on a larger scale, requiring the kind of investment in environmental sciences that qualifies as Big Science. While some investments in controlled ecosystem-level experiments have met with resistance from the ecological community, scientists have joined forces to demand greater support for environmental sciences. Projects like BEACHON illustrate the type of collaborative undertaking that can be mounted to meet the global climate change crisis. Physiological ecology, now linked to the study of broader ecosystem processes in a multidisciplinary scientific enterprise, remains of central importance for addressing the environmental challenges that lie ahead and working toward a predictive ecology.

SEVEN

Exposing the Roots

Nutrient Cycles and the World Underground

THIS CHAPTER EXPLORES A field of research in physiological plant ecology that also focuses on the problem of how species interact, but with particular reference to the question of nutrient cycling in ecosystems. Our story therefore goes underground to consider the roles of soil microorganisms, especially fungi, in nutrient cycling. The cross-disciplinary relationship I examine is between mycologists—specialists in the study of fungi—and ecologists. Both communities studied the relationship between plants and fungi, but for much of the twentieth century they operated separately. Mycologists did much of their experimental work in greenhouses and laboratories, while ecologists emphasized experiments in the field, where it was difficult to study below-ground processes. The challenge was to bring these communities together. From the 1970s through the 1990s different groups of ecologists who had interests in soil ecology and specifically in plant-fungus relationships worked to promote synthesis in these fields of research.

I focus on the study of a symbiotic association between fungi and plant roots that is known as a mycorrhiza (or fungus-root) and on the field-based and laboratory-based efforts to understand the ecological significance of these common associations. Mycorrhizal fungi depend on plants for carbon, and they in turn supply plants with nutrients such as phosphorus and nitrogen. They are therefore important for understanding the mechanisms of nutrient cycling in nature, a problem that was gaining attention in the 1960s as ecologists embraced and applied the new concept of the ecosystem. Ecosystem ecology studies how matter cycles and energy flows through ecological systems. Mycologists at the same time were starting to look closely at symbiotic relationships between plants and fungi, and they were devising laboratory experiments using simplified systems to figure out how these fungi obtained nutrients from the surrounding environment and how those nutrients flowed to plants. Gaining a full understanding of the ecological role of mycorrhizal symbioses required contributions from these two disciplines and therefore prompted efforts to try to bring them into closer communication.

The study of mycorrhizae goes back to the nineteenth century and today is a large and diverse field of research. This chapter focuses on a few key individuals in this community to show how their work and the development of new techniques—such as the use of radioactive tracers and the invention of new laboratories—helped to forge disciplinary connections. Investigations of below-ground processes led to innovative laboratory designs and techniques, including the construction of "rhizotrons," special laboratories that made it possible to observe processes underground, as well as smaller instruments known as "minirhizotrons" used to observe belowground processes. Another approach was to perform experiments on small artificial ecosystems of just a few species, so that a certain level of ecological complexity could be introduced while avoiding the completely uncontrolled conditions of field experiments. (This was an early, simple version of the more elaborate controlled experiments conducted in "ecotrons," discussed in chapter 8.)

The need to forge connections between mycologists and ecologists can be illustrated by charting the circuitous route by which a particular hypothesis about nutrient cycling was tested and evaluated during the 1970s and 1980s. The "direct nutrient cycling theory" was developed by Nellie Beetham Stark in collaboration with Frits Went while they were on an expedition to the Brazilian Amazon. In 1968, Went and Stark proposed that mycorrhizal fungi had a role in transmitting nutrients from plant litter directly to plant roots, and they argued that this role was of particular significance in tropical forests growing on nutrient-poor soils. Ecologists and mycologists understood their core argument differently, however, and the fate of the hypothesis differed depending on which community was evaluating it. These different judgments reveal the fault lines that divided ecologists and mycologists:

ecologists emphasized field work and mycologists concentrated on laboratory studies. But biologists on both sides were also working to bridge these disciplines.

The story starts with an expedition to the tropical rain forests of the Amazon and shifts to a study that began in Venezuela in the 1970s, which was designed in part to test Went and Stark's nutrient cycling hypothesis. This was an important time for the development of ecology in Latin America, as scientists were acutely aware of the need to put a stop to what was referred to as "scientific imperialism," or the practice by foreign scientists of visiting the tropics and taking their findings back home, without contributing in any way to the improvement of local science. The Venezuelan project, which gave rise to a strong program of ecological research and graduate training, countered this "imperialistic" tendency.

Exploring the Amazon Rain Forest

The willingness of the National Science Foundation to fund new kinds of laboratories in the 1960s produced a boom in laboratory design and development, including mobile labs not only on land but also on rivers and seas. In addition to the phytotrons and biotrons discussed in earlier chapters, NSF helped to fund a custom-designed research vessel, the *Alpha Helix*, to facilitate scientific studies on land and in water in remote locations around the world. The vessel was the brainchild of Per (Pete) Scholander, a brilliant physiologist with diverse ecological interests that extended to the physiology of marine mammals, humans, and plants. Born in Sweden and educated in Norway, Scholander had come to the United States during World War II as a postdoctoral fellow and in 1958 was hired as a professor of physiology at the Scripps Institution of Oceanography. There he established a new research unit, the Physiological Research Laboratory, which later included the *Alpha Helix* research vessel.[1]

He was keen to promote what he called "expeditionary physiology," which meant the experimental study of adaptation in natural settings. For such studies, Scholander preferred to work in extreme environments that were often remote.[2] The *Alpha Helix* took groups of scientists from various disciplines into these locations, providing them with well-equipped laboratories that would enable them to do serious experimental research in six-week rotations. The 40.5-meter ship had fully equipped air-conditioned laboratories, a machine shop, and a photographic darkroom. Although the ship was based at Scripps, which covered the travel and subsistence costs of participating scientists, it was a national facility intended for use by scientists across the United States, as well as by foreign scientists.[3] Its first expedition in 1966 was to Australia and the Great Barrier Reef.

Frits Went participated in the *Alpha Helix*'s second expedition, to the Brazilian Amazon's tropical forest in 1967. For this expedition the ship carried two disassembled prefabricated air-conditioned laboratories that could be set up onshore, using power supplied from the ship. In addition to the main laboratories and ship, there were small powered boats and a single-engine amphibious airplane that extended the expedition's operating range. The ship operated in the region of the confluence of two rivers, the Rio Negro and Rio Branco, over 300 kilometers upstream from Manaus, Brazil, and about 1,600 kilometers upriver from the seacoast city of Belem. The expedition got under way in early February, and different scientific programs defined the research agenda at different times of the year. The plant physiology program ran from mid-August through September.[4]

Joining the expedition with Went was Nellie Beetham Stark, an ecologist and soil scientist who had come to Nevada's Desert Research Institute (DRI) in 1964, following six years working for the U.S. Forest Service at its Southwest Forest and Range Experimental Station. During those years she was also a graduate student in botany at Duke University, where she received her M.A. degree in 1958 under W. Dwight Billings and her Ph.D. in 1962 under Henry J. Oosting, two of the leading ecologists of that time. Her doctoral dissertation investigated the ecological tolerance range of seedlings of the Sierran redwood (*Sequoiadendron giganteum*) in environments of three types: natural, logged, and frequented by tourists.[5] Her interest in both natural and disturbed environments was a theme that carried throughout her career. After Went became director of DRI's Laboratory for Desert Biology, she joined the laboratory as an ecologist.

At that time in the mid-1960s, there were relatively few women in forestry or ecology. Jean Langenheim, who published a historical survey of women ecologists in 1996, included Stark among the group of pioneering women who constituted the first modern wave of women ecologists, by which she meant women who got their degrees and established research careers between 1961 and 1975, coinciding with the rise of the environmental and women's movements.[6] Indeed, as a pioneer Stark was determined to forge a career in ecology, then an expanding discipline that was experiencing growth in several sub-fields, notably in physiological ecology, population ecology, and ecosystem ecology.

Stark was not originally going to be part of the *Alpha Helix* expedition to Amazonia, but she was eager to participate, having never traveled abroad, and tried to persuade Went to allow her to go.[7] His initial plan was to study air quality above the jungle canopy, which entailed climbing trees and hauling instruments up to take measurements above the tree tops. Went himself, in his mid-sixties, was not able to perform this task and thought he might hire a local boy to do the work, but Stark argued that she was prepared to do it, since she had been climbing daily to do her own research on transpiration (loss of water) from leaves in tree canopies. Went demurred, but finally agreed to consult Jacob Biale, botanist at the University of California, Los Angeles, who was organizing the botanical program for that portion of the expedition. Stark sent Biale a research proposal, which involved studying transpiration in the rain forest. During the rainy season in the Amazon, parts of the forest would become inundated with several meters of water, making it necessary to move through the jungle by boat or canoe. Stark proposed examining how the trees adjusted to this flooding and whether plants responded by speeding up the rate of transpiration.[8] Her letter to Biale included the humorous suggestion that a lady aboard ship could also help out by administering medicines and applying creams to soothe bug bites. Biale appreciated the humor as well as the research proposal and approved her participation. Stark ended up being the only woman in the group.

Upon arrival at Belem that August, however, Brazilian customs agents confiscated their scientific equipment. Unable to get it released, they were advised to proceed to Manaus, where the *Alpha Helix* was docked, and try to get the equipment released there. The equipment did reach Manaus, but customs agents were still not willing to release it. It appeared that release would depend on making payments of an undetermined amount, which was out of the question. The scientists were at an impasse, but Stark knew a little Portuguese and managed to get through to one person, who advised them to stay on the *Alpha Helix* for a few days to let matters cool down.[9] A few pieces of equipment were released, but the scientists realized quickly that most of the equipment would not be released very soon, leaving many of them unable to do serious research. The expedition leader insisted that they must propose alternative viable research projects within a few days, or they would have to leave the expedition, for the *Alpha Helix* could not afford to stay docked in Manaus if they were unable to do anything.

The day before the equipment reached Manaus, Went, Stark, and three other scientists had visited the Adolfo Ducke Forest Reserve located just northeast of Manaus. This large reserve had been established in 1963; it is now one of the most important sites for rain forest ecology in the world. Went's diary indicates that they spent five hours there but does not record what they did.[10] Stark had been fascinated by that visit and, faced with the urgency of coming up with a new proposal, she hiked back to the reserve on her own, a trek of over twenty kilometers, to spend more time exploring the rain forest. Once off the trails, she noticed something unusual. The ground was soft and roots were growing on top of the ground, mingled with forest litter and exposed to light and air. Picking up the litter and pulling the roots away, she saw tiny white threads breaking away from litter and roots. She recognized that these were the hyphae (thread-like filaments) of mycorrhizal fungi. An idea started to emerge, as she recalled: "I sat beside the trail for a long time and put together the hypothesis that these giant trees, some 160 feet tall, were growing on nutrient depleted soils that were so deficient that the trees were sending roots into the litter to scavenge nutrients from their own recent litter fall!"[11] This fitted with what she knew about nutrient deficiency in Amazonian soils. On the hike back to the ship, she worked out the theory in greater detail and presented it to the other scientists. Went recognized its significance right away, and the two now had an alternative research proposal for a collaborative study that would involve exploration of the role of mycorrhizae in nutrient cycling in the rain forest.

These mycorrhizal studies changed the way Went and Stark thought about the ecological role of these fungi and led to a new hypothesis about direct nutrient cycling, published in 1968. That work generated attention because of its connection to an important problem: how was it possible for lush tropical forests to grow on nutrient-poor soils? Went's biographers commented that the ideas about the ecological role of mycorrhizal associations that Went and Stark proposed as a result of this expedition "energized research in this field."[12] But when the hypothesis was proposed, there was no direct evidence to support it. Finding that evidence would occupy Stark's time for the better part of a decade.

In addition to these studies of mycorrhizae the scientists were also able to resume their original projects after their equipment was finally returned to them. Stark did help Went with his analysis of the air above the canopy, even risking a lightning strike when a storm suddenly blew in while she was high in a tree top.[13] She climbed the trees using a local method that involved placing a hoop (which could be made from an ordinary belt) around the ankles, wrapping one's feet around the tree while pressing the hoop against the trunk, and then shifting upward by alternating the use of hands, knees, and feet.[14] She also worked on her transpiration project as planned. In addition to the Brazilian expedition, Went and Stark extended their stay with visits to sites in eastern Peru, funded by the Smithsonian Institution. They had no difficulty with customs agents in Peru, since Stark knew Spanish and was able to explain the scientific purpose of the equipment.[15] On their return to the United States, following publication of the direct nutrient cycling theory, Stark took up this problem in earnest, returning to the tropics and teaming up with other scientists who also wanted to explain how lush rain forests could grow on poor soils.

How Can Lush Forests Grow on Poor Soils?

Albert Bernhard Frank (1839–1900), a German botanist, coined the term "mycorrhiza" in 1885 to describe the symbiotic associations between fungi and roots. His meticulous observations of truffle-producing fungi revealed that the fungus was an "inalienable part of every beech and oak tree" in the region he studied. Yet its significance had been overlooked, in part because botanists who had observed similar fungal mycelia in other locations had assumed they were injurious to plants. Frank agreed that the fungus benefited from the association, since it depended on the photosynthesizing tree for carbon compounds (in the form of sugar). But he concluded that the fungal mycelium inflicted no harm to the tree. On the contrary, the relationship was a true symbiosis, with the fungus providing key nutrients to the tree. Indeed, he granted the fungus a very large role, arguing that "the fungus takes up soil minerals not only for its own nutrition but also for that of the tree, so we must consider that the root-fungus is the sole organ for the uptake of water and soil nutrients by oaks, beech, etc. It functions as a wet-nurse of the tree in this respect."[16]

Frank's hypotheses were controversial during his lifetime, but by the 1950s the importance of mycorrhizal fungi for certain kinds of plant growth (such as growing orchids) was well known. Scientists agreed that this relationship was primarily symbiotic, not parasitic, with both partners benefiting from the exchange of materials essential for growth. However, there were differences of opinion about the exact nature of the exchange in different cases. Most studies had examined temperate forests or orchids, but little was known about tropical forests. Went and Stark's trip to the Amazon focused their attention on the role of mycorrhizae in the tropical forest. In two important articles published in 1968 in *BioScience* and the prestigious *Proceedings of the National Academy of Sciences* on their return, Went and Stark laid the groundwork for a hypothesis that would give these fungi a more important ecological role than had been assumed.[17]

Since the 1930s, scientists had thought that the paradox of luxuriant tropical forests growing on poor soils might be explained by nutrient cycling, but there were no detailed studies of tropical nutrient cycles. In a study of rain forest ecology published in 1952, Paul W. Richards (1908–1995), professor of botany at the University College of North Wales, Bangor, advanced the idea that nutrient cycling was key to understanding how these forests could be maintained.[18] He argued that the capital of plant nutrients was locked up in the living vegetation and the layer of humus on the forest floor, and he posited that any nutrients set free by decaying vegetation were taken up almost immediately by plant roots in the upper layers of soil, producing a "very nearly closed cycle" of nutrients between humus layer and living plants.[19] Although he noted that some scientists viewed the presence of mycorrhizae as significant, the role of mycorrhizal fungi was controversial at that time and he declined to discuss it further.

The theory proposed in Went and Stark's articles of 1968 built on Richards's ideas but asserted that mycorrhizae did indeed play a crucial role in tropical forests where soils were relatively poor in nutrients. They contended that the nutrient cycle in the tropical rain forest on white sands was really closed and that mycorrhizal fungi provided the closing link by cycling nutrients directly from dead organic matter to the living roots. One question, however, concerned exactly what type of mycorrhizal fungi performed this vital link.

Following Frank's lead, scientists had identified two main groups of mycorrhizae, *endo*- and *ecto*trophic mycorrhizae, based on how the fungus connected to the host plant's roots (while also recognizing two other groups with intermediate properties).[20] In endotrophic mycorrhizae, the hyphae of the fungus penetrated directly into the cortical cells of the root, establishing a very close connection between fungus and root. Endotrophic mycorrhizae were in turn divided into three subgroups; Went and Stark's study involved the most common subgroup, known as "vesicular arbuscular mycorrhizae" or simply "arbuscular mycorrhizae," so-named because the fungi penetrating the roots formed structures called vesicles and arbuscules. (Knowledge of these links was modified by later anatomical studies).[21] The two other groups were ericoid and orchid mycorrhizae. In ectotrophic mycorrhizae, there was no direct penetration of the root cells. Instead, the fungal hyphae formed a mantle outside the root and in between the root's cortical cells.

Went and Stark's hypothesis assumed a difference in the roles of mycorrhizae that were dominant in different kinds of forests. Temperate forests were better studied, and in those forests ectotrophic fungi appeared to be unable to digest freshly dropped branches or cellulose but were able to attack decayed wood. Went and Stark suggested that there was no reason to assume that tropical mycorrhizae behaved the same as temperate mycorrhizae. Since the litter of tropical forests had an abundance of endotrophic mycorrhizae, they focused attention on this group. Went's diary recorded his growing conviction that endotrophic mycorrhizae were important, based on observations during a visit to a biological station in the Alto da Serra forest near São Paulo in southern Brazil.[22] Microscopical observations of the root system clearly showed single fungal hyphae growing within the older root hairs.[23] Stark thought that the litter was partly decomposed by bacteria and fungi while the leaves were still in the canopy, and when nutrients reached the forest floor they could be picked up by mycorrhizal fungi; she would return to this idea in later publications.[24]

Because the fungal hyphae penetrated the plant root cells, nutrients released from the forest floor could be given directly back to the plant. This process was known for orchids but was not thought to be important for trees. The 1968 articles argued that tropical forests appeared to have "largely endotrophic mycorrhiza" and suggested that these might "function more like orchid mycorrhiza than like temperate tree mycorrhiza" and could possibly digest cellulose and lignin.[25] This "direct" nutrient cycling theory gave the fungus an ecological role of great significance, if it could be shown that the fungi did pass minerals and other food substances to living root cells through their hyphae. Nutrients therefore did not pass into the soil, where they would be leached away by heavy rains, but instead were directly recycled back to the plant.[26] However, the articles did not include anatomical evidence identifying the type of fungus involved in the association, and the announcement of the theory in *BioScience* was partly a call for more research.

"If the direct nutrient cycling idea proves correct," Went and Stark stated, "we have the beginnings of an explanation of the enigma of lush forests on poor Amazonian soils."[27] Understanding the key role of mycorrhizae also explained why slash and burn techniques used to clear land for agriculture quickly depleted the land, forcing people to abandon those lands after two or three years. To achieve less damaging agricultural development in the Amazon, Went and Stark suggested, knowledge was needed of the ecological relationships involved in nutrient cycling. These conclusions coincided with growing concern about the Amazon, seen to be on the verge of an ecological crisis as population growth and development of the Amazon caused rapid clearing of the jungle. In January 1969, the Association for Tropical Biology (formed in 1963) held an international symposium and round-table discussion in Colombia to address the impending crisis. Went and Stark's theory was a prominent part of these discussions.[28]

Back in Nevada, Went and Stark pursued work on mycorrhizae in desert plants, helped by the mobile laboratory units of the Desert Research Institute. In the desert, too, they found fungi commonly associated with roots and theorized that the mycelium—the mass of thread-like, branching hyphae—also had a mechanical role helping to bind the soil. Went thought that the plant distributions he had observed in earlier studies of desert ecology, notably the lusher growth of annual plants near shrubs, could also be explained by the presence of fungi that were feeding on the dead organic material produced by the shrubs. He concluded that more attention needed to be paid to the mechanical as well as biological role of fungi in soils.[29]

James M. Trappe, a mycologist and expert on North American truffle species, now retired from Oregon State University, noted that Albert Bernhard Frank's hypotheses in the mid-1890s on the symbiotic associations between fungi and roots had met with considerable opposition. Trappe commented that Frank's particular genius lay in his "intuitive ability to interpret the meaning of what he discovered in light of the evidence and draw the logical conclusions, disregarding conventional thinking. Moreover, once convinced he was on the right track, he was bold enough to put his reputation on the line by announcing it to his peers."[30] The collaborative work of Went and Stark, building on Stark's initial eureka moment in the forest reserve, had this same quality of intuition and boldness. When first proposed, the hypothesis had no direct evidence in support of it, and many years of research were required to find the evidence.

It was Stark, rather than Went, who conducted the sustained research over the next decade that would help to provide more evidence for the hypothesis. In fact, comments by Went a decade later suggest that Stark was responsible for shaping most of the direct nutrient cycling theory in their two co-authored articles. At a conference in 1977, Went explained that he did not wish to discuss the environment of roots and soil in any depth because he had only become aware of the subject recently and still felt himself "too much a 'babe in the woods.' "[31] In that talk, he took credit not for the direct nutrient cycling thesis but for the discovery that, in the desert, soil fungi helped to bind the soil and fix dunes. Went did little further research on the subject after 1968.

Stark, on the other hand, had come up with the hypothesis of direct nutrient cycling and devoted the next decade to finding evidence to support it. She returned to the Amazon three times to follow up on aspects of the hypothesis, publishing her results in a series of single-author articles in the early 1970s. She made comparative studies of Brazilian forests and the white sand savannas of Surinam, as well as detailed studies of several sites in Central and South America, where she measured the nutrient content of wood, leaves, bark, roots, litter, and litter fall. She found that litter was rich in biologically important elements and therefore was a likely source of nutrients if the soils were very poor.[32]

The mycorrhizal research community in North America was at that time relatively small, although studies of mycorrhizae in North America had been going on for more than half a century. The first North American conference on mycorrhizae was held in 1969 and was jointly sponsored by the University of Illinois and the U.S. Department of Agriculture. It involved a small group of fifty-six participants, including Went and Stark. Edward Hacskaylo, a specialist in the study of ectomycorrhizae at the Plant Industry Station of the USDA's main research arm in Beltsville, Maryland, served as co-chair of the conference. His paper, a review of new research on ectomycorrhizae, revealed that scientists were starting to appreciate the "indispensable" roles that these fungi played for the survival of associated organisms.[33] Evidence pointed toward the ubiquity and ecological importance of mycorrhizal associations, which in Hacskaylo's opinion were "among the best examples of balanced reciprocal parasitism in existence."[34] He acknowledged Went and Stark's ideas about the role of these fungi in degrading and transporting nutrients but thought that too little was known about the contribution of other microorganisms to this process. Paul Richards also accepted the idea that fungi "played an essential role in the process of decay" and helped in the transport of nutrients in tropical rain forests, carrying them efficiently "door to door" from decomposing matter to plant roots.[35] However, he did not explicitly endorse the idea that mycorrhizal fungi were themselves decomposing plant litter.

While there was immediate interest in the hypothesis, there were many missing pieces of the puzzle, and research would continue for several more years. The history of this hypothesis was not at all straightforward. It bifurcated along two main paths, each emphasizing different claims within the original publications of 1968. On the one hand was the claim that fungi *retained and transmitted nutrients* efficiently, thereby preventing loss of nutrients to the soil. On the other hand was the claim that some fungi, specifically *endo*mycorrhizae, were also *decomposers of plant litter*. The scientific judgments of the fate of the hypothesis—that is, conclusions about whether it was proven, disproven, or neither proven nor disproven by later research—differed, depending on how scientists understood what the hypothesis was proposing and which part they chose to examine.

Proving That Mycorrhizae Transmit Nutrients

Stark herself approached the problem from an ecological perspective that focused on the question of whether fungi transmitted nutrients. She found that fungi acted as nutrient sinks, holding on to important nutrients that would otherwise be washed away and lost.[36] This finding was part of a broader interest in the role of litter and litter organisms in nutrient cycling and in the maintenance of what would later be called "ecosystem services." Stark explicitly drew out the ecological significance of this research for land management: to maintain high water quality in managed lands, for instance, it was essential to understand the nutrient retention mechanisms of soil and litter and the cycling pathways on land.[37] She was always interested in comparisons between undisturbed ecosystems and systems altered by different land uses, such as agriculture or logging. Stark's follow-up studies were based on more intensive research involving comparisons of twelve study sites in Brazil and Peru, and they included comparisons of virgin and second-growth forests.

Her later statements of the "direct nutrient cycling hypothesis," which she developed in two articles published in 1971, gave a longer version of the hypothesis that emphasized the transportation of nutrients from dead organic litter to living tree roots by mycorrhizal fungi but set aside the question of whether the fungi were directly capable of decomposing organic matter. She envisioned fungi as forming part of a larger ecological system whose details remained to be worked out: "The fungal partners in mycorrhiza, together with help from bacteria, animals, and other members of the rhizosphere are capable of breaking down dead organic litter and can pass nutrients in some form to the living tree root cells."[38] She took for granted that many organisms contributed to decomposition and also suggested that decomposition could begin up in the canopy, where "some forces must be at work digesting the leaves which were caught in the crown before falling to the ground."[39]

The idea that something was going on in the canopy was prescient. A decade later in 1981, Nalini Nadkarni, then engaged in doctoral research on the effects of epiphytes on nutrient cycles in forest canopies, published her important discovery that direct nutrient cycling was actually going on high up in the canopy of both temperate and tropical forests.[40] Her work drew on the studies of Richards, Stark, and other ecologists who were studying nutrient cycles in the rain forest. The canopy held dense mats composed of living and dead epiphytes, along with tree foliage and decomposing bark. Trees were capable of tapping into these nutrient reservoirs by putting out "adventitious roots" that penetrated the mats. She also found endomycorrhizal hyphae and vesicles inside the roots. These discoveries showed there were mechanisms operating in the canopy to conserve and retain nutrients, all critically important for the support of rain forests growing on heavily leached soils.

Stark's plan following the publication of the direct nutrient cycling hypothesis in 1968 was to use radioactive tracers to follow the decomposition products of litter on the forest floor.[41] Radioactive tracers had been used since the 1950s to track how nutrients were being passed between fungi and plants. Elias Melin, a Swedish scientist who was one of the pioneers of mycorrhizal research, published a series of studies starting in 1950 that used radioactive isotopes to track how products of photosynthesis moved from plants to ectomycorrhizal fungi and how the fungi were passing nitrogen, phosphorus, and calcium to pine seedlings.[42] Use of radioactive isotopes quickly became standard in mycorrhizal research. In the 1970s,

Stark teamed up with other scientists working in the tropical forests of Venezuela, and with radioactive tracers they found the first supporting evidence for the direct nutrient cycling theory.

Concern about destruction of the Amazonian rain forest had prompted interest in knowing more about the mechanisms that conserved nutrients in undisturbed forests, and how cut and burn practices might damage them. Scientists interested in tropical research were also sensitive to charges of scientific imperialism, or the practice by foreign scientists of studying the tropics but taking their knowledge back home without contributing in any way to the development of local scientific expertise. A workshop on the Neotropics in 1973 identified pressing research priorities but at the same time emphasized the need to eschew scientific imperialism and instead to reinforce local institutions, share information openly with host institutions, and provide training opportunities for young professionals, technicians, and students from the local region.[43]

Ernesto Medina, one of the scientists at this workshop, was the head of the laboratory for physiological plant ecology at the Centro de Ecologia (Center for Ecology) of the Instituto Venezolano de Investigaciones Cientificas (IVIC, the Venezuela Institute for Scientific Research). On his initiative the center organized an international course on tropical ecology at IVIC. Invited participants included Hans Klinge, a forest expert at the Max Planck Institute for Limnology in Plön, Frank Golley, ecologist at the University of Georgia, Carl Jordan, also affiliated with the University of Georgia, and Ariel Lugo, a tropical ecologist from the University of Puerto Rico. Rafael Herrera, a soil scientist who had been teaching and doing research at IVIC since 1965, also participated. Just after the course, as Herrera recalled, members of that group decided to start a collaborative project to study the structure and function of Amazonian forests and soils.[44] They made an expedition to the Amazonas state in Venezuela to find a suitable location where undisturbed forests thrived on very poor soils, and they selected a forested area near San Carlos de Rio Negro, in the north-central drainage basin of the Amazon River in Venezuela.

The study involved cooperation among institutions in Venezuela, the United States, and Germany.[45] In line with the objective of combating scientific imperialism, the study was headquartered at the Center for Ecology at IVIC in Caracas. However, a multiyear interdisciplinary project of this scope was beyond the resources of Venezuela. The Venezuelan institute invited the participation of scientists from the United States and Germany. In addition to Medina and Herrera from IVIC, the project included Frank Golley and Carl Jordan from the University of Georgia, Hans Klinge from the Max Planck Institute for Limnology, and Eberhard Brünig, a tropical forester from the University of Hamburg.[46]

In 1976 the Venezuelan government signed an agreement with UNESCO to create an International Center for Tropical Ecology within IVIC, and in 1977 the Amazon project was named as a pilot project of UNESCO's Man and the Biosphere Program. To ensure that the project did benefit the scientific and technical infrastructure of Venezuela, all processing of samples and most of the data analysis were done in Venezuela. An international steering committee consisting of Medina, Herrera, Jordan, and Klinge coordinated the endeavor.[47] This project established a series of permanent research sites available to many researchers and students; these sites are now part of a vast international array of permanent plots in a partnership called ForestPlots.net, which monitors and analyzes the world's tropical forests.[48]

The scientific studies inaugurated in the 1970s were designed in part to test hypotheses that had never been tested and to identify what the natural nutrient-conserving mechanisms were.[49] That included testing the direct nutrient cycling hypothesis, and Stark became involved in the project. She had left Reno in 1972 to take a position at the School of Forestry at the University of Montana in Missoula. She started as a research associate and was then a professor of forest ecology from 1979 to 1992, when she retired. She was the first woman faculty member in the school. During the 1970s she continued her research in tropical ecology and joined the Venezuelan project later in the 1970s, working with Herrera and Jordan.

Herrera was at that time also completing a Ph.D. at the University of Reading, England, where he had enrolled in 1974. He completed his degree in 1979, based on his research in the Venezuelan rain forest. He had read some of the classical works on the rain forest, including Paul Richards's book, but he was also influenced by the work of his adviser at Reading, Dennis J. Greenland, an outstanding soil scientist who in the mid-1970s was deputy director general at the International Institute for Tropical Agriculture in Nigeria. In 1960 he and Peter H. Nye had published a study, *The Soil Under Shifting Cultivation*, which interested Herrera because it was a pioneering analysis of nutrients in tropical forests and the role of soils in nutrient cycling in Ghana.[50] Herrera's thesis took up the problem of nutrient cycling in the Amazon forests, especially the idea that forests growing on extremely poor soils would show very efficient nutrient conserving mechanisms.

Carl Jordan had come to the Amazon project with several years of experience studying tropical forest mineral cycling. After obtaining his Ph.D. in plant ecology from Rutgers University in 1966, where he first became interested in mineral cycling, he joined an ongoing project on the tropical rain forest of Puerto Rico. That project had started in 1963 under the leadership of Howard T. Odum and under the auspices of the U.S. Atomic Energy Commission and the Puerto Rico Nuclear Center (later renamed the Center for Energy and Environmental Research). In the early 1960s there was much discussion of how atomic bombs could be used in civil engineering projects, an idea that Edward Teller, nuclear physicist and father of the hydrogen bomb, was promoting under the name Project Plowshare. One of the most ambitious proposals was to build a sea-level canal across Panama, using nuclear bombs to excavate the canal. The Puerto Rican study was meant in part to obtain ecological data that would help to evaluate the effects of using nuclear bombs in Panama. It therefore included studies of radiation on the rain forest ecosystem and studies of mineral cycling that were connected to the Atomic Energy Commission's interests in fallout, earth moving, and waste disposal. The project also sought to understand the basic features and processes of the rain forest ecosystem, especially in relation to problems involving radiation.[51]

Although he remained involved in the project, Odum left Puerto Rico in 1966, and Jerry R. Kline, a soil scientist and ecologist, was head of the project when Jordan arrived in July 1966. By then, the emphasis was shifting from the immediate effects caused by radiation to long-term recovery from radiation, and Kline and Jordan focused on mineral cycling in the tropical ecosystem.[52] For this research Jordan received the George Mercer Award from the Ecological Society of America in 1973 (recognizing an outstanding publication by a young scientist).[53] The study of mycorrhizae was not a major part of the Puerto Rican project, but it was not totally absent. In the 1960s, Odum had been one of the main advocates of a holistic ecosystem approach, which emphasized cycling of matter and flow of energy through ecosystems, and he recognized the importance of diverse microbial processes for nutrient cycling. In his final report on the project, in which he cited Went and Stark's *BioScience* article, he noted that most of the roots of the main trees were found to be mycorrhizal, and he accepted the idea that these fungi were important for conserving nutrients during drenching tropical rains.[54]

In 1969, Jordan moved to the Radiological and Environmental Research Division of the Argonne National Laboratory in Illinois, where he continued research on mineral cycling until the opportunity arose in 1974 to help coordinate the Venezuelan Amazon project. Research at San Carlos de Rio Negro examined two main forest types in the north-central drainage basin of the Amazon; they provided an opportunity for a comparative study of forest ecosystems.[55] The two forest types grew under similar climate conditions but had different soil environments and produced litters with contrasting nutritional quality.[56] The Tierra Firme forest was mixed tropical rain forest growing on top of low hills with well-drained soils (called oxisols) that were highly weathered, rich in iron and aluminum, but low in fertility. This forest was characterized by a dense superficial root mat that could be "peeled back from the soil like a carpet," which made it relatively easy to collect water passing through the root mat.[57] On the flat bottom of the valleys between the hills, forests grew on white sands that were often waterlogged during the rainy periods. These forests were of two types called "tall Amazon Caatinga" and "low Amazon Caatinga" or Bana, the latter being isolated stunted woodlands on bleached sands.[58] In Caatinga forests the roots were embedded in a humus matrix, making it more difficult to measure water percolating through the root mat.

A working hypothesis of this project was that, given the limited amount of nutrients available in these forests, there must be "enhanced structural and physiological mechanisms for absorbing and recycling nutrients."[59] As

Jordan and Medina explained in an overview of the San Carlos project, in addition to gaining valuable knowledge for forest management, an overarching goal was to discover how Amazonian ecosystems differed from temperate ecosystems and to interpret these differences in terms of general ecosystem theory.[60] Nutrient cycling, being an important function of ecosystems, was a central component of this study, and testing the Went-Stark hypothesis about the role of mycorrhiza became one aspect of the larger project. As Stark and Jordan wrote in 1978, "Despite the importance of the idea of direct cycling of nutrients in the Amazon rain forests, there has been little positive evidence that it occurs."[61]

Stark and Jordan examined the role of the root mat in conserving nutrients at two sites with different soil types. They used radioactive tracers to demonstrate conclusively that the root mat did take up dissolved nutrients before they could leach into the soil.[62] They found, for example, that 99.9 percent of all radioactive calcium and phosphorus sprinkled on the root mat was captured by the mat, and only 0.1 percent leached through the mat. They suggested that mycorrhizal fungi, as well as soil microorganisms such as algae and bacteria, had roles in nutrient transfers in the Amazon rain forest. However, their study did not deal with the mechanisms in the root mat that were responsible for this recycling.

Herrera's research, part of his doctoral dissertation at Reading, looked more closely at these mechanisms. By this time, international visitors from many countries were joining excursions to the experimental sites. "For me," Herrera remembers, "the visit of Nellie Stark from the University of Montana and the lively discussion we all had every evening after field work, were perhaps the most inspiring."[63] Jordan, Stark, and Herrera discussed ways to design experiments to study the close contact between decaying litter and root mat. Stark then returned to Montana, and Herrera returned to Caracas. There at IVIC he found plastic capsules resembling Petri dishes but with an opening on one side. Jordan ordered a radioactive phosphorus tracer, and Herrera brought in Tatiana Mérida, an expert in electron microscopy from the Universidad Central de Venezuela. Herrera and Mérida headed back to San Carlos to deploy these capsules on the soil. Each was baited with leaf tissue marked with a radioactive tracer. Eight weeks later the two returned to San Carlos and found that roots had grown into the capsules, attaching themselves to the leaf litter. Returning to Caracas for closer analysis, they could see mycorrhizal hyphae bridging the decomposing leaves and the roots, while radioactivity was detected in leaves, fungal hyphae, and root tissues. This suggested that phosphorus had transferred from the radioactive leaf to the living roots through the fungal hyphae.[64]

Herrera called Stark to convey the good news and immediately started to write a manuscript that they thought could be submitted to *Nature* or a similar high-impact journal. But *Nature* rejected the paper and, at Klinge's suggestion, it went instead to the German journal *Naturwissenschaften*. The study, co-authored by Herrera, Mérida, Stark, and Jordan, concluded that "the direct cycling mechanism here reported can be considered of importance" in nutrient-poor environments.[65] These mechanisms worked in concert with other means of conserving nutrients. But these experiments did not establish which kind of mycorrhizal fungus (endo- or ecto-) was forming the bridge. Supporting evidence would come later from Herrera's research group.

For Herrera this exciting result opened new research opportunities in soil ecology at IVIC, where he trained a series of students who made many contributions to mycorrhizal ecology. These included Gisela Cuenca, who later formed her own group at IVIC; Marcia Toro, who went on to a Ph.D. at the University of Granada, Spain; Alicia Cáceres, who continues her research at the Universidad Central de Venezuela; Tania de la Rosa, now a researcher in Finland; and Bernard Moyersoen, who came to Herrera's lab from Belgium. It was Moyersoen's research in the 1990s that documented the predominance of vesicular-arbuscular mycorrhizae in Amazonian forests and showed that many species formed double symbioses with both ecto- and endomycorrhizae.[66]

Jordan and his colleagues also found that lichens and mosses growing above the root mass fixed nitrogen from the air and scavenged nutrients from rain. A *Science News* report in 1977 on this growing body of research on the capture and conservation of nutrients described these new insights as "startling."[67] Jordan and Herrera recognized that several nutrient-conserving mechanisms were likely to be at work in these Amazon rain forests, and they argued that nutrient-poor (oligotrophic) and nutrient-rich (eutrophic) forests represented extremes on a spectrum of forest types: it was essential to recognize these differences when drawing comparisons across different ecosystems.[68] As Jordan and Medina wrote in 1983, "mechanisms such as the above-ground root mat and direct recycling by mycorrhiza are adaptations to help the ecosystem survive in the nutrient-poor conditions. The implication is that destruction of the Amazon forest on a large scale will cause an irretrievable loss of nutrients and consequently of the ecosystem, because large scale clearing destroys the nutrient conserving mechanisms."[69] Jordan's work on tropical forests continued through the 1980s, while he was on the faculty of the University of Georgia (where he is now emeritus professor). He studied problems of conservation and rehabilitation of tropical forests in regions of intensive pulp production and mining, and later in the 1990s he developed an interest in agroecology and sustainable agriculture in the American South.[70]

Stark also related her research on direct nutrient cycling to sustainable forest management. By the late 1970s she had done several comparative studies of different forests and types of soil in temperate, tropical, and desert environments, and these comparisons led her to think more deeply about how long a soil would be capable of supporting a forest.[71] She proposed that nutrient cycling mechanisms changed during the long lifetime of a forest, as soil and plants evolved together. She argued that during the earlier stages in the development of a forest, indirect cycling occurred: nutrients were taken from inorganic soils. As soils became older, direct nutrient cycling became more important, and nutrients passed from organic to organic, largely bypassing inorganic soil. In cases of direct cycling, many organisms were involved in releasing nutrients, including bacteria, fungi, and litter animals, and some of those nutrients were passed to roots by mycorrhizal fungi. Any form of sustainable forest management had to take account of the nutrient-supplying and nutrient-retaining power of the soil.

During the 1980s, Medina and his graduate students made detailed investigations of the nutrient dynamics of the Amazonian forests at San Carlos. Elvira Cuevas (now Medina's wife and professor of biology at the University of Puerto Rico) was the first graduate from IVIC in the field of ecology in 1984. She used mesh bags filled with fresh litter to study litter decomposition and nutrient fluxes in different ecosystem types. Some bags were placed on the forest floor and left undisturbed to be colonized by roots. Others were placed on the soil but were lifted weekly to prevent root penetration and break hyphal connections between roots and litter. A third group was suspended in baskets above the soil surface, so that colonization only by airborne microorganisms was possible. In the Tierra Firme forest, calcium and magnesium were released faster when the litter was in continuous contact with roots, an effect not observed in the Caatinga or Bana forest. Cuevas concluded that different nutrient-release mechanisms were operating, mediated by roots and/or associated microorganisms. This in turn implied that mycorrhizal roots were operating differently in these ecosystems.[72]

In the light of all these investigations, I return to the question of the status of the Went-Stark hypothesis in the 1980s. How did scientists perceive it? Had it been confirmed? This turns out to be a complicated question not amenable to a yes or no answer, because the hypothesis had several components and biologists chose to focus on some while ignoring others. In a review of the field of tropical physiological ecology of plants published in 1980, a group of ecologists, including Medina, agreed that experiments such as those done by Herrera, Mérida, Stark, and Jordan confirmed that the roots of tropical forest plants could recapture phosphorus from leaf litter through mycorrhizal root connections. They in turn took these findings as confirmation also of Went and Stark's 1968 proposal. However, they also acknowledged that there was "a wide array of mechanisms by which tropical plants acquire and distribute nutrients" in environments where nutrients were not stored in soil.[73] In a review of tropical nutrient cycling published in 1986, ecologists Peter Vitousek and Robert L. Sanford, Jr. (who was one of the many doctoral students at the Amazon project in San Carlos) also supported the conclusion that mycorrhizal fungi could transfer nutrients such as phosphorus from decomposing leaves to roots. However, they noted that there was "no evidence that the mycorrhizae of trees actually decompose litter, and evidence is accumulating that the most abundant tropical mycorrhizae (vesicular-arbuscular mycorrhizae, or VAM) lack the enzymes necessary for decomposition."[74]

The study by Herrera, Stark, and others had not identified which mycorrhizae were involved, nor had it directly addressed the proposal that *endo*mycorrhizae (that is, vesicular-arbuscular mycorrhizae) were capable of decomposing fresh litter. In the 1970s, this part of the hypothesis was not Stark's main concern. For her, as well as others involved in the field study in Venezuela, the central claim of the hypothesis was that there were mechanisms for the conservation and recycling of nutrients, preventing their loss in nutrient-poor environments, and that mycorrhizal fungi had a role in that process of conservation and recycling. Ecologists agreed that in this respect the hypothesis was confirmed, but there remained the problem of whether mycorrhizal fungi could be active decomposers. That part of the puzzle would fall to mycologists doing laboratory and greenhouse experiments.

Fungal Links: Toward the "Wood-Wide Web"

The 1980s saw growing interest in mutualistic interactions among species in general. Robert May, a leading mathematical ecologist who wrote a regular column for *Nature,* commented in 1982 on this interest and predicted that "empirical and theoretical studies of mutualistic associations are likely to be one of the growth industries of the 1980s."[75] But his examples were confined to associations involving animals, and in reply David H. Lewis at the University of Sheffield in England pointed out that missing from May's forecast were mutualistic associations of fungi and bacteria with plants in all terrestrial ecosystems. Lewis further remarked that he and others had long "bemoaned the slow recognition by taxonomists and ecologists of the importance" of these kinds of associations.[76]

Lewis had studied under one of the giants in the field of mycology, John (Jack) Laker Harley (1911–1990), who spent most of his career at Oxford University, except for four years at Sheffield from 1965 to 1969.[77] Harley had enormous influence on the field of mycorrhizal research through his own work and the work of his students. His monograph *The Biology of Mycorrhiza,* first published in 1959 with a second edition in 1969, helped to generate interest in this field of research. He co-authored a textbook, *Mycorrhizal Symbiosis* (1983), with his daughter, Sally E. Smith, who was an expert on mycorrhiza at the Waite Agricultural Research Institute in Adelaide, Australia. While at the University of Sheffield for four years, Harley worked to modernize its research facilities in botany, brought in his former research student, David Lewis, and encouraged the work of David J. Read, who joined the faculty at Sheffield in 1963 and became a world authority on mycorrhizal fungi. When he left Sheffield to return to Oxford in 1969, Harley left a greatly strengthened botanical department.

Harley was exceptional in having very broad interests that included ecology and forestry. In 1948 he published an influential review article that dealt with mycorrhiza and soil ecology.[78] Although he started out in the department of botany at Oxford, support for mycology waned after C. D. Darlington, a geneticist, became head in 1954, and Harley moved in 1958 to the department of agriculture. When he returned to Oxford from Sheffield in 1969, it was to the department of forestry, which he proceeded to transform. Others at Oxford viewed forestry as a weak discipline, not worthy of a place in a research institution, but by strengthening its scientific basis he bolstered its credibility as an academic subject. From 1961 to 1983 he was editor of *The New Phytologist,* which became one of the leading journals for mycorrhizal research, and he served as president of the British Mycological Society as well as the British Ecological Society. His presidential address to the ecological society in 1971, titled "Fungi in Ecosystems," made a strong case for studying fungi in order to answer ecological questions. However, he also pointed out how difficult it was to place fungi in a broader ecological context. In studies of nutrient cycling or energy flow in ecosystems, he cautioned, trying to define the part played by fungi "is a long lifetime's work" entailing a lot of fundamental research "before relevant information can be obtained on some of the most important problems."[79]

At the time that Lewis drew attention to Robert May's omission of plants, fungi, and bacteria in the burgeoning study of symbiosis, the University of Sheffield was one of the centers of fundamental mycorrhizal research, with many eminent mycorrhizal specialists on its faculty. By the mid-1980s, new evidence showed that some mycorrhizal fungi could decompose organic matter and could in consequence directly cycle nutrients from leaf litter to plant roots in the way that Went and Stark had imagined. In the following discussion I focus mainly on the Sheffield group, especially the work of David J. Read, who with his collaborators contributed many important

experimental studies, to show how scientists were trying to integrate laboratory findings and field studies on mycorrhizal research within a broader
ecological and evolutionary framework.

Read joined the University of Sheffield as a junior research fellow in
1963 and spent most of his academic career there; in 2007 he was knighted
for his contributions to biological science and is now emeritus professor of
plant sciences at Sheffield.[80] Read began his extensive research on mycorrhizae in the early 1970s. His work, along with that of his colleagues, collaborators, and students, ranged across all groups of mycorrhizal fungi and
uncovered many interesting features of these underground networks during
the 1980s, often with the help of radioisotope tracers. One set of experiments in the early 1980s used both laboratory and field studies to tackle
earlier hypotheses that had never been resolved, which involved the transfer
of materials such as carbon and phosphorus from one plant to another plant
along a direct hyphal pathway. Some plants appeared to "donate" nutrients
to other plants through the fungal pathway, although the extent of the donation and its ecological significance was not yet clear. This possibility had
not been widely recognized at that time.[81]

The movement of nutrients from donor to receiver plants was potentially of great ecological value. For example, such supplies could nurture
seedlings through their early stages of development, when there might be
intense competition for both nutrients and light from larger established
plants. Read and his collaborators concluded that these attributes of the
mycorrhizal system could "improve the vigour of the individual receiver
plant, optimize the efficiency of resource distribution within the plant community and enhance the conservation of nutrients at the ecosystem level."[82]
These comments were made at a special symposium of the British Ecological Society in April 1984 at York, which focused on ecological interactions
in soil. The editors of the published volume of papers, of whom Read was
one, commented that "it is increasingly realized that the mycorrhizal condition is the norm in natural vegetation, and the implications of this, particularly in terms of nutrient transport through interplant connections . . . must
be profound."[83] Read's paper at the symposium discussed those implications. As one reviewer of the volume noted, the key message was that "the

separation of soil from the rest of a terrestrial ecosystem is a travesty" and that this area of ecology was replete with "new ideas which should be more widely shared."[84]

Laboratory experiments by Read and his collaborators in the mid-1980s uncovered evidence that some mycorrhizal fungi could be very efficient decomposers and could obtain carbon as well as nitrogen from proteins, peptides, and amino acids in the soil, providing plants with key nutrients.[85] These experiments focused on ericoid mycorrhizae, an endotrophic mycorrhiza that occurs in heathland ecosystems in the Arctic and Mediterranean (in association with shrubs in the sub-family *Ericaceae*). The research of Read and his colleagues in Britain during the 1980s and 1990s strongly suggested that ericoid and ectomycorrhizae were operating not only as transmitters but as active foragers for nutrients bound up in organic matter. To study this process, he and his collaborators used transparent root observation chambers filled with peat, to which was added organic material that served as a source of nitrogen and phosphorus.[86] Pine seedlings inoculated with fungus were planted in the chambers, and it was possible to observe directly the growth of the fungal mycelium around and below the seedling roots, as the fungus colonized the nutrient-rich organic matter placed within the chamber. This colonization was associated with increased enzyme activity, which suggested that degradation of the nitrogen- and phosphorus-containing substrate was also enhanced.

Read also collaborated in the 1980s with a leading ecologist at Sheffield, John Philip (Phil) Grime (1935–2021), and through this collaboration strengthened the argument for the significance of mycorrhizal symbioses for community structure. Grime received his Ph.D. from the University of Sheffield in 1960 and except for a postdoctoral year at the Connecticut Agricultural Experiment Station was at Sheffield's botany department for his entire career. In the 1980s he was deputy director of the Natural Environment Research Council's Unit of Comparative Plant Ecology at Sheffield. He had long been interested in understanding why some plant communities showed higher diversity than others, a central question in community ecology. Theoretical models were based on the idea that different species exhibited different adaptive strategies depending on their environmental and ecological contexts. However, some of these theoretical approaches were based on animal species and did not conform well to observations of diversity in plant communities.

In the 1970s Grime proposed both a model and a general evolutionary theory to explain differences in plant species diversity. The model was a graphical representation of how diversity changed along a continuum that ranged from high stress or highly disturbed environments to those of low stress or disturbance.[87] Under conditions of high stress or disturbance, diversity was low, then it increased at moderate intensities of stress or disturbance (creating a hump in the graph), and then decreased again at the opposite extreme under conditions of very low stress or disturbance. He also identified three main evolutionary strategies characteristic of species in different environments. Some adaptations made species better competitors for resources, others made them more tolerant of stressful conditions, and others made species (called ruderal species) adaptable to highly disturbed environments.[88] These ideas gave rise to what became known as the CSR theory, where the letters stood for *competitors* (species that exploit conditions of low stress and low disturbance), *stress-tolerators* (species that exploit environments of high stress and low disturbance), and *ruderals* (species that exploit environments of low stress and high disturbance). These ideas were developed more fully in a book, *Plant Strategies and Vegetation Processes,* published in 1979.[89] That book discussed mycorrhizal fungi relatively briefly, drawing on previous studies suggesting the importance of mycorrhizal fungi for plants growing under low-nutrient conditions.[90]

In the 1980s, Robert Peet, a plant ecologist at the University of North Carolina, proposed a collaboration among several laboratories worldwide to do field experiments to study how species coexisted in species-rich grasslands. Grime was interested in this problem but realized that doing field experiments had serious drawbacks. As he explained, "Manipulations of ancient, complex ecosystems with incompletely known histories have the potential to set in motion chain reactions that may be hard to follow and have scant relevance to the functioning of real ecosystems."[91] "Against this uncertain background," he recalled, "I decided to conduct a radically different experiment in which simplified ecosystems would be allowed to assemble from seed under controlled conditions and we would examine the consequences for community development of '*leaving something out*' in order to assess its importance. The '*something*' in this experiment consisted of arbuscular mycorrhizal fungi."[92]

This study, conducted with Read along with M. L. Mackey and S. H. Hillier from the Unit of Comparative Plant Ecology, examined the effects of specific factors on diversity in laboratory microcosms that received different treatments.[93] The experiment ran for twelve months and assessed the impact of arbuscular mycorrhizal fungi, grazing (simulated by cutting vegetation with scissors), and soil heterogeneity (which affected the depth of roots and soil moisture). The study, published in *Nature* in 1987, revealed a pronounced effect on diversity from grazing and mycorrhizal infection in phosphorus-deficient soil. A striking result was that mycorrhizal infection reduced the yield of the canopy dominant, *Festuca ovina* (sheep fescue), while strongly stimulating yields of other species. Studies of the movement of carbon using radioactive tracers showed that the dominant canopy plants were losing carbon to the mycorrhizal network, while understory plants gained carbon.

This research provided strong evidence in favor of earlier conclusions by Read and his collaborators that the export of carbon from "source" plants to "sink" plants through a common mycorrhizal network "may be an important element of the mechanism maintaining species-rich communities in infertile soils."[94] This project also showed the value of supplementing field studies with experiments on artificial communities of different sizes (microcosms and mesocosms) where certain variables such as nutrient supply or grazing could be controlled, an approach that Grime continued to develop.[95] (Grime recalled that the experiment was also memorable for another reason altogether: "Twenty-four hours after the experiment was harvested on the 7th floor of the Biology Building our growth room was incinerated by an electrical fault, two staff were briefly trapped between floors in an adjacent lift and order was restored by the Sheffield fire department.")[96] The second edition of his book *Plant Strategies, Vegetation Processes, and Ecosystem Properties,* published in 2001, discussed these results and their implications more fully and affirmed the conclusion that arbuscular mycorrhizae increased the species richness of understory plants.[97]

By 1991, Read viewed the field of mycorrhizal research as having reached a crucial turning point. As he argued, "The first century of research on mycorrhizas has provided description of the major types of infection and a careful, largely laboratory-based, analysis of their function. We now enter a new phase in which the challenge is to evaluate the function of the symbiosis in the real world."[98] "Mycorrhizas in Ecosystems," echoing Harley's lecture on "Fungi in Ecosystems" two decades earlier, became the theme of the Third European Symposium on Mycorrhizas, hosted by Sheffield University in August 1991.[99] In a review essay also entitled "Mycorrhizas in Ecosystems," published prior to the symposium in 1991, Read placed the ecological role of different mycorrhizae within a broader ecological and evolutionary framework.[100] His review focused on three groups of fungi, ericoid mycorrhizal fungi, ectomycorrhizae, and vesicular-arbuscular mycorrhizae, each of which predominated in different regions of the world.

He pointed out that limiting nutrients such as nitrogen and phosphorus can be in short supply in ecosystems not because they are present in small quantities but because they are locked up in organic matter and therefore are unavailable to plants. Decomposers must break down complex organic compounds to release those nutrients, which can then be transmitted to higher plants by the mycorrhizal fungi. A key question was whether these fungi also contributed *directly* to this decomposition. Until the mid-1980s most scientists thought they did not, but that belief was now being challenged by experimental evidence. However, how mycorrhizae contributed to nutrient transport depended on the context: one had to understand what else was going on within the ecological community, including in the soil.

Read argued that it was important to focus on soil and on the different conditions that affected the rates of decomposition of organic matter. With changes in latitude or altitude came changes in climate; climate change in turn would affect the number and activity of different decomposers, for instance, the number and activity of bacteria. With shifts in rates of decomposition, one would also expect corresponding shifts in associations with mycorrhizal fungi. Taking an evolutionary perspective, Read proposed that natural selection should favor "associations between plants and those fungal mutualists that were capable of unlocking the two key growth-limiting resources," namely nitrogen and phosphorus.[101] Locations where nitrogen was limiting should favor one type of symbiosis, and locations where phosphorus was limiting should favor a different type. These mutualistic relationships would change as one moved around the globe and as climate and soil conditions also changed. Fungal mutualists should contribute to plant nutrition in different ways, depending on the environmental context. His review drew attention to the growing evidence that the role of mycorrhizae was far more important than had been realized.[102]

But natural systems are messy and dynamic: relationships between plants and fungi were constantly shifting through time and across space. Read recognized that, apart from changes due to climate, local variations would affect the quality of soil and availability of nutrients. As ecological communities aged and went through the processes of ecological succession, there would be changes in soil and in mycorrhizal relationships. The mycorrhizal infection, he concluded, would be important in determining what type of plant community predominated at each stage of the succession. But even within a climax community, different soil conditions would produce distinctive localized communities.[103]

Read did not refer to Went and Stark, and there was no reason to address their original articles directly, since a large amount of research had been done in the intervening two decades. However, within the community of scientists doing laboratory and greenhouse-based experiments, the status of the Went-Stark hypothesis was in fact still unsettled two decades later. As John Dighton remarked in a literature review published in the same issue as Read's 1991 article, "the general hypothesis proposed by Went and Stark for tropical forests has never truly been proved or disproved."[104] But he drew attention to evidence, such as Read's findings, suggesting that some mycorrhizal fungi, especially ectomycorrhizae and ericaceous mycorrhizae, did produce enzymes that allowed them to derive mineral nutrients and carbon from organic sources. These mycorrhizae grew on soils with high levels of organic matter but where climate limited the rate of decomposition. As Dighton concluded, the "direct nutrient cycling system as envisaged by Went and Stark . . . would be of advantage to plants growing under such conditions."[105] Thus the Went-Stark hypothesis evolved into a hypothesis about ectomycorrhizae and ericaceous mycorrhizae rather than the vesicular-arbuscular mycorrhizae of the rain forest, which was the original form of the hypothesis. But Dighton also noted that the subject was far from completely understood, in part because most work had been done under laboratory conditions. He called for more research to demonstrate the degradation of organic substrates by mycorrhizae in the field.

By the late 1990s, scientific studies had provided abundant proof that both ericoid and ectomycorrhizal fungi were able to decompose litter and that plants were not solely dependent on microbial generalists for their nitrogen and phosphorus. The research of Read and others helped to break down a functional distinction that had been made between "decomposers" (such as bacteria) that released nutrients and the "transmitters" of those nutrients, the mycorrhizae. In a review of the field published in 1997, Jonathan Leake and Read pointed out that new research showed that this separation of functions was not correct and that there was "increased interest in, and evidence for, direct recycling of nutrients from organic matter to plants by mycorrhizal fungi."[106] Although Went and Stark's thesis as originally stated was not found to be correct in all its details, their intuition that mycorrhizae did serve a direct role in nutrient cycling was increasingly supported by subsequent experimental work. Their theory in turn supported Paul Richards's ideas, advanced much earlier in 1952, about the importance of nutrient cycling in tropical forests. Many experiments conducted in the field and in laboratories over decades were required to build, in a slow incremental way, an understanding of the ecological role of mycorrhizal fungi.

Read continued to argue for field experiments to extend and test the validity of greenhouse and laboratory experiments, and he drew attention to innovative field research as it was published. In 1997 a key publication appeared in *Nature,* a study by Suzanne Simard and colleagues from British Columbia and Oregon.[107] This work was based on Simard's doctoral dissertation (completed in 1995) at Oregon State University and her later studies as a forester in British Columbia. These experiments used radioactive carbon tracers to determine whether carbon was passing from one plant to another via mycorrhizal fungi in a forest. Although such transfers were known from earlier experiments (including those of Read and his collaborators), it was not known whether they occurred in two directions, whether one plant gained at the expense of another in the field, and whether such transfers affected plant performance in the field. Simard's study found that there was bidirectional transfer of carbon and that one species could gain at the expense of another. In this case, carbon was moving through the mycelial network from birch trees (*Betula papyrifera*) to Douglas fir (*Pseudotsuga menziesii*), resulting in a gain for the Douglas fir.

On the cover of the issue of *Nature* in which Simard's study appeared, a photograph of a forest accompanied the headline, "The wood-wide web," in bold letters, introducing a term now commonly used as an expression of the intricate linkages between plants and fungi within ecosystems. As Read noted in a discussion of Simard's article in the same issue of *Nature*, the implication was that the role of competition should be reassessed, with emphasis instead placed on distribution of resources within the community. He proposed that "if mycorrhizal colonization results in an equalization of resource availability, as suggested by this and a number of microcosm studies, it would be expected to reduce dominance of aggressive species, so promoting co-existence and greater biodiversity." But this was just one step forward, as he wrote: "The challenge now is to quantify further the contribution of these fungal linkages to the maintenance of biodiversity and stability in terrestrial ecosystems."[108]

A Matter of Perspective

This was not the only challenge. Alastair H. Fitter, now emeritus professor at the University of York, England, and his collaborators also pointed out that interpretations of such below-ground interactions tended to be "phytocentric" or focused on what the interactions did for plants. He and his colleagues countered with a different perspective, asking "What's in it for the fungus?"[109] Fitter is a plant ecologist whose interest in soil ecology and plant-root systems went back to his graduate student days. He completed his Ph.D. at the University of Liverpool under Anthony (Tony) David Bradshaw, a plant geneticist who had done important studies on phenotypic plasticity in plants and the evolution of plant resistance to heavy metals. Bradshaw had been on the staff of the department of agricultural botany at

University College of North Wales (now Bangor University), where Paul Richards was on the faculty, before moving to Liverpool in 1968. His research at Bangor had focused on evolutionary biology, but after his move to Liverpool he shifted to restoration ecology and was one of the founders of this field.[110] As Fitter explained, "I was fortunate to be the first of a series of students who focussed on the reclamation of colliery waste, and I learned about the importance of soil (or the lack of it) very quickly. . . . As a scientist, Tony was exceptional in the ease with which he moved between the worlds of what were then called 'pure' and 'applied' science."[111]

During the 1990s Fitter and his collaborators did extensive research on mycorrhizal associations, zeroing in on the question of what the precise costs and benefits of these associations were to both parties.[112] This question was relevant to understanding how mycorrhizal associations could evolve: did they improve the fitness of fungi and plants? Since the fungi needed their plant hosts to obtain carbon, the benefit to the fungus was clear. But what about the plant? Fitter and David Robinson, who was at the Scottish Crop Research Institute in Dundee, evaluated Simard's results on ectomycorrhizal fungi, while also considering how arbuscular mycorrhizae might complicate the picture. They agreed that if plants could share carbon through a common mycorrhizal network, that type of resource-sharing would demand changes in the way ecologists viewed resource competition in plant communities, as Read had suggested. But they argued that Simard's results, as well as earlier experiments on carbon transfer between plants colonized by ectomycorrhizal fungi, were open to alternative explanations, so that more research needed to be done.

Turning to the different case of arbuscular mycorrhizal networks, Robinson and Fitter argued that it was possible that any transfers of carbon from one plant to another remained within the fungus's own storage structures, which were associated with plant roots. In their "mycocentric" view, the fungus could be moving carbon through its extended mycelial network in response to its own carbon demands. It was not responding to demands of its plant hosts, and plants were not "beneficiaries of non-Darwinian fungal altruism."[113] Such transfers of carbon might have no direct significance for plant fitness, therefore, but they would be significant for understanding fungal ecology.[114] This critique introduced an important dimension to this discussion: the question of perspective and the tendency to see the fungus-plant relationship from the perspective of the plant rather than the fungus. Different perspectives could lead to very different conclusions. Subsequent research has developed this argument and advocated explicitly for "myco-centrism" in considering such questions as how plants and mycorrhizal fungi respond to elevated levels of carbon dioxide.[115]

Recycling the Went-Stark Theory

Research by Fitter and his collaborators, published in 2001, also pointed to a different and larger role for arbuscular mycorrhizae in decomposing matter and obtaining nutrients from the decomposition.[116] In the 1980s, experimental evidence had suggested that arbuscular mycorrhizal fungi could not decompose dead organic matter, did not produce the enzymes needed for decomposition, and did not depend on litter for their carbon or energy.[117] Their role in mineral cycling was thought to be limited to transfer of inorganic mineral nutrients, especially of phosphorus (as inorganic phosphate) to plants. The new experiments by Fitter and colleagues challenged that idea, showing that, at least in these experiments, arbuscular mycorrhizae did colonize patches of decomposing organic matter and that the fungi appeared to promote decomposition and then acquire the nitrogen that was released, which increased the growth of fungal hyphae. The mechanism was not understood and they did not know whether other arbuscular mycorrhizae behaved the same way.

Biologists working with David P. Janos, a mycologist and tropical biologist at the University of Miami, Florida, also pointed toward a larger role for arbuscular mycorrhizae but did not conclude that the fungi were active decomposers. Janos had been engaged in research on mycorrhizal symbioses in tropical forests since the 1970s, with particular attention to vesicular-arbuscular mycorrhizae.[118] Many of the plant species he studied in the rain forests of Costa Rica depended on mycorrhizae for their growth, and his experiments on pot-grown plants confirmed that mycorrhizal infection improved their growth and survival. He joined the faculty at the University of Miami in 1979 and continued research on tropical forests, examining, among other problems, how mycorrhizae might influence ecological succession in the tropics.[119] In a study co-authored by Catalina Aristizábal, Emma Lucía Rivera, and Janos, published in 2004, the colonization of decomposing litter by arbuscular mycorrhizae was examined closely. The study's analysis confirmed that these fungi were indeed "very well positioned to scavenge mineral nutrients, even if mineralization is carried out by other microorganisms," and the authors concluded that arbuscular mycorrhizal fungi were "more important for retention and recycling of mineral nutrients in montane tropical ecosystems than previously recognized."[120]

A follow-up study by Janos and his colleagues published in 2019 came to a similar conclusion, but this time the researchers also revisited the Went-Stark hypothesis from a half century earlier.[121] This study was co-authored by Rebecca A. Bunn, Dylan T. Simpson, Lorinda S. Bullington, Ylva Lekberg, and Janos. Coincidentally, Bullington and Lekberg were associated with the University of Montana in Missoula, where Stark had established her career in soil science and forestry after leaving Reno. This study sought to explain the success of arbuscular mycorrhizal hosts in tropical forests. What matters for our purposes is the way this question brought the group back to the Went-Stark theory of 1968, which they described as languishing "in the bin of unsupported hypotheses" but which they now proposed to revisit and reframe. The way they referred to the Went-Stark theory suggests that it had acquired a status comparable to a scientific folk memory, like a persistent legend that continued to be passed down through generations, although its details were lost in the mists of time. In their account, the Went-Stark thesis was taken to be arguing that arbuscular fungi "alone" were accessing litter-bound nutrients. They did not cite later studies by Stark that would have shown her attention focused on the fungal role in transmitting nutrients and did not propose that mycorrhizal fungi acted alone. The field experiments on direct nutrient cycling from the 1970s also were not cited. The lack of reference to the line of research described earlier in this chapter suggests that the research communities had indeed been divided and that different constituencies within these communities had different perceptions of the fate of the original hypothesis.

The 2019 study noted that arbuscular mycorrhizae actively colonized leaf litter, and it applied sophisticated microscopic observations and molecular techniques to identify the fungi accurately, which Went and Stark had not been able to do in 1968. The study concluded that the effects of these fungi on decomposition "must be mediated by other organisms," particularly microbes, and that there was a "tight coupling" between arbuscular mycorrhizae and "decomposer groups that have historically been considered separate." The fungi, the authors hypothesized, might stimulate decomposition by other organisms and the nutrients released could then be transferred by fungi to host plants. They did not know how general this process was but concluded that there were good reasons to reconsider the direct nutrient cycling hypothesis, "not because we believe that these fungi can directly mobilize organically bound nutrients, but because of ample evidence that AM [arbuscular mycorrhizal] fungi can influence degradation of organic matter and subsequently acquire and transfer a portion of released nutrients to their associated host plants."[122] This study affirmed what Read had also pointed out: that the historical distinction made between "decomposer" groups and mycorrhizal fungi that had taken hold by the 1980s was not valid.

If one were to revisit not just the 1968 articles but also the later work by Stark and her colleagues in the 1970s, then the 2019 study could be taken as compatible with earlier conclusions, but with new details added. It is fitting, therefore, that these later authors thought to revisit the Went-Stark hypothesis, when they might have ignored it entirely. But the fact that for this later group the Went-Stark hypothesis had acquired something like the status of a folk memory, that is, something lost in time and only vaguely remembered, suggests that different groups of researchers had been operating in separate spheres. This chapter has aimed to convey a sense of those divisions and yet highlight efforts to bring these communities together. The next sections also examine the convergence of these disciplines from the perspective of different bridge-builders.

Connecting Disciplines

A common theme in the literature on mycorrhizal symbiosis has been the neglect of mycorrhizae in the broader ecological literature and the long-standing lack of communication between mycologists and ecologists. James

Trappe, a mycologist with ecological interests who since the 1960s has been working to close the gap between these communities, also remarked on how long it took to test and confirm A. B. Frank's prescient ideas.[123] He raised the important question of why ideas can be slow to catch on, even though the intuitions that lay behind them were found to be correct. Trappe thought one problem was the strong emphasis on competition in neo-Darwinian theory and the corresponding lack of interest in symbiosis, cooperation, or mutualism. That argument, which would apply to Darwinian theory as it developed in the mid-twentieth century, agrees with Robert May's comment that mutualism and symbiosis were becoming trendy subjects only in the early 1980s. Another difficulty that Trappe identified was lack of communication across disciplines. What was known and accepted within the community of mycologists was slow to be picked up by evolutionary biologists and ecologists. In a survey of references to mycorrhizae in the *Annual Review of Ecology and Systematics* between 1970 and 2003, Trappe found only twenty-three references to mycorrhiza, many made just in passing. However, he noted that the reluctance to "consider the importance of the below-ground ecosystem" was fading by the early twenty-first century.[124]

Such observations, published in 2005, echo similar complaints a decade and a half earlier. Michael F. Allen, a fungal ecologist (now a distinguished professor emeritus at the University of California, Riverside, and formerly director of the Center for Conservation Biology), published a review of the field in 1991, *The Ecology of Mycorrhizae*.[125] His goal was to draw ecologists' attention to the way mycorrhizae must be considered at all levels of ecology—including physiological ecology, population and community ecology, and ecosystem ecology—and factored into such problems as the study of ecological succession. He noted that the ecological literature, including textbooks, tended to slight the role of mycorrhizae and failed to engage with the extensive literature on mycorrhizal dynamics.

Yet, as Allen pointed out, new hypotheses in the 1970s and 1980s were drawing attention to mycorrhizal fungi. For instance, K. A. Pirozynski and D. W. Malloch proposed in the 1970s that mycorrhizal symbioses enabled the ancestor of vascular plants to invade the hostile terrestrial environment. Malloch, Pirozynski, and Peter Raven extended this idea to the longer history of plant evolution, arguing that ectotrophic mycorrhizae conferred a selective advantage on their plant symbionts, enabling them to grow in pure or nearly pure stands in extreme environments.[126] They used this idea to explain the expansion of ectomycorrhizal forests from the Middle Cretaceous onward: these forests are low in plant species diversity but have high diversity of fungal symbionts. Hypotheses like these pointed to the centrality of mycorrhizal symbioses for understanding evolutionary history. As Allen noted, the research potential on mycorrhiza and practical application of that research was "almost limitless."[127] Sally Smith and David Read, in their 2008 revision of the textbook *Mycorrhizal Symbiosis,* credited Allen and his wife, Edith Allen, with broadening ecological thought in the United States to include the role of mycorrhizal associations.[128]

Why were some ecologists exceptional in their attention to mycorrhizal symbioses? Allen, for instance, completed his Ph.D. dissertation in 1980 on arbuscular mycorrhizae at the University of Wyoming, where his adviser was Martha Christensen (1932–2017), who was trained in mycological taxonomy but also considered herself a community ecologist. She obtained her Ph.D. at the University of Wisconsin, working with mycologists Myron P. Backus and W. F. Whittingham, as well as the eminent ecologist John T. Curtis, whose own dissertation had been on orchid mycorrhizae. When Christensen was hired at Wyoming in 1963, she replaced a mycologist who had become a dean, and it was he who had invited her to apply for the position. Receiving graduate training that cut across these disciplinary boundaries was clearly important, as was being accepted into a botany department with strong traditions in natural history and ecology. Christensen also mentioned being influenced by the ecological study of fungi in British sand dune soils published by Juliet C. Brown (later Juliet Frankland) in 1958.[129] Christensen was unusual in covering subjects in taxonomy, plant pathology, and applied botany, as well as ecology, but this breadth was common to people with interests in the ecology of mycorrhizae and in soil ecology, as illustrated by several of the people profiled in this chapter.

Michael Allen, trained in this tradition, similarly had broad interests and a conviction that mycorrhizae were of great ecological importance. He also had a strong collaborator in his wife, Edith Bach Allen. Her Ph.D.

dissertation in ecology at the University of Wyoming, under the direction of Dennis Knight, was not focused on mycorrhizae, but she had married Michael while a graduate student and she became interested in mycorrhizae and their ecological roles. In addition to community ecology, soil, and mycorrhizae, her interests included restoration ecology. She and Michael started a long collaboration in the 1980s. One of their first collaborative studies, published in 1980 when they were still in Wyoming, examined the reestablishment of vesicular-arbuscular mycorrhizae following strip-mine reclamation in Wyoming.[130] Again, we see the same breadth of interest, with particular emphasis on restoration ecology, that Alastair Fitter noted in his adviser, Tony Bradshaw.

Christensen herself, in a review of a book on mycorrhizae that was published in 1982, noted several reasons for the renaissance of the subject at that time.[131] She identified four causes: the increasing and overwhelming evidence of enhanced growth in mycorrhizal plants; the development of precise techniques for extracting and observing the fungal symbionts and plant responses; general acceptance of James Trappe's much-repeated assertion that mycorrhizae were the "nutritional norm" for 90 percent of the plant species of the world; and an "applied sciences stimulus" from practical needs presented by outside interests such as forestry, agriculture, commercial fruit production, phytopathology, arid-land management, greenhouse gardening, and reclamation of mined lands. We might add another reason for the renewed interest: a growing understanding of the importance of a broad ecosystem framework, with its emphasis on the cycling of matter. Decomposers are on the same footing as primary producers, herbivores, and carnivores in the cycle. One characteristic common to many of the biologists discussed in this chapter is their interest in ecosystem structure and function and their recognition that mycorrhizae and other microorganisms have an important role in these functions.

By the 1990s, soil scientists were becoming more vocal, insisting that greater attention should be directed toward what was happening underground. The Ecological Society of America (ESA) released a major strategic planning report in 1991 called the Sustainable Biosphere Initiative, which was meant to define research priorities for ecology in the 1990s.[132]

The report identified three priorities—global change, biological diversity, and sustainable ecological systems—that should have embraced the whole of ecology. However, the community of soil scientists in the United States complained that the study of below-ground processes was only *implicitly* included in the report and needed to be *explicitly* identified as a research priority. Soil scientists were concerned that large-scale research efforts had for too long ignored below-ground processes or treated them as a "black box," where inputs and outputs were measured but detailed knowledge of ecological processes was lacking.[133] ESA approved the formation of a section on Soil Ecology soon after, in 1993. Scientists expressed similar concerns at international meetings. In 1997 a major symposium convened at Montpellier, France, to consider how the terrestrial biosphere would respond to and influence climate change also brought the question of soil processes into view, with scientists repeatedly stressing the importance of below-ground productivity.[194]

Rhizotrons, Soil Biotrons, and Minirhizotrons

Christensen's identification of the stimulus from forestry, agriculture, fruit production, and other applied sciences as drivers of advances in this field brings us to another laboratory innovation of the 1960s and later decades: the *rhizotron,* a laboratory that enabled scientists to go underground to observe the rhizosphere. As discussed in previous chapters, the need to improve agriculture drove phytotron construction worldwide. In parallel fashion, practical work in forestry and fruit production stimulated technical improvements in the study of the rhizosphere. Agricultural scientists led the way in designing the underground laboratories that became known as rhizotrons.

A rhizotron is an underground laboratory that permits direct observation of the rhizosphere. The scientists who designed the first rhizotrons were working on fruit cultivation. Agricultural scientists had long been interested in root systems, but the idea of creating special underground observatories to view the rhizosphere in situ was a relatively recent innovation dating from the early 1960s. W. S. Rogers, who was working at the East Malling Research

Station in Kent, England, which specialized in the study of fruit crops, designed what became the model for subsequent laboratories. Frits Went had visited East Malling during his European tour in 1952 and found the biologists very eager to get controlled chambers, but they could not afford a phytotron.[135] Instead, Rogers pursued the idea of an underground laboratory. He had been using special boxes with glass panels to observe roots as early as the 1930s, but these offered limited ability for direct observation of the rhizosphere. Rogers started experimenting with digging underground trenches beneath orchards; placing windows in the walls of these underground corridors made it possible to observe roots and soil organisms. In 1960–1961 he and G. C. Head designed and built an improved laboratory based on this idea, and it became famous as the first "rhizotron," although they did not use that term. They also built a second root-observation laboratory at East Malling in 1966.

These underground laboratories were built into trenches dug about seven feet deep and seven feet wide, allowing for relatively comfortable observation chambers. The viewing windows along the sides were divided into panes, and some had small detachable panels for experiments. Microscopes and cameras could be mounted on the windows to make observations and take pictures. Weekly observations could include full-scale drawings, photographs, and time-lapse films. The purpose was to observe the life history of the roots of fruit trees and to study the growth of roots in relation to shoot growth and fruiting. Two other observatories based on the East Malling model were built in South Africa (for studies of sugar cane) and Kenya (for studies of coffee). The term "rhizotron" seems to have been first used by Walter H. Lyford and Brayton F. Wilson, scientists working at the Harvard Forest in Petersham, Massachusetts. In 1966 they used the term to describe a different set-up consisting of a shed-like laboratory built on top of a bulldozed trench, one meter deep, exposing the roots of trees. The term "rhizotron" caught on, and in the 1960s and 1970s several rhizotrons designed for in situ studies of root systems were built at agricultural research centers. Most copied the East Malling design with various modifications.[136]

Robert D. Fogel, a mycologist at the University of Michigan, visited the East Malling laboratory in 1984 while attending the symposium of the

British Ecological Society on ecological interactions in soil. He was impressed by the potential of these laboratories for research beyond the problems of agricultural interest. His Ph.D. dissertation at Oregon State
University in 1978 had been on the systematics of a mycorrhizal fungus. He
then joined the University of Michigan faculty and served as curator of
fungi in the herbarium. In the late 1970s and early 1980s he was studying the
role of roots and mycorrhizae in primary production and mineral nutrition
in forests. This subject was gaining more attention, but mycorrhizae were
still relatively neglected, in part, he thought, because of the enormous
amount of labor needed to study these fungal associations.[137]

Fogel's methods included taking core samples of soils and using what
were called "minirhizotrons," or long glass cylinders inserted at an angle
into the soil and equipped with a light source and camera for making observations. The concept behind the minirhizotron went back to the 1930s; the
device was refined in the early 1970s and the term "minirhizotron" was introduced in 1974.[138] This method allowed only simple experiments in small
areas of soil, and the photographic quality was often poor. Fogel's visit to
East Malling opened his eyes to the possibilities of a larger underground
laboratory, especially for doing manipulative experiments. He teamed up
with John Lussenhop, an invertebrate biologist at the University of Illinois
at Chicago, and Kurt Pregitzer, a soil scientist at Michigan State University,
to explore the possibility of building a rhizotron at the University of Michigan's Biological Station in northern Michigan. Pregitzer had been developing software for digitizing and analyzing video images of roots. Fogel
preferred the name "soil biotron" rather than "rhizotron" because he imagined a laboratory that would go well beyond the study of root growth and be
used to explore interactions among soil organisms, roots, and mycorrhizae
in situ. He also hoped it "would provide a focus and forum for soil scientists
from different disciplines and institutions to interact and hopefully synthesize new approaches to soils research."[139]

Fogel obtained a grant of $70,000 from NSF's division of environmental
biology to build the Michigan Soil Biotron in 1987. The observation equipment cost much more, an estimated $100,000 by the early 1990s for micro-
scopes, video cameras, computers and GIS software, infrared gas analyzers (to

measure and control carbon dioxide levels), probes for detecting soil moisture and temperature, light meters, and other equipment. Unlike the East Malling laboratory, the Michigan Soil Biotron was not located in an agricultural setting but at a biological station that had a seventy-five-year history of research and that had been designated an International Biosphere Site. Most of the land was relatively undisturbed: logging had ceased ninety years earlier, and fire suppression had been imposed sixty years earlier. It was close to the Stockard Lakeside Laboratory, a modern, well-equipped laboratory.

The soil biotron was at the time an exceptional laboratory for the analysis of the rhizosphere of a natural system. As James Teeri, the director of the Michigan Biological Station, remarked, its software for computer-aided analysis of video records, which had been specially written for analysis of root data, provided new and unique capabilities. He also noted the multidisciplinary and cross-disciplinary nature of the work done there by the 1990s, ranging across the disciplines of "soil science, agriculture, forestry, plant and animal ecology, meteorology, climatology, physics, plant physiology, biochemistry, geography, molecular biology, genetics, systematics, and microbiology."[140] Lussenhop, Fogel, and Pregitzer declared in 1991 that the opportunities for soil research opened up by the new techniques and laboratory innovations of the 1980s, aided by the new methods of video and image analysis, represented "a new dawn" for soil research.[141] (In 2005–2006 another rhizotron was constructed by the U.S. Forest Service in Houghton, Michigan, northwest of the Michigan Biological Station near Lake Superior.)

Minirhizotron technology also has continued to develop, to overcome the limitations involved in taking field samples or extrapolating from laboratory and glasshouse experiments. These newer designs have benefited from more sophisticated technologies. As described by Michael Allen and his colleagues, automated robotic minirhizotrons sealed within buried tubes were developed in the early twenty-first century, and data and instructions are communicated through a wireless network. Automated minirhizotrons can be coupled with a networked sensor system that measures soil temperature, moisture, oxygen, and carbon dioxide. These "Soil Ecosystem Observatories" are similar to technologies used for satellite and airborne remote sensing, but they observe changes in space and time at scales appropriate to soil processes. These innovations allow for direct observation of the growth and mortality of mycorrhizal associations and make it possible to get better assessments of carbon fluxes in soils, which are important for understanding ecosystem functions. Such devices are helping finally to pry open the black box of the underground world. But they also produce a lot of data, requiring development of tools for the analysis of data.[142]

Conclusion

These episodes from the physiological ecology of mycorrhizal symbioses illustrate several problems that are inherent to ecological research. One is the great difficulty of making observations beneath the earth's surface; therefore new technologies, including the use of radioisotope tracers and the invention of new kinds of laboratories and instruments, have been essential in moving the science forward. The second is the challenge of breaking through disciplinary barriers; communication and synthesis across disciplines have always been important for advancing physiological ecology. The circuitous history of the Went-Stark hypothesis reveals these disciplinary divisions. Nonetheless, biologists have constantly striven to bridge disciplines, a process that can be slow and frustrating. In addition to the usual cross-disciplinary mechanisms of symposia, international meetings, and publications promoting disciplinary links, the construction of workspaces that bring scientists from different disciplines together, as Fogel anticipated, has helped to facilitate this kind of synthesis. Such workspaces include research vessels like the *Alpha Helix* as well as rhizotrons and phytotrons. As the individuals profiled in this chapter illustrate, a broad interest in biological problems and exceptional willingness to move between different fields of knowledge have been characteristic of the scientists who have contributed to this field.

The difficulty in bridging these divisions has sometimes been exacerbated by international inequities. In the development of Latin American ecology, as we saw in this chapter's focus on the Venezuelan programs at IVIC led by Ernesto Medina and Rafael Herrera, North American and European researchers had to confront the accusation of scientific imperialism and find new strategies that would promote and strengthen Latin American scientific institutions. This subject deserves much more study.[143] The educational and research programs developed by Medina and Herrera were critically important for the growth and improvement of Venezuelan science, while also fostering collaborations with scientists from abroad.

Herrera's career illustrates how this kind of local development of science fed into and supported broader international projects that were important for understanding global processes. He continued his research on tropical ecology and in 1987 became involved in the global change research agenda that was then emerging.[144] He served as vice-chair of the International Geosphere-Biosphere Program that had been endorsed by the International Council for Science in 1986. After retiring from IVIC Herrera moved to Europe in 2003 and has continued research in association with several universities and research institutes, with particular attention to the global carbon cycle and climate change. The long-term international research that was made possible by the establishment of permanent forest plots in the Amazon has been especially important for understanding Amazon ecosystem dynamics and global change.

An ongoing challenge has been to relate studies under controlled conditions to the more complex environment of the field. In a review of the ecological role of the mycorrhizal mycelium in 2004, scientists from Sheffield (including Read) and Cardiff University stressed how hard it was to observe and study mycelial systems without disturbing or destroying them. That problem, they pointed out, was hard to overcome outside the laboratory, which meant that "knowledge of the functioning of these networks has been strongly based on reductionist studies, the relevance of which to the field situation is often difficult to confirm."[145] This review identified key methodological advances that have helped scientists to understand the "hidden half" of the symbiosis, the mycelial network. Moving back and forth between laboratory and field studies—seeking ways to link findings from both sides and fostering communication between disciplines—has been a persistent challenge over many decades. Long-term commitment to research is important, and that commitment requires us to invest in the kinds of laboratories or other infrastructure that can support long-term studies.

EIGHT

The Drive for Synthesis

MODERN DARWINIAN THEORY, known as the "Modern Evolutionary Synthesis," is said to have taken shape from the late 1920s to 1950, beginning with the recognition that Darwin's theory of natural selection was compatible with Mendelian genetics. The term "modern synthesis" is taken from Julian Huxley's book *Evolution: The Modern Synthesis,* published in 1942, which surveyed the vast field of evolutionary biology to show that Darwinism, modified in the light of developments in genetics, was alive and well.[1] The main claim of the modern synthesis was that natural selection was the sole directing force in evolution and that "soft" inheritance, or the inheritance of acquired characteristics, often referred to as Lamarckian inheritance, was not possible. An additional claim was that microevolution and macroevolution were the result of the same general causes. Across several disciplines, scientists formulated new interpretations of how natural selection operated, linking their understanding to knowledge gained by the science of genetics, a science that did not exist in Darwin's time. The history of the modern synthesis has focused on certain canonical texts published over three decades and on scientists who are viewed as the leaders of this synthesis. Ernst Mayr, one of the chief architects of the modern synthesis in the discipline of systematics, was influential in defining what the modern synthesis involved and encouraging historical study of its origins, its leaders, and its central texts.[2]

Contributions from physiological ecology are largely missing from the historical reconstructions of the modern synthesis. Part of the reason for the neglect is that these contributions lagged behind those of other disciplines. However, the field of physiological plant ecology offered novel perspectives on plant adaptation and microevolution, the processes causing the separation of a species into two distinct subspecies. These perspectives began to be articulated in the 1920s and continued to the 1980s, with some of the most important work being done after the 1950s, the decade when physiological plant ecology emerged as a discipline within ecology. Instead of seeing these advances as events occurring after the modern synthesis, I propose that we extend our time frame and consider the modern synthesis to be an ongoing process of reformulation of Darwinian theory that continued into the 1980s, rather than a process that was largely completed by the 1950s. Far from being completed, from the 1950s on the modern synthesis expanded to embrace new disciplines such as physiological ecology. We should therefore follow its progress into the next generation. By taking a longer view we can appreciate the unfinished nature of the synthesis in 1950 and realize that much more scientific research and more than one generation were required to construct modern Darwinism.

My discussion of the evolutionary synthesis focuses on these later decades (1950s to 1980s) and on central developments that helped to define the field of physiological ecology, not just as a subdiscipline of ecology but as a branch of evolutionary biology. I also look at the role of new instruments and laboratory innovations in shaping this scientific field. I explore this development through the program in "experimental taxonomy" at the Stanford botanical laboratory of the Carnegie Institution of Washington. In 1970 the name of that program was changed to "physiological ecology," signaling recognition that a new discipline had emerged, partly as a result of the creativity and leadership of the Carnegie team. I look especially at the cooperative relationship between the Carnegie program and Frits Went's activity at Caltech in the 1950s, as well as other collaborations important for the development of evolutionary ecology. In focusing on the Carnegie group and its collaborators, I wish to show how, in this particular case, the new field of physiological plant ecology came to be defined as an evolutionary science.

One important and far-reaching advance made during this later time period was the discovery of what became known as "C_4" photosynthesis, a distinct biochemical pathway that differed from the more common C_3 pathway. This discovery has rightly been called one of the most exciting developments in this field in modern times, and it stimulated a vast amount of scientific work that encompassed diverse disciplines. C_4 photosynthesis came to be understood as an evolutionary adaptation that had arisen in C_3 plants in response to environmental conditions prevailing between fifteen and thirty million years ago. I will use the example of C_4 photosynthesis to explore how scientists from different fields, but especially in physiological ecology, carried on the integrative spirit of the modern synthesis.

The chapter ends by considering the expansion of ecology that occurred in the 1990s in recognition of the global consequences of human population growth. By this time there could be no doubt that human activities were not only affecting every ecosystem on the planet but were also changing the composition of the planet's atmosphere and altering its climate. The need to meet these new challenges ushered in a laboratory innovation that was a reinvention of an older idea from the 1960s: the ecotron, or a climate-controlled laboratory designed for research on ecosystems. These new laboratories also served as vehicles of integration, bringing together ecologists from different disciplines, especially physiological ecologists, population and community ecologists, and ecosystem ecologists.

From Experimental Taxonomy to Physiological Ecology

In chapter 6 I wrote in general terms about the rise of physiological plant ecology as a synthesis of plant physiology and ecology. In this chapter I examine the transition toward physiological plant ecology within a single long-term program at the Carnegie Institution of Washington's department of plant biology, which from 1929 on was based at its laboratory on the campus of Stanford University. The program had begun a few years earlier under the title of "experimental taxonomy," a subject that grew from European botanical traditions extending back to the nineteenth century. This research centered on transplant experiments, which entailed moving plants to different locations to test the effect of climate and soil on the plant.[3] Many of the early experiments seemed to suggest that characters acquired in the new environments could be inherited, leading to the belief that varieties or species were actually being transformed by the process. In America, Frederic Clements, a leading ecologist and a longtime member of the Carnegie staff, was enthusiastic about transplantation experiments. He was a late holdout for the inheritance of acquired characteristics; most American biologists were rejecting such explanations by the 1920s, largely because experimental tests failed to demonstrate a clear environmental inherited effect.

Better understanding of genetic variability in species made it possible to reinterpret transplant experiments in terms of Darwinian natural selection. An important insight was to realize that climatic and soil conditions were operating as selective agents on genetic races (called "ecotypes") within these populations. Göte Turesson (1892–1970), a Swedish ecologist, advanced this subject in the 1920s by emphasizing the role of climate and soil conditions in promoting evolutionary divergence.[4] Turesson had lived in the United States for a few years as a young man and received his bachelor of science and master of science degrees from the University of Washington in Seattle. He returned to Sweden in 1915, where he completed a doctorate at the University of Lund in 1922 and then became a lecturer and research docent there. He was unable to obtain a faculty position at Lund, and in 1935 he became a professor of botany at the Agricultural College of Uppsala University, from which he retired in 1959.

Turesson argued that both genetic and ecological causes should be considered to explain why certain varieties of plants were favored in given locations. He emphasized the environmental causes that explained the distribution of varieties and species, while firmly rejecting the Lamarckian idea that the environment could produce adaptive changes that were inherited.[5] He also proposed a new field of research that he called "genecology," meaning the study of genetic variation within plant species as it affected their relations to the environment. He believed that the species itself should be regarded as an ecological unit, not just the individual (the subject of autecology) or the community (the subject of synecology). He coined the term "ecotype" to describe ecological races that formed within a species: these different ecotypes were adapted to different environmental conditions within the species' range.

The Carnegie program in experimental taxonomy came to mirror what Turesson was trying to do in Sweden. The program began under the direction of Harvey Monroe Hall, who had been doing transplant experiments since 1921. He had been engaged in debate with Clements over the interpretation of these experiments, and the program in experimental taxonomy was created to give Hall independence from Clements. Experimental taxonomy's main goal was to solve the problem of heredity versus environment, or to clarify what features of a plant were genetic in nature, and therefore inheritable, and what were caused by environmental effects and were not inherited. For taxonomists this distinction would help to determine which plant characters were most useful for classification. But Hall realized that experimental taxonomy had broader implications for understanding microevolution, or how species became separated geographically and morphologically into two subspecies.[6]

Hall died prematurely in 1932 but other Carnegie scientists carried on this line of research. Jens Clausen (1891–1969), who had joined the Carnegie group from Denmark in 1931, took over as head of the program after Hall's death. Collaborating with him were David D. Keck (1903–1995), a taxonomist, and William M. Hiesey (1903–1998), a physiologist. Experimental taxonomy became a long-term program, and it continued even after people retired or left. In 1950, Keck left to become head curator at the New York Botanical Garden, and Malcolm Nobs, then a graduate student in botany at Berkeley, joined the group in January 1951 as a research assistant to help with the taxonomic and experimental work. He finished his Ph.D. in 1957 and became a permanent staff member in 1960. Hiesey became head of the program when Clausen retired in 1956, although Clausen remained actively involved in the program after retirement (figure 10). In 1960 Olle Björkman from Uppsala University, Sweden, arrived on a postdoctoral fellowship and then joined the group permanently in 1964. He became head of the group after Hiesey retired in 1969. Other scientists also worked with the group for shorter periods of time.

During the time that Clausen headed the program from the early 1930s to mid-1950s, the emphasis was on problems of adaptation and evolutionary divergence, with a focus on the idea of ecotypes or ecological races. Clausen's research on species of wild pansies (*Viola tricolor*) in Denmark in the early 1920s had independently led him to agree with Turesson that

Figure 10. William Hiesey (*left*) and Jens Clausen (*right*) in the greenhouse of the Carnegie Institution's department of plant biology. (Courtesy of the Carnegie Institution for Science Archives)

species were composed of genetically distinct races, each one fitted to survive in its own environment. By transplanting species into different environments, one could test whether their growth characteristics changed in the new environments and therefore determine whether they were genetically-based characteristics. The Carnegie department had three transplant stations for this work, one on the Stanford campus (just above sea level), one at Mather, on the edge of Yosemite National Park at an elevation of 1,400 meters, and a third at Timberline, just beyond the eastern edge of the park boundary at 3,000 meters. But laboratory and greenhouse experiments were also needed to get a clearer picture of how these ecotypes differed from each other. Their work, which was meticulous in its detailed study of ecological races and hybrids of different races, constituted an important contribution to the emerging modern synthesis in evolutionary biology.

We associate the modern synthesis mainly with a group of canonical texts published from 1930 to 1950, such as Ronald A. Fisher's *The Genetical Theory of Natural Selection* (1930), which helped to found the field of population genetics, T. D. Dobzhansky's *Genetics and the Origin of Species* (1937 and subsequent editions), Ernst Mayr's *Systematics and the Origin of Species* (1942), and George G. Simpson's *Tempo and Mode in Evolution* (1944). In botany the canonical text was G. Ledyard Stebbins's *Variation and Evolution in Plants* (1950), which was based on lectures given in 1948.

Stebbins's book was a broad survey of botanical research and included discussion of the results that Clausen, Keck, and Hiesey were obtaining by the late 1940s. The relevance of their work to the modern synthesis has also been recognized in more recent scholarship, notably by J. Núñez-Farfán and C. D. Schlichting in a study published in 2001.[7] They characterized the Carnegie research done from the 1930s to 1950s as significant for providing empirical support for the emerging synthesis and for defining the new field of ecological genetics, which was conceived more broadly than the ecological genetics being advanced in Britain by E. B. Ford. The Carnegie scientists themselves understood their work to be a contribution to ecological genetics by the 1950s. They noted that the principles for this field had "not been formally stated" and that the field's goals were to study "the hereditary structure that controls the mechanisms and processes through which ecological races and species have become adjusted to their environments."[8]

Because of the long-term nature of the work by the Carnegie group, formal publication in scientific journals was often delayed, and in some years the only available accounts were those appearing in the Carnegie Institution's annual reports. Stebbins later commented on how this slow publication rate delayed recognition of the value of their work as the modern synthesis was being constructed. For instance, they had made careful studies of hybrids between six species in the genus *Layia* (tidy lips, a daisy-like flowering plant common in California), which Stebbins considered to be one of the best demonstrations either in plants or animals of the geographic theory of speciation. Although the work was done in the 1930s, it did not appear until Clausen's book of 1951, *Stages in the Evolution of Plant Species*, a short book based on a lecture series at Cornell University in 1950. Similarly, he remarked, "they had a fine example of a carefully analyzed natural allopolyploid [plants having chromosomes derived from different species]. The tragedy, I think, was that this didn't come out in a series of papers during the 1930s and early 1940s."[9] Betty Smocovitis has commented on the lack of a major synthetic book from the Carnegie team by the 1950s: although Clausen's book was influential, it was not a comprehensive monograph on the level of Stebbins's work.[10] The delays had various causes: World War II interrupted some of their research, but Stebbins thought the main problem was that Clausen, unlike Turesson, "was reluctant to publish until every detail was understood."[11] This delayed publication in part accounts for the lack of recognition that Núñez-Farfán and Schlichting later sought to rectify.

I agree with Núñez-Farfán and Schlichting that this research was indeed in the spirit of the modern synthesis and was a contribution to the construction of modern Darwinian theory and to ecological genetics. But their analysis focused mostly on the early part of the Carnegie program, during the years coinciding with the emergence of the modern synthesis. I take a longer view, because the Carnegie program did not stop in the 1950s, when Clausen retired, but extended for more than half a century from the early 1930s to the 1980s, covering two or perhaps three scientific generations. The full sequence is represented in a six-part monograph series called *Experimental Studies on the Nature of Species,* published by the Carnegie Institution in 1940, 1945, 1948, 1958, 1971, and 1982. I will highlight in particular the fifth volume of the series, published in 1971, because it illustrates exceptionally well how the Carnegie group blended laboratory and field studies and brought this hybrid approach to bear on the study of adaptation and evolution.

In addition to journal articles from the group, there were also discussions of new research directions and summaries of results in the annual reports of the Carnegie Institution. These reports allow us to chart how this research program evolved and how the availability of new laboratories and instruments contributed to that evolution. If we include this group's longer research trajectory and if we expand our definition of the modern synthesis to embrace developments in physiological and ecological disciplines, we will see that this synthetic enterprise continues to unfold after the 1950s. I also argue that, if we take this longer view, we discover that the Carnegie group's long-term studies did indeed produce a major synthetic work beyond the *Experimental Studies* series mentioned above. This work did not appear until 1980, but I regard it as being in the same class as the much earlier canonical texts of the modern synthesis.

Over the several decades of this group's work, as it passed from one generation to the next, the goal remained fairly constant—to learn how different races were formed in adaptation to changing environmental conditions. Its overall goals were constant, but its focus and methods changed and became over time more analytical and laboratory-based, although field research was always important. Experimental taxonomy was a multidisciplinary enterprise studying the kinships between plants, the relations between plants and their environments, and the evolutionary processes that produced plant diversity and adaptations of plants to their environments. It therefore embraced, in addition to taxonomy, the subjects of genetics, cytology, comparative morphology, and ecology, which together were said to comprise *biosystematics* in the 1940s.[12] But it also included physiology and biochemistry, especially in later years.

Rather than cutting off the story at its midpoint in the 1950s, I include the middle and later stages of this decades-long research program. If we restore the emphasis on physiology that was a main focus of this program in the second half, we can appreciate the way it contributed not only to ecological genetics but also to another emerging field, physiological ecology. This contribution was clearly signaled by 1970, when the program in "experimental taxonomy" changed its name to "physiological ecology." The availability of new kinds of laboratories and instruments for physiological experiments had a role in the evolution of this program and in the emergence of physiological ecology as a new field within ecology.

New Directions

As Patricia Craig has suggested in her centennial history of the Carnegie's department of plant biology, around 1950 the experimental taxonomy program shifted more decisively toward laboratory experiments and a greater emphasis on physiological research.[13] She also noted the significance of Hiesey's physiological interests, which led him to do experimental work at Caltech in the 1940s and 1950s, using both the Clark greenhouses and the Earhart phytotron. She sees the transition as happening gradually during the 1940s, but I suggest instead that the opening of the phytotron in 1949 was a catalyst for change, exciting new interest in physiology and experiments in controlled environments.

When Went built the Clark greenhouses at Caltech in the late 1930s, knowledge of how regional races or climatic races originated was rudimentary.[14] Climatic races appeared to differ in their physiological properties, and plant reactions to the environment appeared to be under genetic control. But how genes governed the physiological processes was not known. As Hiesey pointed out, incorrect interpretations stemmed from failure to distinguish between hereditary and environmental factors.[15] Experimental work, both in the field and in laboratories, would help to clarify these relationships over the next three decades.

Went was keen to relate experimental results to the natural ecology of plants, thereby bringing the results of modern physiological discoveries to bear on ecological problems of plant adaptation and species distribution. His cooperation with the Carnegie plant biologists at Stanford began before the phytotron was built in the 1940s, and it coincided with growing recognition by evolutionary biologists of the value of the Carnegie's field stations for studies of speciation. The Mather station was strategically located for research on plant evolution because it combined plants from higher and lower elevations and from northern and southern distributions, a number of which could hybridize. After the war, biologists began to appreciate the opportunities that these stations offered for their evolutionary studies.

Theodosius Dobzhansky spent several summers at Mather studying native populations of the fruit fly *Drosophila,* his main experimental animal. Ledyard Stebbins worked between Mather and Timberline, studying genetic races of blue wild rye (*Elymus glaucus*) and California brome (*Bromus carinatus*). Went also visited Mather to gather seeds to grow under controlled conditions. These three, along with a group of American and foreign biologists, attended a roundtable hosted at Mather during the winter season of 1946–1947 to discuss the role of genetics, physiology, and the environment in the evolution of natural local populations, climatic races, and species. A decision was then made to search for a group of plants with ideal characteristics for "co-operative investigations on genetics, cytology, physiology, ecology, and evolution."[16] This cooperative, multidisciplinary enterprise defined the Carnegie group's work in "experimental taxonomy" for the next couple of decades, and it involved cooperation not only within the Carnegie staff but with outside scientists and institutions.

William Hiesey provided the link between the Carnegie and Caltech laboratories in his experiments on how climatic or ecological races were formed when plants adapted to changing environmental conditions. He had worked in the Clark greenhouses in the 1940s, but the opening of the Earhart laboratory in 1949 to showcase the world's first phytotron was a giant step from air-conditioned greenhouses, and it came at a critical time when biologists at Carnegie and elsewhere had decided to pursue cooperative multidisciplinary investigations on the nature of species. The Carnegie project, which started in January 1950 and ran for six months, was one of the first to use the phytotron. The project involved species from three genera of plants, *Poa* (grasses), *Achillea* (yarrow), and *Mimulus* (monkey flower), which were first cloned at Stanford and then brought to the phytotron to grow under different temperatures.[17] The Carnegie group had become interested in plants in the *Mimulus guttatus* complex because it was highly variable and had evolved races growing all over western North America, ranging from Alaska to Baja California, and from sea level to 3,400 meters. Most races were perennial, cloned easily, and crossed easily. They looked like good candidates in the search for a model organism to study experimental taxonomy.

Achillea milleflorium, the common yarrow, was a hardy and abundant perennial that grew in diverse climates and had evolved an exceptionally complete set of races, making it a "unique experimental subject."[18] Hiesey thought that experiments in controlled environments could bridge the long-standing gap between physiology and genetics by clarifying how genes as well as external influences affected physiological and biochemical processes. We can see these investigations in the 1950s as promoting interdisciplinary synthesis or integration, that is, continuing the momentum toward the modern synthesis in evolutionary biology that scientists had been advocating since the 1930s at least. In the mid-1950s, Hiesey did not think that this synthesis had been accomplished, but he did think that experiments in controlled environments were key ingredients for success.

The Carnegie-sponsored studies on grasses had practical applications. During the 1940s, Carnegie scientists became interested in the problem of creating new species of range grass, with new qualities and suited to new environments, by hybridizing species from contrasting environments. This problem would grow into a lengthy research program involving extensive field work, and it was the subject of the sixth volume of *Experimental Studies on the Nature of Species,* published in 1982. Figuring out how to reseed depleted ranges and establish grass cover on burned areas required information about how the germination of seeds was affected by temperature. The phytotron provided a way to test some of the key hybrids under controlled conditions. The value of laboratory studies for estimating how new strains would adapt in the field seemed self-evident, but in fact it was quite difficult at this time to judge a plant's field success from laboratory studies. It was clear that distinct climate races had quite different growth responses when subjected to the same temperatures. This result suggested that ecological races had different internal functions, but closer work on the comparative physiology of ecological races was needed.[19]

Experiences in Caltech's phytotron helped to sharpen the group's research questions and pushed them more in the direction of physiological experiments under controlled conditions. In the annual report for 1950–1951, Hiesey noted that the laboratory studies looked promising for understanding the genetic and physiological mechanisms that accounted for the plant's external characteristics. This area of plant biology, he observed, was "still mostly unexplored."[20] In 1951–1952 he announced that work had started on a program "designed to investigate the basic physiological characteristics of ecologically different races."[21] By this time he was working with Harold W. Milner, a chemist on the Carnegie staff in the biochemistry program. These groups had previously worked separately: the biochemical group was distinct from the experimental taxonomy group at that time.

The next year the group's work began to focus more closely on the study of rates of respiration and photosynthesis under controlled conditions of temperature, light intensity, and atmospheric composition. To support this project the Carnegie team began in 1952–1953 to construct its own apparatus. The next year's work was devoted exclusively to the development of apparatus for the measurement of photosynthesis and respiration under controlled conditions. By the mid-1950s, when Hiesey took over as head of the group after Clausen retired, the research strategy had evolved to encompass three approaches: field studies at the different altitudinal stations, controlled laboratory experiments such as the ones conducted at Caltech's phytotron, and development of apparatus for quantitative comparison of rates of respiration and photosynthesis under controlled conditions. By 1958 the Carnegie scientists were constructing their first controlled growth chamber, with plans to develop more as the research advanced.[22]

There was still "much confusion," as Hiesey wrote in 1957, regarding the interaction between genetic and physiological processes.[23] In his view, that meant that the "time was ripe" for extensive experiments under controlled conditions. The synthesis they were working toward in the 1950s involved study of the interaction among genes, physiological processes, and environment. They were convinced that "these fields are not separate, but that the causal chain reaches from the gene through biochemical and physiological processes to the morphology of the plant and its adjustment to environment."[24] There is no question that the Carnegie group was thinking in a synthetic way about evolutionary problems in a Darwinian framework and that it was trying to knit together the fields of genetics, physiology, ecology, and taxonomy to get at the very difficult question of how the expression of genes operated in variable environments.

The point I wish to emphasize is that their evolving strategy was stimulated by experience in the Caltech phytotron. This work had three components: tests at the three Carnegie field stations of old and newly acquired plant materials from different environments, quantitative measurement of respiration and photosynthesis using specially designed apparatus, and growth studies of plants in specially designed controlled-growth chambers.[25]

Ecology was being brought into a closer relationship with experimental sciences, especially physiology and biochemistry.

As late as 1971 they noted that recent progress had depended on new instruments and methods and "the extensive use of controlled growth facilities."[26] Laboratories such as phytotrons and their smaller-scale variations provided essential tools for working out some of the most difficult problems of modern Darwinian biology relating to ecological genetics. In the 1950s progress on these problems still lay largely in the future, but the potential contribution of new laboratory technologies was already well appreciated, as shown by Hiesey's eager exploitation of the possibilities of the phytotron. Nonetheless, progress was slow on these complex problems. The group admitted it was hard to connect a physiological response to an environmental factor when "the observed response is actually the resultant of interaction between the genotype as a whole and an entire complex of environmental variables."[27] In short, unscrambling all the variables was very hard.

When Olle Björkman from the University of Uppsala joined the Carnegie group, initially as a postdoctoral fellow in 1960 and then as a staff member in 1964, he brought great experimental skills—both in the laboratory and in the field—to this program. Björkman's training had been in chemistry and genetics, though not in botany, but at Uppsala he was in a department of genetics and plant breeding where Turesson's legacy remained strong, and so he turned to botanical subjects to explore the problems that interested him.[28] The mechanism of photosynthesis became his research focus. Shortly after Björkman's arrival at Uppsala a scout from the Rockefeller Foundation happened to visit and invited the group to submit a brief proposal on the spot. A few months later the Rockefeller Foundation started funding Uppsala's research, giving the group ample laboratory resources for the pursuit of its inquiries. The foundation was interested in supporting biological research relevant to agriculture and was investing in Swedish research; it also donated funds for a phytotron at the Royal College of Forestry in Stockholm, constructed between 1958 and 1964, which opened in January 1965.[29]

In Sweden, Björkman and his collaborators developed *Solidago* (goldenrod) as a model organism to study the difference between sun-adapted and shade-adapted ecotypes. Goldenrod was not an opportunistic species—that is, it always grew in the same place without spreading, but it had very wide distribution. Turesson had used *Solidago virgaurea* as an example of a species possessing several ecotypes that lived in contrasting environments across Europe. Björkman's studies looked at the photosynthetic rates of these contrasting types in relation to temperature and then examined differences between plants living in sunny and shady habitats. There were dramatic differences in these ecotypes. The races living in sunny habitats or exposed alpine tundra thrived under high light intensities. But the same intensities damaged plants accustomed to deeply shaded habitats. Sunny races could use high light intensities efficiently for photosynthesis, but shady plants lacked this ability. In 1960 Björkman brought these research strengths to the Carnegie group. His move to California coincided with major breakthroughs in the understanding of photosynthesis, and Björkman quickly took command of the subject from an ecological and evolutionary perspective. In the words of his colleague Joseph Berry, "Under Björkman's leadership, progress was spectacular."[30]

Photosynthesis from an Ecological Perspective

During the 1950s, Melvin Calvin and his collaborators at the University of California, Berkeley, had worked out the biochemical pathway of photosynthesis using radioactive carbon-14 to trace the path of carbon from atmospheric carbon dioxide through the plant during photosynthesis. The pathway they discovered is called the photosynthetic carbon reduction cycle or Calvin cycle (or the Calvin-Benson cycle to recognize the contributions of Andrew A. Benson to this research).[31] For his research Calvin received the Nobel Prize in Chemistry in 1961. Most studies of this process were done on the alga *Chlorella,* and only a few higher plants had been investigated, but it was assumed that this biochemical process was the same in all plants. However, there were indications in the botanical literature that a different process for assimilating carbon dioxide might operate in tropical grasses, including crop plants like maize and sorghum. Some plants had ring-like structures in their leaves, first observed by Gottlieb Haberlandt in the nineteenth century and named "Kranz" (wreath) anatomy, and they showed physiological differences related to photosynthesis. The significance of these differences for how plants photosynthesized was not realized at first.

A breakthrough came in the mid-1960s with the work of two Australian scientists, Marshall (Hal) D. Hatch and C. Roger Slack, who were studying sugar cane for the Colonial Sugar Refining Company in Brisbane, Australia. They had superb facilities for their work, because the company had built a phytotron, the David North Plant Research Centre, which opened in 1961, even before the larger phytotron in Canberra opened.[32] It had a direct link to the Caltech phytotron, for the plans were initiated in 1958 by Harry Highkin and Pret Keyes of Caltech and by Dov Koller, who was then visiting Caltech from Israel. Work at the David North Plant Research Centre focused exclusively on all aspects of the physiology and biochemistry of sugar cane.

Hatch and Slack learned of sugar cane research in Hawaii, where scientists were trying to repeat Calvin's experiments with *Chlorella* but were getting results at odds with Calvin's work.[33] They took up this problem in 1965, after the Hawaiian group published its findings. Hatch and Slack concluded that the photosynthetic pathway in sugar cane was indeed different, and they proposed a new model in 1966, observing that it was highly unlikely that the pathway they proposed was peculiar to sugar cane.[34] The new model was not a completely different pathway but rather an additional pathway for the fixation of carbon dioxide added to the Calvin-Benson cycle common to all plants. They proposed calling it the C_4 dicarboxylic acid pathway of photosynthesis (also referred to as β-carboxylation photosynthesis), which was shortened to C_4 pathway or C_4 photosynthesis, so-named because the first stable intermediate products formed were 4-carbon acids. The pathway was found in a wide variety of plants, which came to be known as C_4 plants. Plants with just the Calvin-Benson cycle came to be known as C_3 plants, because the first stable intermediate product of the reactions was the 3-carbon compound phosphoglyceric acid. The key difference was that C_4 plants separated the initial fixation of carbon dioxide from the Calvin cycle, so that these steps were performed in different types of cells.

C₄ plants also differed from C₃ plants in their leaf anatomy and their physiological responses, a cluster of differences called the "C₄ syndrome." Scientists from several countries gradually sorted out the details of the C₄ syndrome. One important center for this investigation was the Canberra phytotron, where research helped to clarify the adaptive role of the C₄ pathway.[35] In fact, C₄ species were common in Australia. They had a competitive advantage in open and dry habitats where light was stronger, and they also used water more efficiently. C₃ plants did better in cool environments with less intense light. The differences in leaf structure between the two types of plants were understood to be consequences of these different photosynthetic pathways.

The discovery of the C₄ pathway stimulated interest in studying photosynthesis in a much wider variety of species than was the case before the 1960s. As Harold Mooney explained, when all plants were thought to have a single photosynthetic pathway (the Calvin-Benson cycle), relatively few species were studied and "the choice of experimental organism was of little importance."[36] But that changed dramatically with the discovery of the C₄ pathway, which prompted a search for more C₄ plants, in part because of the potential agricultural implications of these new discoveries. The discovery was also important for understanding basic questions of plant distribution and evolution. Herbert Baker, a botanist and president of the Society for the Study of Evolution in 1969, called the discovery of two biochemical pathways for photosynthesis "one of the most exciting developments" of the 1960s.[37]

Coinciding with the discovery of the C₄ pathway was another discovery that was given the name "photorespiration" by E. Bruce Tregunna, then working at Queen's University in Kingston, Ontario.[38] Photorespiration is different from normal respiration in plants and is linked to photosynthesis, but with the opposite effect. In photosynthesis the initial step is to "grab" a carbon dioxide molecule from the atmosphere, which is then added onto an organic compound called ribulose-1,5-bisphosphate (shortened to RuBP). This process is known as *carboxylation*, which is catalyzed by a protein enzyme in the leaf. Samuel Wildman at Caltech, the person said to have coined the term "phytotron," had first characterized this enzyme in the

1950s, and soon after it was identified as ribulose-1,5-diphosphate carboxylase (diphosphate was later changed to bisphosphate). The next step is to produce 3-phosphoglycerate, which is then converted to sugars. Oxygen is given off as a waste product during photosynthesis.

Photorespiration appeared to be the result of an error during photosynthesis. Instead of grabbing a molecule of carbon dioxide from the air, an oxygen molecule is captured from the atmosphere and fixed, and a carbon dioxide molecule is given off. Photorespiration was considered a wasteful pathway that competed with photosynthesis and had no benefit to the plant. By the early 1970s scientists realized that the same enzyme was catalyzing both reactions, and therefore the enzyme was also an *oxygenase*. It could take on two functions depending on conditions. After its role in photorespiration was discovered and following a change in nomenclature from diphosphate to bisphosphate, its name became ribulose-1,5-bisphosphate carboxylase-oxygenase in recognition of this dual function. In 1979, at a retirement party for Samuel Wildman, David Eisenberg as a joke coined the name *rubisco* (or RuBisCO) for the enzyme; the name stuck, and it is still commonly called rubisco.[39] Unraveling the mystery of C_4 photosynthesis— the "why" question, or its adaptive significance—would hinge on this dual role of rubisco as a catalyst both for photosynthesis and photorespiration.

Coinciding with Tregunna's discovery of photorespiration in the 1960s, Björkman also made a discovery related to photorespiration; in fact, Tregunna and Björkman published their findings in the same issue of the journal *Physiologia Plantarum*. Björkman was making comparative studies of photosynthesis in species and races of higher plants from diverse ecological habitats. He found that oxygen inhibited the fixation of carbon dioxide in certain higher plants, an inhibitory effect first observed in algae by Otto Warburg in 1920 and called the "Warburg effect." Björkman suggested that these plants had evolved at a time when the oxygen content of the atmosphere was lower than it is today.[40] As recalled by Joseph Berry, who had been Tregunna's student and who joined the Carnegie group as a postdoctoral fellow in 1970 (becoming a staff scientist in 1972), these and other studies "made it clear that oxygen inhibition of photosynthesis occurred in many species and that it resulted in significant reduction of photosynthesis by these plants under normal ambient conditions."[41] These discoveries, coming rapidly on top of each other, catalyzed cross-disciplinary interactions that were exceptional for the time. As Berry described, the result was to foster a cooperative culture:

> We were something like the proverbial group of blind men encountering an elephant for the first time, each beginning to touch different parts of the mysterious beast's anatomy. In the original Indian fable, this led to bickering and arguments between the blind observers; in the case of C_4 photosynthesis, there seemed to be a general recognition among researchers that they were dealing with something that was bigger than their respective areas of competence. This led to rapid and (for the most part) generous sharing of information among research groups, and to a sense of adventure. In my opinion, this episode had an important and beneficial impact on the way plant biologists work and think across disciplinary lines.[42]

An international conference held in Canberra in November 1970, jointly sponsored by the American and Australian governments, brought together scientists in the rapidly developing fields of photosynthesis and photorespiration.[43] Marshall Hatch was also in Canberra, having been appointed chief research scientist in the CSIRO's Division of Plant Industry in 1970. By this time there was overall agreement about carbon dioxide fixation by the C_4 pathway, although many details still needed to be worked out.[44]

The Carnegie team (Björkman, Malcolm Nobs, Robert Pearcy, John Boynton, and Berry) presented a paper on hybrids between C_3 and C_4 species, and Björkman in a separate paper spoke about the ecological significance of the C_4 pathway.[45] As he explained, the coinciding discoveries of the C_4 pathway and rediscovery of the Warburg effect and of photorespiration generated new interest in the significance of these photosynthetic pathways. "It is particularly encouraging to those of us whose primary interest lies in ecological aspects of photosynthesis," he wrote, "that experts on the photochemical and biochemical mechanism of photosynthesis have also become interested in this field." He noted that "a workshop with investigators representing such different specialties in the study of photosynthesis as those gathered in this room . . . would have been an unlikely event only a few years ago."[46] The C_4 pathway discovery was quickly recognized as ecologically significant, for the two pathways represented adaptations to different environments. C_4 plants were adapted to more extreme environments having high light intensities, high temperatures, and often severe drought. Björkman's experiments involved hybrids between species of the same genus, *Atriplex* (saltbush), which had different photosynthetic pathways. How was this cluster of C_4 traits inherited? What was their significance? As he concluded in his Australian talk, evidence that the C_4 pathway had evolved independently in several unrelated taxa "strongly suggests that this pathway represents an adaptation that has evolved in response to selective forces operating in specific ecological habitats."[47]

This subject perfectly fitted the Carnegie group's long-standing goal of synthesizing genetics, physiology, and ecology in the study of microevolution. The team also published its fifth contribution to its ongoing series on experimental studies on the nature of species in 1971. Written by Hiesey, Nobs, and Björkman, this volume focused on photosynthesis and the comparative study of ecological races and species in the genus *Mimulus,* which they had identified as a good model organism a couple of decades earlier. The results of these and other studies showed the importance of physiological divergence in the pathways of photosynthesis. These were adaptive changes and therefore "an important aspect of natural selection and plant evolution."[48] The publication of this research report coincided also with a shift in how the Carnegie styled this group's work. For decades called "experimental taxonomy," that designation changed after 1969, when Hiesey retired and Björkman took over leadership of the group. In 1970 he replaced the term "experimental taxonomy" with "physiological ecology." Hence the fifth volume, published in 1971, was subtitled *Biosystematics, Genetics, and Physiological Ecology of the Erythranthe Section of Mimulus.*

This volume made frequent reference to the need for controlled-environment facilities and precision instrumentation as the key to furthering this subject. The idea that this line of research was dependent on, or limited by, technological developments in laboratories was a running theme. Because the work focused on species growing in extreme climates, which simply would not survive under uncontrolled conditions in an unfavorable climate, the researchers depended on controlled-climate growth facilities to maintain stock clones of races and species from these extreme environments. Technical limitations also restricted how many species and races could be studied: "To date only a few ecologic races have been studied critically under controlled laboratory conditions," the authors wrote.[49] As they concluded, "current interest in this relatively new field has been stimulated by the development of new technical advances that make possible precise quantitative measurements combined with excellent control of external variables with reasonable speed."[50]

Berry's recollections emphasize the high degree of skill as well as the synthetic thought process that Björkman brought to his experimental studies. Although Berry maintained his focus on biochemistry after joining the team and achieved important results on the mechanisms of photosynthesis, he credited Bjorkman with introducing him to a new way of thinking about science: "I also began to see how Olle was linking our work to genetics, ecology, and evolutionary biology. Olle did experiments on another scale than I had been accustomed to. . . . He had a self-built gas exchange system that was the equal of any in the world at the time; it was one of the first to be able to estimate intercellular CO_2 concentration during photosynthesis." Berry confessed to feeling "truly overwhelmed" in his first few years after coming to Carnegie but also acknowledged learning "a whole new interdisciplinary research approach to science."[51]

After the Canberra conference Björkman spent a sabbatical year in Canberra in 1971–1972 and started a long association with Australian scientists. Charles Barry Osmond was one of the Australian botanists who formed part of this collaborative group. He was then at the department of environmental biology, led by Ralph Slatyer, located within the new Research School of Biological Sciences at the Australian National University in Canberra. The school had been created in 1967 as one of six research schools that formed the university's Institute of Advanced Studies. Osmond was also influenced by work being done at the CSIRO and the Canberra phytotron. He recalled that Bjorkman's sabbatical visit was a "game-changer" for photosynthesis research in Canberra, and later he also worked with Björkman in California.[52] Slatyer's department was similar to the Carnegie group in size and in its freedom of operation; these similarities enabled a strong partnership between the two groups that lasted many years, forming what Berry referred to as the "Carnegie-Canberra pipeline."[53] Also in the 1970s interactions increased with Stanford's department of biological science, making it possible for Björkman to co-advise doctoral students at Stanford. James Ehleringer, his first student, worked with Björkman and with Hal Mooney on adaptations of desert plants, including the differences between C_3 and C_4 plants; Ehleringer went on to a long and stellar career at the University of Utah.[54]

Björkman remembered the period from the mid-1960s to the mid-1970s as particularly good years at the Carnegie department. There was complete freedom to do research and very few duties outside of research. The Carnegie's approach, he recalled, was to encourage "Blue Sky research" that was high risk: "As long as it was new and innovative stuff, then go ahead. You did not have to get your proposals reviewed by anyone; you could just go ahead and do it." Carnegie also had generous funds for post-doctoral fellowships, without administrative strings, making it easy to bring in people. This focus on research, combined with constant easy grouping and regrouping with international colleagues from Australia and Europe, created a vibrant research environment. Björkman also preferred to collaborate with small groups of people, sometimes just one or two, because he liked the experience of discovering things: "If you have a small group working closely, you're on top of it immediately and share the joy [of discovery]. If the group becomes too large, you become an administrator. Also, it becomes more costly and you have to rely more and more on outside grants, and you spend more time on writing grant proposals than on doing research."[55] Berry also commented on other aspects of the social environment: "Björkman's co-workers experienced some of the best meals and drank some of the best wines of their lives in his home."[56]

Although Björkman considered himself mainly a bench scientist, he had strong ecological interests and approached the problems of physiology from an ecological perspective. This involved penetrating the question of biological adaptation with ever deepening analysis and experimental rigor, which was not something all ecologists practiced. It also entailed constantly moving back and forth between laboratory and field: both were places of discovery and insight. This principle can be seen at work in the next section, which looks at experiments that Björkman and his colleagues conducted in Death Valley, California, in an attempt to get at a very old Darwinian question: how do plants cope?

How Do Plants Cope? Experiments in Extreme Environments

Even in studies focused on single species and their adaptations to the abiotic environment, research could not be confined to controlled laboratory environments when species lived in climates so extreme that they were hard to duplicate in a laboratory. This problem led to what Berry has called "the most audacious physiological experiment of all time," one that in his view "redefined the field of physiological ecology."[57] This field study, conducted in the summer of 1970, was a collaborative project between Björkman and Robert Pearcy from the Carnegie team and ecologists at Stanford University, Tyrone Harrison and Hal Mooney.[58] The link between Mooney and the Carnegie group was a natural match. Mooney, after graduating from the University of California, Santa Barbara, went to Duke University in 1957 for a Ph.D. (completed in 1960) under the direction of Dwight Billings. Billings's interests in the physiological basis of ecotypic differentiation built directly on the work of Clausen, Keck, and Hiesey. He was also interested in controlled experiments; as Mooney recalled, "We designed and brought into the field one of the first gas exchange systems based on an infrared gas analyzer. We also designed and built growth chambers (out of deep freeze chests) for the subsequent controlled environment work."[59] Duke did not have the funds to build a full phytotron when Mooney was there, but the idea was in the air. Paul Kramer, a plant physiologist at Duke, helped to bring not one but two phytotrons to Duke and North Carolina State later in the 1960s.[60]

After completing his Ph.D., Mooney took a faculty position at the University of California, Los Angeles, and in 1968 he moved to Stanford University, where he was able to work with the Carnegie group. This team originally consisted of Mooney, Björkman, and Berry, each bringing different strengths in ecology, physiology, and biochemistry respectively to the team. As Mooney recalled in a later memoir, there were four main centers for plant physiological ecology at that time: Duke University, the University of Würzburg in Germany (under the leadership of Otto Lange), the Canberra group led by Ralph Slatyer, and the group at Carnegie/Stanford.[61] There were considerable interactions among these groups, and Mooney himself spent a sabbatical leave working with Slatyer's group in Canberra, as did Berry in 1976–1977.

One of the "high points" in his research, Mooney recalled, involved studies on plant adaptations to stress done in collaboration with Björkman and Berry. To study temperature adaptations they selected a hot environment, Death Valley, and a cold environment, Bodega Bay, California. They examined metabolic responses first in the field and then in experiments in controlled environments and experiments in reciprocal environments, with plants from Death Valley grown in Bodega Bay and vice versa. These studies, which also included collaborations with other scientists, helped to reveal the different adaptive pathways used by plants in extreme environments. The Death Valley studies, conducted in one of the most inhospitable environments on the earth, were especially revealing. In *Tidestromia oblongifolia,* a herbaceous annual in Death Valley, they found that its photosynthetic optimum was at 47 degrees C, a striking contrast to most plants, which had temperature optima close to 25 degrees C. "One of the interesting things," Mooney later wrote, "was that there were so many different ways for plants to cope with this extreme environment."[62] They attributed this extraordinary performance to the presence of the C_4 pathway, plus an unusually high thermal stability of the photosynthetic apparatus.[63]

To carry the instruments needed for these field studies, they used an air-conditioned mobile laboratory that was funded by an NSF grant to Mooney. Berry remarked how Björkman "adjusted, calibrated and redesigned each instrument until it met his standards" (figure 11).[64] The labora-

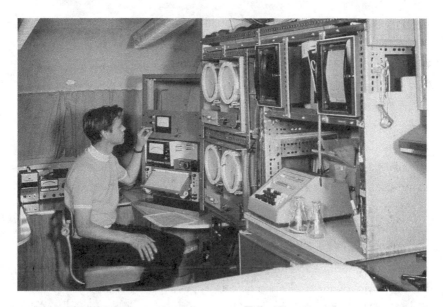

Figure 11. Olle Björkman inside the mobile laboratory about 1970, when he was working on plant adaptations to desert conditions. (Courtesy of the Carnegie Institution for Science Archives)

tory was a Dodge motor home modified to accommodate instruments and bench space, with additional changes made to withstand the heat of Death Valley.[65] As Mooney explained, the research on *Tidestromia* really had to be conducted in the field, for these plants did not have such high temperature optima when grown in normal growth chambers, probably because of limitation of light.[66] For some types of physiological research, the inability to duplicate the natural environment might not be important, but for ecological research involving adaptive strategies to extreme conditions it was crucial to study the plant in its natural environment. The mobile laboratory demonstrated that at least some measure of control could be brought to the field, satisfying the requirements of an ecological experiment.

The first mobile laboratory was relatively primitive by modern standards. "Initially in the late sixties," Björkman recalled, "they were not computer based. I had to get rather clunky kinds of equipment: recording systems with lots of paper, and the equipment was also a lot more heat dissipating . . .

which is hard when it's already extremely hot outside. That means the generators have to be larger and suffer more. So it was a big deal: you had to construct a mobile laboratory that could take these conditions and still not be too large, so you could get into the sites where the plants were without causing a lot of destruction of the habitat."[67] Björkman used the same laboratory to study *Oxalis oregana* (redwood sorrel) in 1980 in the coastal redwood forest, noting that "by that time, ten years later, it did have a computer the size of a washing machine, and we could do a lot more." This plant, related to the shamrock of Ireland, was interesting because it could quickly adjust its leaflet angle to maximize light interception when the light was low and to completely avoid it when the light was high.

Björkman's later comments on the Death Valley work reveal how important it was for him to move back and forth between laboratory and field. Ecology began with field observations to get some idea about what plant responses might be or ought to be, but then the plants were brought to the lab and grown under controlled conditions to see if those ideas made sense. If they did, it was back to the field to examine how the plants were growing in natural surroundings. In the Death Valley studies, it was obvious that plants were there but not obvious that they were active or could photosynthesize. To find out, one had to go to the field and "share the misery with the plants." This was why the mobile laboratory was needed. Once in the field, the researchers discovered new things and got a new perspective on the interaction between high temperature, high light, and water relations, but it was also important to approach the problem in an analytical way. This approach led them to consider the ecological aspects of C_4 photosynthesis, whereas the earlier work had been on the physiological aspects. In Death Valley that led to new research using transplants, to examine light interactions with temperature. As Björkman explained, photosynthesis could not be understood in a vacuum; instead, the plant had to be studied in its environmental context. "On the whole," he said, "I probably didn't spend more than ten percent of my time in the field. But that time was crucial to how we would approach and design the experiments that we would be doing in the lab." Planning the field research, he explained, took much longer than the actual time spent in the field.[68]

By the mid-1980s the understanding of the different photosynthetic pathways was much deeper, and it was clear that the discovery of the C_4 pathway was indeed a very important and exciting discovery.[69] In an article reviewing the latest research, co-authors Pearcy, Björkman, Martyn Caldwell, Jon Keeley, Russell Monson, and Boyd Strain developed the argument that looking at how plants assimilate carbon was a useful conceptual tool as well as a practical approach for studying low plants function in different environments. This placed the study of photosynthetic pathways (where carbon was gained) as well as respiration (where carbon was lost) at the center of physiological ecology. The discovery of the C_4 pathway had also opened up study of another pathway called the crassulacean acid metabolism (or CAM) pathway, which was common in succulent plants as well as more primitive plants such as ferns and their allies. Both of these pathways were mechanisms for maintaining a high efficiency of photosynthesis while preventing water loss. They enabled plants to thrive in hot and dry environments.

In the introduction to a special issue of *BioScience* devoted to the question of "how plants cope," Mooney, Pearcy, and James Ehleringer noted that scientists from many countries had contributed to the development of physiological plant ecology. They briefly reviewed the diverse contributions of biologists from Germany, Scandinavia, Britain, France, and Australia. In highlighting U.S. contributions, they drew attention to the work of the Carnegie team, including the strong evolutionary approach stemming from the work of Clausen and others as well as what they called the "unique working philosophy" illustrated by the research of Björkman and his colleagues on photosynthetic pathways. "This philosophy," they argued, "brought a vertical integration to the study of plant adaptive traits by leading investigators to seek the biochemical and physiological mechanisms underlying adaptive features and to demonstrate the relevance of these mechanisms to performance under natural conditions."[70] What they refer to as "vertical integration" involved the bridges formed between physiological ecologists and metabolic plant biologists, bridges that also linked laboratory and field studies. This type of integrative philosophy would also be important for connecting physiological ecology to evolutionary and ecosystem ecology.

The extremely productive research environment created by the Carnegie Institution did produce a major synthesis volume, which I would place alongside the volumes that make up the main canon of the modern synthesis published from the 1930s to the 1950s, although we do not associate it with the modern synthesis. Titled *Physiological Processes in Plant Ecology: Toward a Synthesis with Atriplex,* it was authored by Barry Osmond, Björkman, and Derek John Anderson (from the University of New South Wales, Australia) and published in the Springer series *Ecological Studies: Analysis and Synthesis* in 1980. Unlike most of the earlier modern synthesis volumes, it did not attempt to cover a wide field of research involving diverse organisms. Instead it focused on one genus, *Atriplex* (saltbush), a genus with worldwide distribution, hundreds of species, and thousands of taxa at the subspecies level. There were C_3 species and C_4 species within the genus. They had become adapted to an exceptionally wide range of conditions, making this genus ideally suited to serve as a model organism for understanding microevolution. It had been studied by a diverse group of biologists for many years, going all the way back to the work of Hall and Clements at the Carnegie Institution and to the field of genecology that Turesson developed in Sweden.

This synthetic volume brought together research from several disciplines, especially physiology and ecology, to analyze the mechanisms of adaptation to a range of environmental stresses, in the hope of providing insight into the general processes of speciation and adaptation of plants to diverse habitats. Its subject was a classic Darwinian problem: to understand adaptation, or to understand what makes an organism fit to live in an environment relative to its competitors. It was a plea for physiological ecologists to pay attention to the processes that determine and show evidence of fitness, and to do so by means of carefully designed experiments.

The reason I characterize this work as a Darwinian synthesis that is comparable to the earlier texts of Dobzhansky, Mayr, Stebbins, Simpson, and others is not because I wish to diminish the accomplishments of those earlier biologists. They certainly articulated a modern version of Darwinism and refocused scientific attention on the importance of natural selection. Their synthesis was largely aimed at the integration of genetics with other disciplines. I would not go so far as to say the modern synthesis as understood by most historians did not exist or was not a real movement. But I do think that it was an unfinished synthesis in 1950, that certain fields were left out and could not be included until they had amassed a lot more research.

The literature shows that additional synthetic undertakings, such as those of genecology, were starting already in the 1920s and the goal of achieving synthesis was there at the outset. It just took a lot longer to reach the point where synthesis could be attempted because of the complexity of the problems and the need for sophisticated experimental techniques. The Darwinian mystery of understanding adaptation had by no means been fully solved during the early period of the modern synthesis. With the development of new techniques and with the willingness of scientists from different fields to combine forces and work out problems of adaptation in detail, the grounds for further development of the modern synthesis were laid. These problems were so complex that they required long-term commitments along with partnerships across many disciplines, which meant the project became a multi-generational undertaking. I would argue that the work of the Carnegie group and its many collaborators was not a "post-Modern Synthesis" phenomenon but was in fact a continuation of the goals and ambitions of earlier endeavors. Formulating the modern synthesis was an ongoing quest that was only at the halfway point in the 1950s. We can better appreciate how difficult a task it was, and what kinds of resources were needed to support the task, if we take the story forward in time.

The Spirit of Canberra: Continuing the Integrative Enterprise

The synthetic enterprise that culminated in the Canberra conference in 1970 carried forward for the next three decades. The next generation of scientists, too young to have attended the conference, nonetheless recognized how the "integrative spirit" of that first group had been "seminal to the wave of understanding of C_4 photosynthesis that has been achieved in the almost 30 years since."[71] These comments were made by Rowan F. Sage of the University of Toronto and Russell K. Monson of the University of Colorado, who in 1999 edited a volume devoted to an overview of the subject of C_4 photosynthesis, with the express aim of further catalyzing the integration begun at the Canberra conference. In addition to chapters by Sage and Monson, the other twenty-two contributors provided a broad review of a subject that had grown into a major field of research. In this section I examine how the "spirit of Canberra" carried over to the next generation as exemplified by Monson's research in the 1980s and 1990s. I draw on a recent review of Monson's contributions to the evolution of C_4 photosynthesis by Rowan Sage, who noted that in about two dozen papers published between 1982 and 2003, Monson helped to "lay the foundation for decades of evolutionary investigations into origins of C_4 plants and their ecological interactions."[72] Much of this research built on the earlier foundations laid by the Carnegie team and its collaborators.

The 1970s saw increased interest in understanding the ecological significance of photosynthetic pathways in different kinds of ecosystems. Extra funding became available from NSF for studies of major biomes in the United States in connection with the International Biological Program, one of these being a study of the structure and function of grassland ecosystems. By the mid-1970s it was known that C_4 grasses in North America decreased in higher latitudes where summer temperatures were lower but that they were more common in warmer climates. In the North American shortgrass prairie, however, the two types of grasses lived together. How could this be explained in ecological terms? George J. Williams III, an ecologist at Washington State University in Pullman, Washington, explored these adaptive differences, focusing on two dominant species in Colorado, the C_3 species *Agropyron smithii* and the C_4 species *Bouteloua gracilis*. *A. smithii* grew during the cooler, wetter months of spring and early summer, while *B. gracilis* grew better during the hotter summer months. These variations were caused by the different photosynthetic pathways. In a follow-up study, Paul R. Kemp and Williams grew the plants in controlled-environment chambers, which easily confirmed that the two species responded dramatically differently to different temperatures. These functional variations meant that the two species occupied different ecological niches—in other words they were partitioning resources in a way that reduced competition. The ecologists studying these differences concluded that niche separation enabled the two species to reduce their competition for limited water resources.[73]

Russell Monson arrived at Washington State University in 1979 and took up this problem for his doctoral research, completed in 1982. His research looked at the same two grass species but added a third species, *Carex stenophylla,* a C_3 sedge that grows throughout cool and warm seasons. Using sophisticated experimental techniques that drew in part on the work of Björkman and others on the Carnegie team, Monson's research, which included both laboratory and field studies, looked closely at carbon dioxide fixation in relation to temperature and light intensity and measured changes in chlorophyll fluorescence (emission of light by chlorophyll). Chlorophyll absorbs light energy to drive the process of photosynthesis, but a small amount of excess energy is re-emitted either as heat or as light (fluorescence). When fluorescence increases (for instance, as temperature rises), it indicates that photosynthesis is being inhibited or that there is damage to the photosynthetic apparatus. The study involved a comparative analysis of the different adaptations to heat in the three species, with *C. stenophylla* representing an intermediate, less specialized plant that could acclimate to a broader range of environmental conditions than the other two species.[74] The defining feature of this work was to place the plant in its natural context. Sage's review noted that "by placing physiology into the field context, Monson et al. (1986b) demonstrate how simple physiological responses take on new significance in dynamic field settings, thus improving understanding of how species gain fitness in their environmental niche."[75]

After completing his thesis in 1982 and before joining the faculty of the University of Colorado at Boulder later that year, Monson was a postdoctoral fellow with Gerald E. (Gerry) Edwards and Maurice S. B. Ku, also at Washington State University, and began a long collaboration with that group. This research focused on species that had traits intermediate between C_3 and C_4 plants. One genus that had many intermediate phenotypes was *Flaveria* (a genus in the marigold family). The Edwards lab happened to acquire a nearly complete set of these intermediates about the time that Monson joined the group, creating, in Sage's words, a "synergy of talent, novel techniques, and unique plant material that produced major contributions to C_4 plant biology."[76] Their close examination of the intermediate types led to proposals about the evolutionary transition from C_3 to C_4 plants.

C_4 plants occur in diverse families of the more highly evolved orders of flowering plants but not in the more primitive orders. The pathway was thought to have evolved independently several times. But what was the exact evolutionary transition? Were the intermediate forms simply hybrids, or were they truly intermediates in the evolutionary stages leading from C_3 to C_4? Monson, Edwards, and Ku proposed that species in the genus *Flaveria* did appear to be true intermediate forms, and they developed a model to explain what the steps of this transition might entail.[77] This model outlined what later became known as C_2 photosynthesis, which Sage explains is "now recognized as a major intermediate phase in the evolution of C_4 photosynthesis."[78] C_2 photosynthesis involves capturing, concentrating, and then reassimilating the carbon dioxide that is given off during photorespiration.

It was important to place the evolution of C_4 species in the context of the paleo-environments in which those lineages had evolved: a time of relatively low carbon dioxide levels but higher temperatures at low and mid-latitudes. These created conditions favoring elevated photorespiration. A critical development in the evolutionary sequence, then, was to evolve a mechanism that could trap and refix the carbon dioxide that would normally be lost in photorespiration. Building on this idea, Monson and his colleagues developed a fuller picture of the transitional steps leading from C_3 to C_4 photosynthesis. The key insight was to appreciate how a fitness deficit could become an opportunity, such that "photorespired CO_2 became a valuable resource that selection could work with instead of an inhibitory waste product." Photorespiration, as Sage explained, "became the bridge to C_4 photosynthesis."[79]

By 1999, when the volume on C_4 biology edited by Sage and Monson was published, the importance of the C_4 pathway was much better understood, even though many specific questions were still being debated. C_4 plants had evolved independently from C_3 plants at least thirty-one times, and this pathway was present in 8,000 to 10,000 species in several families of flowering plants. C_4 plants had arisen and diversified within the past 15 to 30 million years, at a time when oxygen levels were relatively high and carbon dioxide levels were low. Under such conditions, photorespiration would pose a problem, competing with photosynthesis. The anatomical and physiological changes that led to the evolution of the C_4 pathway were an effective way to overcome that problem because the waste product of photorespiration, carbon dioxide, could be captured, stored, and refixed rather than lost.

The spread of C_4-dominated savannas during the late Miocene and early Pliocene also promoted the evolutionary success of grazing mammals and altered the environment in ways that presented new challenges for the hominid lines that were evolving in Africa at that time.[80] The animals grazing on the expanding grasslands would have provided rich sources of protein for any hominid line that learned how to live on the savanna and hunt those herbivores. The expansion of C_4 grasslands was therefore the backdrop for human evolution, creating a new niche "that favored many of the traits (large brains, complex social organization, sophisticated weaponry) that distinguishes [sic] our evolutionary line from all others."[81] Modern humans have literally reaped the benefits of C_4 evolution, for C_4 crops are major contributors to our food and their cultivation has enabled the rise and expansion of modern civilizations.

The chapters of C_4 Plant Biology unfolded this grand scenario, at the center of which lay an enzyme nicknamed rubisco with the unusual ability to catalyze reactions that fixed carbon (photosynthesis) and oxygen (photorespiration), the former essential to plant growth and the latter wasteful. In the early evolution of plants photorespiration was not a problem because oxygen levels in the atmosphere were low. Environmental changes occurring over millions of years turned this dual function into a liability, triggering evolutionary solutions that had major ecological consequences. These ecological changes in turn affected the evolution and success of humans. C_4 Plant Biology was indeed an expression of the "spirit of Canberra," but it was expanded now to extend from the level of the molecular sciences to ecological and evolutionary biology and finally to all of human history. I also see works like this as continuations of the spirit of the modern synthesis and of the Atriplex synthesis volume of 1980 that resulted from the "Carnegie-

Canberra pipeline": synthetic and integrative works bridging disciplines that dealt with processes on very different scales of time and space.

Understanding Ecosystem Functions

The study of the genus *Atriplex* and of such problems as the evolution of different photosynthetic pathways provided a good way to integrate physiology (and molecular sciences such as biochemistry), ecology, and evolutionary biology. However, there was also a need to integrate the different disciplines within ecology. Physiological ecology, community ecology, population ecology, and ecosystem ecology had all developed along different lines, and as late as the 1970s there was relatively little interaction between them. Ecologists saw this lack of integration as a problem, especially in view of the major global changes that humans were causing in the earth's ecosystems, which were also threatening to human well-being. Some of these crises were new, or newly recognized, such as the accumulation of greenhouse gases in the atmosphere and their effects on climate change. Such problems only exacerbated older problems connected to feeding the world's growing population: how would changes in climate affect human migrations, agriculture, pasturelands, forests, and water supplies? What kind of world were we heading into, and how could we adapt?

Ecological science experienced a major shift in the late 1980s and early 1990s in response to new appreciation of the global impacts of humans on biotic systems. Studies began to consider much larger dimensions than were customary in the past. One driver of this broadening perspective was better understanding of the exchange of materials between terrestrial ecosystems and the atmosphere. Modern challenges demanded a new kind of global ecology that was part of an interdisciplinary earth system science. Harold Mooney, Peter Vitousek, and Pamela Matson drew attention to the need for new global initiatives, pointing out that "an understanding of atmosphere-biosphere interactions and their effects is a problem vast in scope, one that encompasses the entire planetary ecosystem." They argued that experiments performed at the level of whole ecosystems would be important to "develop the capacity to predict the consequences of a changing atmosphere and climate."[82] Mooney referred to these changes in perspective as the "glo-balization of ecological thought" and described how this expanded vision in the 1980s and 1990s was accompanied by international cooperative research efforts, in which he took a leading role.[83] Scientists were establishing new coalitions and looking for ways to link disciplines to study global change.

These initiatives also influenced Carnegie's program in plant biology. There, Christopher Field's laboratory had been working to scale up physi-ological studies in tropical rain forests from leaf to whole plant to the forest canopy. Field, a physiological ecologist, completed his Ph.D. under Mooney's direction in 1981, and he joined the Carnegie department in 1984. In the mid-1980s the problem of scaling up from the level of the leaf to whole organism and to the canopy was considered an important but still formidable problem. In 1987, writing in a special issue of *BioScience* de-voted to physiological plant ecology, a group of ecologists (including Björk-man and Monson) reminded readers that the synthesis volume by Osmond, Björkman, and Anderson had argued that "the concerns of physiological ecology, which are shared with fields ranging from biochemistry to evolu-tionary ecology, cover a size scale from molecules to ecosystems and a time scale from milliseconds to thousands (if not millions) of years." A great deal had been learned about the mechanisms underlying the functions of the leaf, but "scaling up to understand the significance of these mechanisms for whole-plant performance in communities is still a difficult challenge." Moreover, the authors continued, "merely extending results from single leaves to whole canopies requires a formidable sampling program."[84]

Plant physiologists seized this challenge. The director of the Carne-gie department of plant biology, Winslow Briggs, noted that in 1990 the range of scale of Field's laboratory "increased dramatically to include re-gional and global as well as plant and ecosystem levels. The challenge of understanding global change is clearly emerging as one of the central scien-tific problems of this era."[85] In December 1990 a workshop at Snowbird, Utah, brought scientists together to discuss biological scaling. The subse-quent volume of papers, edited by James Ehleringer and Field, was consid-ered groundbreaking in its efforts to bring plant physiological ecology into the orbit of earth system science.[86] Field continued to lead the efforts to forge these links: he was the founding director of the Carnegie Institution's department of global ecology, created in 2002, and in 2016 he became director of the Stanford Woods Institute for the Environment.

Plant physiology, Field pointed out, "is deeply entwined with climate change."[87] Plants are not only sensitive to climate change but are also major regulators of climate because of their role in removing carbon dioxide from the atmosphere (by photosynthesis) and putting it back into the atmosphere (by respiration as well as combustion). Plant physiology and physiological ecology were key disciplines for the study of global change. By the year 2000, one of the main goals of global change research was to understand how plants and ecosystems regulated atmospheric carbon dioxide.[88] In a talk given to the Ecological Society of America in 2020, Field developed the point that physiological plant ecologists were important leaders in this shift toward global ecology.[89]

This subject—the leadership of physiological plant ecologists in broadening the scope of ecology in the 1990s and meeting the challenges of earth system science—deserves a lot more study and analysis. That topic is beyond the scope of this book. My purpose in drawing attention to these events and discussions is to emphasize that at this critical time around 1990 ecologists were becoming acutely aware that they needed not just to form links with physical scientists who studied climate or atmospheric processes but also to improve communications within ecology itself. Ecology was not a unified science, and ecologists often had strong disagreements about the value of certain approaches. However, new global crises such as climate change helped to drive efforts to improve communication across the different sub-disciplines within ecology. The next section examines how this drive to unify and communicate was expressed through novel laboratory designs, a development that brings us back to our main theme of laboratory innovation and the integration of disciplines.

From Phytotrons to Ecotrons

As it became clear that humans were transforming the face of the earth, ecologists revisited an idea proposed in the 1960s—namely, to build specially designed laboratories suited to ecological experiments, including experiments analyzing population dynamics, community structure, and ecosystem functions. These laboratories are known as "ecotrons." The movement to construct ecotrons began in the 1980s in response to global changes. Coinciding with these developments, environmental scientists were also trying to evaluate the effects of air pollutants as well as increased levels of carbon dioxide on plants and especially on forests. In Munich a new phytotron was commissioned in 1985 for forest research, because large forested areas in Germany and other countries were showing declines in response to air pollution, acid precipitation, and climatic changes. In Britain, outdoor hemispherical sunlit glasshouses with forced-air ventilation systems, called "Solardomes," were built to study the effects of low concentrations of atmospheric pollutants on plants. Scientists at the University of Lancaster had been using such domes since 1976 and had gradually improved the designs to give more precise control and make experiments repeatable. In the Netherlands experiments were being conducted on outdoor plots to measure the effects of air pollutants on plants.[90]

Building on these initiatives, new types of facilities for outdoor research were designed in the 1980s and 1990s to test the effects of enhanced carbon dioxide concentration on outdoor plots, including forested areas. These were called Free-Air Carbon dioxide Enrichment (or FACE) facilities; they did not involve enclosed chambers and were meant to test plant reactions under realistic microclimatic conditions. FACE experiments had no walls. Instead, a system of vents was constructed in a circular area that defined a plot of vegetation, which could be an agricultural field or even a forested area of moderately sized trees. Carbon dioxide was added from vents and carefully monitored by computer to control the concentration within the plot area. In the United States the Department of Energy funded these initiatives, originally undertaken by the Brookhaven National Laboratory, which worked on improving the design and performance of these experiments in the late 1980s. The first experiments, jointly supported by the departments of Energy and Agriculture, were large, cooperative ventures employing scientists from different disciplines and institutions. At Brookhaven's facility, located at the University of Arizona's Maricopa Agricultural Center, more than thirty scientists and engineers from fifteen government and university groups participated. The experiments were meant first of all to evaluate the effect of increased carbon dioxide on plants and ecosystems, a problem that had implications for the security of food and fiber production in the future. But they also had the longer-term goal of evaluating how terrestrial biota regulated atmospheric carbon dioxide concentration. Such experiments, which were necessarily expensive to set up and maintain, helped to move ecology in the direction of Big Science.[91]

The ecotron concept, which was emerging at the same time, fell in between traditional experiments in growth chambers and these types of FACE experiments. Although ecotron experiments could be used to test the effects of carbon dioxide concentration on ecosystems, they were designed to have a much broader range of uses. The first laboratory formally dubbed an "Ecotron" opened in January 1990 at Imperial College London's postgraduate campus at Silwood Park, about twenty-five miles west of London near Ascot, Berkshire. Operational trials began in July 1991 and full-scale experiments started in 1993. The ecotron was part of a new Centre for Population Biology, which had been created in 1989 by the Natural Environment Research Council as part of a larger program on the ecology of communities and populations. Its director was John H. Lawton, an eminent population and community ecologist (knighted in 2005 for his contributions to ecological science). He was the driving force behind the Ecotron Controlled Environment Facility, and in the ensuing discussions about the place of these facilities in ecology he became the spokesman for a multi-pronged approach to ecology that would include these kinds of laboratories. However, the building cost was high: about $1.5 million in U.S. dollars, with operating costs around U.S. $113,000 per year, excluding salaries. Lawton made the same point that had been made earlier about phytotrons: "By ecological norms this is an expensive and complicated exercise, but it is still cheap and straightforward compared with the ocean-going ships, satellites, radiotelescopes, and accelerators upon which oceanographers, astronomers, and particle physicists routinely depend."[92] Ecology's version of Big Science was still a bargain.

Just as phytotrons brought disciplines together, so did ecotrons. As Lawton and his colleagues explained, the ecotron was designed to create better links between ecosystem ecology and population and community ecology.[93] The ecotron was unusual in allowing scientists to build *model ecosystems* that were purely laboratory constructs, not replicas of any natural ecosystem. Its chambers allowed for the creation of up to sixteen small-sized ecosystems (a square meter each) with plants and animals representing four trophic levels, that is, levels in the food chain including primary producers (plants), herbivores, carnivores, and decomposers. Although less complex than natural ecosystems, the artificial systems were modeled on real, early successional ecosystems—weedy meadows—and embodied the essential features of such communities. Ecologists could also examine population dynamics—that is, changes in relative species abundance over a few generations. A wide variety of manipulative and comparative experiments could be done, and such experiments could be replicated.

These experiments were mainly run without seasons; all the organisms reproduced continuously, allowing for more generations in a given span of time. Such experiments were not possible in the field. The ecotron therefore functioned "like a biological accelerator."[94] The theme of the laboratory as an "accelerator," which harkened back to the comparison of the phytotron to a cyclotron, carried forward into ecotrons. The ecotron was meant to provide a way to tackle ecological theories that could not easily be tested in the field. But the results from the ecotron were not meant to replace field work. Instead, its proponents clarified, ecotron experiments provide "guidance for future field work, may provide insights into the mechanisms explaining past results from field work, and are an essential halfway house between the relative simplicity of mathematical models and the full complexity of the field."[95]

One of the earliest projects undertaken at Silwood Park's ecotron was to study the relationship between biodiversity and ecosystem functioning. This topic had recently risen to the top of the ecological agenda and was of interest to an international group of ecologists, all worried about how human activity was altering diversity in ecosystems of all types, causing changes that were not well understood or even documented. An international scientific advisory group called the Scientific Committee on Problems of the Environment (SCOPE) took up this problem in 1989 at a joint meeting with the International Union of Biological Sciences held in Washington, D.C. SCOPE's subsequent program, which included support from UNESCO's Man and the Biosphere Program, was launched at a meeting in Bayreuth, Germany, in October 1991. Lawton participated in that meeting, as did David Read, who discussed plant-microbe mutualisms and community structure. Hal Mooney served as chair of the steering committee for the portion of the joint project that involved biodiversity and ecosystem function. The committee included ecological experts from the United Kingdom (David Hawksworth), South Africa (Brian Huntley), France (Pierre Lasserre), Argentina (Osvaldo Sala), Venezuela (Ernesto Medina), Russia (Valerie Neronov), Germany (Ernst-Detlev Schulze), and the United States (Otto Solbrig). The meeting brought together population ecologists and ecosystem ecologists and led to a volume published in 1993 on *Biodiversity and Ecosystem Function,* which included contributions by many of the leading ecologists of the time.[96]

These discussions coincided with the United Nations Conference on Environment and Development, held in Rio de Janeiro, Brazil, in June 1992. The Rio meeting, known as the Earth Summit, focused on the goal of sustainable development, meaning a level of development that would promote healthy, productive lives for humans but still be in harmony with nature. Maintaining ecosystem health and biodiversity were linked to the concept of sustainable development at that meeting.[97] The United Nations Environment Program raised the idea of making a global assessment of biodiversity, and the SCOPE program became linked to this larger assessment.

Given the intense interest in understanding why it was important to conserve biodiversity, it seemed a natural subject to take up in the unique facilities of the ecotron. At least, this was the view of Shahid Naeem, then a postdoctoral fellow in Lawton's lab (and now a professor of ecology at Columbia University). He proposed the idea to Lawton, who was initially skeptical of the plan, because he believed that most ecosystems had redundant species—more than one species that performed the same functional roles, so that the loss of a few such redundant species would not affect ecosystem processes.[98] However, two other postdocs in the lab, Sharon Lawler and Lindsey Thompson, were enthusiastic about Naeem's proposal, and

Lawton gave the project his full backing. Naeem's idea was to construct three kinds of ecosystems having lower, medium, and higher levels of plant and animal diversity and assess how several ecosystem properties were affected by these differences. Initial results were reported in 1994 in *Nature,* and the five authors of the article (Naeem, Thompson, Lawler, Lawton, and Richard Woodfin) expressed their conclusions very cautiously.[99] Higher diversity did result in higher levels of productivity and community respiration. But the study also found that different ecosystem processes responded differently to loss of biodiversity, allowing for different interpretations of the underlying causes. The five authors published a more detailed discussion of the experiments in 1995, but again there was the same note of caution: a connection "may" exist between biodiversity and ecosystem functioning. Yet they also suggested that their arguments could be added to the growing list of reasons to conserve biodiversity.[100]

The ecotron experiments coincided with publication of results from long-term field experiments conducted in Minnesota, also announced in *Nature* in 1994.[101] The authors, David Tilman (University of Minnesota) and John A. Downing (University of Montreal) reported on an eleven-year study beginning in 1982 in Minnesota that assessed how ecosystem functions, especially primary productivity, responded to environmental stresses such as drought. The study period included a severe drought lasting two years, and the species-rich plots survived better, mainly because those plots included species that were resistant to drought. Tilman and Downing argued that ecosystem functioning was sensitive to biodiversity and that diversity promoted ecosystem stability.

As Mooney later remarked, these studies stimulated a great deal of discussion and new research even though they "left considerable room for further analyses and experimentation."[102] But coming as they did right in the middle of a wide-ranging international discussion of biodiversity and ecosystem function, they inevitably garnered attention, and in the context of the wider discussions they received a lot of publicity.[103] The studies were discussed at a final SCOPE conference at Asilomar, California, in 1994. The committee's report, published in 1996, broadly endorsed the idea that changes in biodiversity would affect ecosystem functions and declared that understanding the functional role of biodiversity had "crucial implications for the management of the Earth System."[104] But these studies had also generated controversy, and the controversies in turn affected ecologists' judgments not only about artificial communities but about such facilities as the ecotron.[105]

Some ecologists appreciated how ecotron experiments could help to address questions that were difficult to study in the field. Peter Kareiva, an ecologist at the University of Washington, Seattle, threw his support behind such experiments and hoped that the vision behind the ecotron would prove to be "contagious" and that a dozen more would be built.[106] But Lawton recognized that some ecologists were "suspicious of, if not downright hostile to" artificial communities of any kind and "to facilities like the Ecotron in particular."[107] He took up the defense of such "big bottle" experiments in 1996 and again pointed out that the ecotron "would never replace field observations and experiments, analyses of long-term data bases, and the search for large-scale patterns and processes in ecology."[108]

The ecotron was simply one tool and therefore had advantages as well as disadvantages. One advantage, as Lawton explained, came from the knowledge gained by trying to build an ecosystem from scratch: "A bench mark of the ability of ecologists to understand and predict the behavior of ecological systems is to build our own, and see if they work as expected! We should do more of it, more often."[109] Lawton's defense also touched on a sore point, the idea that the expense of ecotrons would divert funds away from other kinds of ecological investigations. He pointed out that the money used to build the ecotron was new money that came into science from the Natural Environment Research Council and had not been diverted from other areas of ecology. In fact, he asserted, none of those funds "would have been made available to ecologists without the vision of the Ecotron facility."[110] From the halfway house of the ecotron, the next step was into the field. The Centre for Population Biology took the lead here as well. In the mid-1990s it coordinated a large-scale comparative study called Biodiversity and Ecological Processes in Terrestrial Herbaceous Ecosystems (shortened to BIODEPTH). This project conducted biodiversity and ecosystem function experiments at field sites in the United Kingdom (at Silwood Park and Sheffield) and six European countries.[111]

A related approach was to do comparative studies of ecosystems that could be viewed as natural experiments. Mangrove ecosystems had the potential to serve as natural experiments because they showed a clear diversity gradient: thirty mangrove species grew across Australia, New Guinea, and Malaysia, but there were progressively fewer species as one traveled eastward. The Carnegie botanists, along with Hal Mooney, were involved in discussions about how to use mangrove forests to study how biodiversity affected ecosystem processes.[112] Björkman had studied Australian mangroves in the mid-1980s, as part of his broader interest in how plants coped with stressful environments. Mangroves grew where light was strong and temperatures high, and they often stood in full-strength seawater, conditions that would kill most other plants. After spending six months studying mangroves in Australia's coastal salt marshes, Björkman brought the plants into Carnegie's greenhouses and growth chambers in 1984 to study the physiological and biochemical mechanisms behind these plants' adaptations. These experiments were conducted with postdoctoral fellow Barbara Demmig from the University of Würzburg (now Barbara Demmig–Adams, professor of distinction in ecology and evolutionary biology at the University of Colorado, Boulder). In 1996, Björkman (then three years shy of retirement), along with Chris Field, initiated a project to do long-term studies of mangrove ecosystems as a way to understand the ecological role of biodiversity.

These shifts into the field with larger-scale and longer-term projects directed by international teams by no means signaled the end of ecotrons. They are now very much a part of the landscape of ecological research, although Silwood Park's ecotron closed in 2013 as research priorities shifted. In the United States, the first ecotron to be built was the Frits Went Laboratory, which opened in 1995 at the Desert Research Institute in Nevada. The centerpiece of the laboratory is four EcoCELLs (Ecologically Controlled Enclosed Lysimeter Laboratories), which are living-room-sized mesocosms that are used for research on the response of ecosystems to global change. (Lysimeters measure percolation of water, along with dissolved materials, through soil.) The ecosystems studied in Nevada were not assembled from scratch, as had been done at Silwood Park, but were taken from natural ecosystems. In one four-year study led by John (Jay) Arnone and Paul Verburg of the Desert Research Institute, twelve-ton plots of prairie grassland and soil from Oklahoma were dug up and transported to the EcoCELLs to measure how changes in temperature affected the uptake of carbon dioxide in these ecosystems. That study, reported in *Nature* in 2008, demonstrated that just one unusually warm year led to decreased carbon dioxide uptake lasting for two years, an unexpected result that revealed the sensitivity of these ecosystems to warmer temperatures.[113]

The European Ecotron of Montpellier (located north of Montpellier, France) opened in 2010, a half century after Frode Eckardt proposed the idea, but it represents a continuation of Eckardt's vision and legacy. It is owned by the Centre National de la Recherche Scientifique (CNRS) and is a laboratory of CNRS's Institute of Ecology and Environment. (The ecotron's website is www.ecotron.cnrs.fr/en/.) The director of the Montpellier ecotron, Jacques Roy, was also trained at Montpellier. Its adjunct director, Alexandru Milcu, had been the ecotron project leader at Silwood Park and moved to Montpellier when that facility closed. The Montpellier ecotron currently includes three experimental platforms of different sizes (referred to as macrocosms, mesocosms, and microcosms) and attracts researchers from many countries. CNRS also opened a second laboratory in 2017, the Ile-de-France Ecotron, located at Saint-Pierre-lès-Nemours at the Centre de Recherche en Ecologie Expérimentale et Prédictive.

Details about these ecotrons and related laboratories can be found in a recent review article by Jacques Roy and colleagues from several countries.[114] The authors identified thirteen facilities that qualify as ecotrons: eleven in Europe, one in North America, and one in Australia. Although a few were operating in the 1980s and 1990s, most have opened since 2010, and two were under construction in 2021. About half the experiments conducted in ecotrons and similar facilities deal with global changes, about a quarter investigate the effects of loss of biodiversity on ecosystem functioning, and a quarter center on physiological ecology. The authors assess the advantages and limitations of using ecotrons for ecological research, while stressing how modern work in such laboratories can pull together teams of scientists from a wide range of fields.

These newer ecotrons work with more realistic ecosystems and are therefore closer to experimental field conditions, rather than serving as halfway houses. But ecotron experiments are relatively short-term, generally running about a year or up to three years, so for long-term studies they have to be paired with field experiments. However, the review's authors emphasize that the significance of such laboratories, in addition to their capacity to provide realistic ecosystems, is their ability to project into the world of the future, a world with elevated carbon dioxide, warmer temperatures, and extensive drought. The investments in Big Science are demanded by the fact that the world is changing, and we should find out sooner rather than later how to adapt to those changes. The main question is how to translate that research—and the future scenarios it can create for us—into sound policy decisions.

Conclusion

There is no question that today's environmental and ecological sciences have become "big." The remarkable thing is that the founding visions and aspirations appeared very early, even if they could not all be realized at the time. Some projects ended when they fell victim to shifting scientific priorities. Nonetheless, the vision persisted and biologists continued to innovate in the quest to understand that most basic but also complicated question: how do organisms become adapted to such a wide range of environments? Physiologists and ecologists have continued to insist on the centrality of these questions because they are the foundation for understanding the global processes that affect us, the structure, function, and stability of ecosystems on which we depend, and even our own evolutionary history and success as a species. Physiological ecology shows us what evolution has accomplished over millions of years, creating so many solutions to the problem of living on the earth. Physiological ecology is a science of high ambition: it seeks to understand at a mechanistic level, but it also seeks to scale up to the global level.

Developing the science has required a tremendous amount of creativity—in experimental design, invention of equipment and laboratories, and the type of synthetic thinking that pulls information from different fields and disciplines to form novel explanations. In the story of the phytotron, as well as later initiatives using other kinds of laboratories, scientists have formed coalitions, collaborations, and cross-disciplinary links in their campaigns to bring resources into plant science. We commonly think of science as a highly competitive activity, but the stories I have told foreground the importance of cooperative relationships and multidisciplinary team efforts. Creating a scientific culture that emphasizes cooperation and sharing of data and that actively seeks to dissolve disciplinary barriers has been important in these synthetic sciences.

These efforts were successful to the extent that nations understood that investment in science was required to secure food and other resources to meet the needs of a growing population. In the face of climate change the urgency is now that much greater. A Big Science approach was needed, and the research community stepped up to persuade governments around the world to support big investments in biology. This book illustrates how much these fields of science have benefited from visionary thinkers, those who have had broad conceptions of science and a willingness to move across different fields and different subjects. We must analyze and understand how these sciences operate because they have been, and will be, the source of ideas about how we can solve our problems now and in the future. The initial building of phytotrons, intended in part to bring resources into what botanists considered to be basic or fundamental science, was never divorced from practical needs in agriculture and forestry. In these laboratories it did not make sense to distinguish between basic and applied science; the two were woven together.

What these places of inquiry collectively stand for is the idea that solutions to problems (from population expansion to climate change) require extensive investment in science, in this case, plant science. That is, the solutions are not going to be solely political, educational, cultural, or economic, although of course environmental problems include all those aspects. Solutions stemming from investment in science will entail some form of engineering or intervention, whether it be an effort to protect habitat, restore a damaged system, conserve the genetic diversity of wild and domesticated species, or develop new varieties of crops that can survive better in our altered systems. Scientific knowledge is needed to guide the next steps. If we care about finding solutions in this way, as I believe we should, then we should also be interested in understanding the scientific process and the kinds of challenges that scientists must address in order to figure out what the appropriate interventions might be.

Epilogue

THROUGHOUT THIS BOOK I have been interested in how, over the long development of plant physiology and physiological ecology, biologists have dealt with problems that required synthetic thinking, or thinking beyond the confines of an established discipline, method, or level of analysis. One challenge they faced was to combine the approaches or perspectives of different disciplines, a second was to integrate laboratory-based and field-based research, and a third was to scale up from the level of the leaf and individual organism to ecosystems and biomes. Such synthetic challenges are a particular feature of ecological and evolutionary sciences, and I have suggested that they deserve scrutiny beyond the few historical episodes that have received close attention, such as the "Modern Synthesis" in evolutionary biology. Biological literature contains frequent references to synthesis—or synonyms such as "fusion" or "integration"—and these references should prompt us to examine this process more closely.

Today it is common to use the term "integrative biology" to characterize twenty-first-century trends toward research that cuts across several disciplines and taxa; in the past few years programs in integrative biology have proliferated.[1] Since I am not specifically charting the historical development of such programs, I prefer to use "synthesis" rather than "integrative biology." By the term "synthesis" I mean a general type of scientific activity that involves combining perspectives; it would include what is meant by "integrative biology" but is also broader. I also agree with Massimo Pugliucci's argument that efforts to integrate different biological disciplines should not imply "all-encompassing reductionism," or the primacy of one particular field over others.[2] Seeking a synthetic interpretation does not mean reducing all of biology to a single level or discipline.

I would like to conclude with some recommendations for future research, beyond topics I have already mentioned in the chapters. These suggestions are tied to the themes of this book but carry the narrative further
into the twenty-first century and into the age of genomics. As discussed in
chapter 4, some plant physiologists worried about the rise of molecular biology in the 1950s and 1960s because they perceived molecular biologists as
determined to remake all of biology into a molecular science. A counter-
argument was that the goal of biology was to understand higher levels of
organization, and a strictly reductionist program could not succeed because
the whole was greater than the sum of the parts. However, even if all of biology was not reducible to the molecular level, it was still important to understand the molecular basis of an organism's development and organization.
The integration of molecular biology with other biological disciplines has
been the great synthetic achievement of the past thirty years. In 2005, Hans
Mohr, then at the end of a long and distinguished career in plant physiology, admitted that although he had initially feared the molecular collapse of
quantitative physiology, he recognized that the important challenge for the
next decade was to construct "an interface between genomics and whole
plant and animal physiology."[3] Creating such an interface is essential to the
task of linking genotype and phenotype, which has been the main challenge
of biological science since the rise of genetics in the early twentieth century.
It took nearly a century to fulfill this long-held aspiration.

Starting in the 1980s, but especially from the mid-1990s, the construction of these interfaces transformed a variety of disciplines. The third edition of Sally Smith and David Read's authoritative textbook on mycorrhizal
symbiosis, which appeared in 2008, spoke of the "explosion of molecular
biology, now applied to almost all aspects of biological investigation" that
had occurred over the previous decade. Molecular genetic techniques, the
adoption of new model plant species (in this case, legumes), and their inclu-
sion in genome sequencing projects all "revolutionized our understanding
of genetic programming and control" of arbuscular-mycorrhizal colonization.[4] Ecology was among the disciplines that felt the impact of these developments. A new journal titled *Molecular Ecology* started publication in
1992; its editors wisely did not attempt to define this field but welcomed any

subject that addressed the link between molecular biology and ecology. More recent textbooks on the subject also refer to molecular biology's "revolutionary" effects on ecological sciences.[5]

The advance of this revolution and the widespread application of molecular biological techniques is so recent that it has barely been studied within the history, philosophy, and social studies of science. How, and under what circumstances, did molecular biology infiltrate other biological disciplines, and what effect did these changes have on the balance between disciplines and on the way science is conducted? These are the transcendent problems that those who wish to explain the development of modern life sciences must address. The subject invites collaboration between scholars who focus on molecular biology and those who study other biological disciplines ranging from cell biology to ecology, whether they are historians, philosophers, or social scientists. Collaborative projects with biologists would also be valuable for such recent events, since many biologists are interested in the history of their fields. Each of the suggestions that follow offers fertile grounds for collaborative projects. The first centers on the introduction of a new model organism into botany, *Arabidopsis thaliana* (mouse-ear cress), a weed with no economic value that turned out to be an excellent vehicle for synthesizing molecular genetics, physiology, and development. Research on *Arabidopsis* has also contributed to our understanding of how plants defend themselves against pathogens, and so from this problem I segue to my second topic, the ecological study of plant chemical defenses against herbivores, a core problem of molecular ecology that also has important implications for control of agricultural pests.

Creating a Useful Weed: *Arabidopsis* as Model Organism

We would benefit greatly from a full scholarly study of the adoption of *Arabidopsis thaliana,* mouse-ear cress, as a model organism for the study of flowering plants and their development. *Arabidopsis* is a small plant of Eurasian origin in the mustard family and is found in Europe, Asia, Africa, Australia, and North America. Having no economic value, it seemed an unlikely prospect as a model organism compared to such well-studied plants as

wheat and maize. By the mid-1980s, however, it was well on its way to becoming as important to plant genetics as the fruit fly *Drosophila* had become to animal genetics. The tiny plant is now famous because it was the
first to have its genome sequenced (with most of the genome published by
December 2000); within five years this achievement was hailed as having
had a "catalytic effect" both on the research community and on the conduct
of plant science.[6]

Jim Endersby offered a brief account in his history of model organisms, while Sabina Leonelli opened the path to a broader analysis of *Arabidopsis,* but the secondary literature on the subject is still sparse. Leonelli
raised interesting questions about what it took for *Arabidopsis* to become a
model organism and highlighted the process of standardization that enabled the mass production and distribution of specimens. She and Rachel
Ankeny have recently published a philosophical discussion of the concept
of the model organism in general, which provides a good starting place for
further analysis of particular model organisms. Clearly the history of *Arabidopsis,* and of other plant model organisms that followed in its wake, is a
large and complex story that merits more study.[7]

Most of the history of *Arabidopsis* has been published by the biologists who had leading roles in convincing an initially skeptical community
of the plant's great potential as a model organism. Their accounts are valuable for giving an inside view of both the science and the social processes
that fostered the growth of the *Arabidopsis* community. But these relatively
condensed accounts deserve expansion and in-depth analysis, for *Arabidopsis* is a story of many breakthroughs that have influenced other areas of
plant science. The outline that follows, which points out some of the themes
that this history could explore, comes from these sources.[8]

The idea that *Arabidopsis* would be a good model organism goes back
to the German botanist Friedrich Laibach, who had worked with the plant
as early as 1907. It was a convenient plant for experiments because it was tiny,
had only five pairs of chromosomes, a short life cycle of about six weeks, was
prolific in seed production, and could self-fertilize. It appeared to be a good
candidate for the role of "botanical *Drosophila,*" or the botanical equivalent
of the fly that became one of the first model organisms in genetics. Starting

in the 1930s, Laibach gathered a large collection of what he called "ecotypes" of the plant, or varieties collected at specific locations and times (now called "accessions"), and in 1943 he recommended its suitability as a model organism both for genetics and for the study of microevolution.

Other botanists were similarly searching for botanical *Drosophilas* in mid-century. Frits Went favored California flowering plants as candidates because *Arabidopsis,* though able to grow over a wide geographical range, does not exist in the wild in southern California. Relatively few botanists heeded Laibach; two who did were George Rédei, a Hungarian botanist who emigrated to the United States in 1957 and obtained a position at the University of Missouri in Columbia, and John Langridge, who used the Canberra phytotron for experimental studies on *Arabidopsis* mutants. Although the number of scientists interested in *Arabidopsis* was small, they began to organize in the 1960s in the hope of generating interest in the plant. However, in the 1970s interest declined as biologists switched to other species for which funding was available. It was close to impossible to get funding for the study of a "useless" plant.

It was not until the 1980s that the use of *Arabidopsis* as a model organism for flowering plants took off. A few younger biologists, initially working independently, began to realize that *Arabidopsis* might be well suited to the new techniques of molecular genetics. They were attracted to the subject in part by an influential review article that Rédei had published in 1975. Another factor was the recent discovery, in 1977, that a soil bacterium, *Agrobacterium tumifaciens,* which causes crown gall tumors, could transfer DNA to the nuclear genome of higher plants. This finding suggested that the bacterium could be used to make transgenic plants. This discovery, along with Rédei's comprehensive review, caught the imagination of a small group of scientists early in their careers.[9]

This younger group included Chris and Shauna Somerville, Elliot Meyerowitz, and David Meinke in the United States and Maarten Koornneef in the Netherlands. It is from this group that we have many of the historical accounts documenting the expansion of *Arabidopsis* research from the 1980s and explaining its significance for current biological thought. The Somervilles (first at the University of Illinois and then at Michigan

State University in East Lansing) were initially attracted by the idea of applying the techniques of molecular biology to plants; their work with mutant plants resolved several problems relating to the biochemistry of
photosynthesis and photorespiration. Meinke (at Oklahoma State University in Stillwater) used *Arabidopsis* as a model system for the study of plant
embryo development. Koornneef and his colleagues at Wageningen University identified a series of mutants, a starting point for analyzing many basic
processes, and created the first genetic linkage maps showing the estimated
orders and positions of genes along the chromosomes.

Meyerowitz's dissertation at Yale University dealt with a problem in
developmental genetics in *Drosophila,* but he was also interested in plants
and took courses from plant scientists at Yale, including Arthur Galston.
After a postdoc at Stanford he was hired by Caltech in the early 1980s, and
shortly after that he and others in his lab turned to *Arabidopsis.* His group
worked out the size of the plant's genome (the smallest of any flowering plant
and about five times larger than yeast) and showed that the genome had relatively little repetitive DNA compared with other plants. This finding gave
Arabidopsis unique advantages, setting it apart from all other species used in
plant genetics. It suggested that *Arabidopsis* would be useful for cloning
genes relatively quickly and efficiently, which was hard to do in species with
large genomes. Meyerowitz's group at Caltech was the first to clone and sequence an *Arabidopsis* gene in 1986 and would go on to develop *Arabidopsis*
into a model for studying plant development, especially floral development.

These biologists, along with several others and with the help of a
few distinguished geneticists who were not plant scientists (Gerald Fink,
Ronald Davis, and Howard Goodman), strongly promoted *Arabidopsis* as a
model organism in the 1980s. The National Science Foundation, which had
forward-thinking program officers and directors, stepped in to fund much
of the *Arabidopsis* research and later supported the sequencing of the genome in the 1990s.[10] To attract more people to this field it was critically
important to create a welcoming culture of cooperation and openness, to
establish stock centers and databases, and to make data and materials available to all. When the molecular biology laboratory at Cold Spring Harbor
organized a course on plant molecular biology, led by Frederick Ausubel

and John Bedbrook, it soon became a course on *Arabidopsis*. Another course in Cologne, Germany, also helped attract people. In 1994, Cold Spring Harbor published a reference work, edited by Meyerowitz and Chris Somerville and written by a consortium of authors, that contained all available information about *Arabidopsis*.[11] All of these activities brought biologists into this field.

Chris Somerville in 1994 commented on how growth in the field of molecular genetics was causing shifts in plant biology.[12] The disciplines of plant physiology and biochemistry, he thought, had lost people, while plant morphology, a discipline that had become moribund, had been invigorated by new interest in developmental biology. Somerville was writing as the new director of the Carnegie Institution's department of plant biology at Stanford. Both he and Shauna had joined the department in January 1994, bringing in fourteen students, postdocs, and associates, which was a relatively large group. By that time about two thousand scientists were working with *Arabidopsis* worldwide, and thousands of genes were being identified in *Arabidopsis* as well as in rice, an important crop that has a compact genome compared to other cereals.

The decision to sequence the genome was taken in the mid-1990s after a few years of discussion, and the sequence was mostly completed by December 2000, a landmark achievement by an international team. One surprise was the discovery that there was more genetic redundancy, or duplicated genes, in *Arabidopsis* than was originally thought. The next stage was to determine the function of the thousands of genes identified. For their role in establishing *Arabidopsis* as a model organism for plant molecular genetics, Meyerowitz and Chris Somerville received the prestigious Balzan Prize, sometimes referred to as the "Italian Nobel," in 2006. *Arabidopsis* as a model organism for the study of flowering plants changed plant biology in fundamental ways, providing deep insight into basic processes of growth and development. This sketch merely hints at the many themes that biologists have identified in their historical accounts, themes that warrant much greater attention.

Meyerowitz's historical account, published in 2001, emphasized the conceptual advances that *Arabidopsis* brought to biology through the synthesis of classical and molecular genetics with plant development, plant physiology, and plant pathology. He used the word "fusion" rather than "synthesis," but I take the meaning to be the same. Thomas Hunt Morgan had sought to bring genetics and physiology closer together in the 1930s, and to this end he had brought a strong group of young plant physiologists to Caltech, among them Frits Went. But at that time it was premature to synthesize these disciplines, for the discovery of the double helix was still two decades in the future. It was the advent of molecular genetics and then the development of *Arabidopsis* as a model organism that made this synthesis possible in plant sciences many decades later.

Meyerowitz explicated the conceptual breakthrough that was involved. Plant physiologists, he pointed out, were mainly interested in the flow and movement of substances in plants, but genetics introduced a new idea: that there was a flow of information as well as substances. Information flowed not just from one generation to the next through the act of reproduction. It also flowed within the organism: it flowed into cells, from cell surface to nucleus and from nucleus to cytoplasm. These ideas led to different ways of approaching such problems as how hormones acted: a hormone became a "carrier of information that transmutes through a series of different biochemical forms, . . . a rather different view than that before genetics came to plant physiology."[13] Meyerowitz emphasized that the fusion of molecular genetics and plant physiology was not just a methodological change but a basic conceptual change: the concept of plant life was shifting as a result of these disciplinary fusions. Moreover, as a result of these syntheses experiments could be imagined and performed that were not even conceivable in earlier times. Discoveries made possible using *Arabidopsis* were relevant at all levels of plant science. As Meyerowitz noted, the subjects that the new knowledge could address included "such fundamental questions as how plants grow, how plant cells function and communicate with their neighbors, how plants sense and respond to their environments, and how plants change over evolutionary time."[14]

Shauna Somerville's research program, which continues now at the University of California, Berkeley, illustrates how the model could be applied to problems that were also important in ecology, evolutionary biology, pathology, and agronomy. Her group used *Arabidopsis* to study the mechanisms of disease resistance and the relationship between plant hosts and disease organisms like powdery mildew (caused by a fungus) and black rot (caused by a bacterium). The central question was how plants have evolved ways to sense and respond to various pathogens. These responses, which have a genetic basis, confer resistance to specific pathogens and even to specific races of pathogens. While the general goal of this research was to understand disease resistance in important cereal crops such as barley, several ecotypes of *Arabidopsis* also showed either resistance to the same diseases or tolerance of disease organisms. Identifying and isolating resistance genes in *Arabidopsis* could therefore serve as a bridge to understanding homologous genes (that is, genes derived from a common ancestor) in cereal crops. Moreover, *Arabidopsis* provided a way to compare resistance in a weedy species to that of cereal crops that had been cultivated for thousands of years.[15] Apart from using *Arabidopsis* to understand genetic mechanisms in species of economic importance, ecologists and evolutionary biologists also began to see *Arabidopsis* as a means of studying plant adaptation to a broad range of environmental conditions. Therefore, the study of *A. thaliana* and its close relatives in the wild also benefited from the tools and resources created by molecular genetics.[16] The story of *Arabidopsis* extends to many disciplines, including those that mainly study plants in natural or agricultural contexts.

The history of *Arabidopsis* addresses some of the most central problems of modern biology, including the Holy Grail of the past century, namely, the synthesis of genetics, physiology, and developmental biology. Another important synthetic achievement was to link this research to applied problems in agriculture and plant pathology. One would also wish to investigate, as part of the broader impact of *Arabidopsis,* the development of other model organisms that followed in its wake.[17] Finally, interest in using *Arabidopsis* to study problems in ecology and evolutionary biology carries the synthetic enterprise to the field sciences, while also returning to some of the questions that interested Laibach. The disciplinary, institutional, social, and conceptual themes that this history raises are all central to our understanding of how biology has developed in the past four decades. The quest for synthesis is integral to the analysis of these themes.

Plant Signaling and the Emergence of Molecular Ecology

My second suggestion regarding a productive direction for collaborative projects is to study the transition from chemical ecology to molecular ecology through the history of plant communication by chemical signaling. This topic, which deals with the coevolution of plants and animals, cuts across physiological, population, and community ecology. A central problem is the analysis of how plants mount defenses against their enemies, including herbivores, parasites, and pathogens, and how these complex relationships drive the coevolution of plants and animals. Because this is a large topic, I will not attempt an overview here but rely on the work of two leaders in the development of chemical ecology, Jack C. Schultz and Ian T. Baldwin, to suggest some of the themes that should interest historians, philosophers, and social scientists. The genomics era has led to major changes in the field of chemical ecology, has stimulated new kinds of syntheses between disciplines, and has produced new thinking about the training of graduate students. Since this topic emphasizes the ecological perspective, it underscores the importance of field research and the enduring value of natural history, or becoming intimately acquainted with the organism in its natural habitat.

The study of how plants responded chemically to attacks by herbivores and pathogens was a relatively new field of research in ecology in the 1970s. Some plant defenses are always expressed; these are called "constitutive" defenses. They operate whether or not a plant is being attacked. Many defenses, however, are called into action by an attack, which stimulates a response that helps protect the plant from further damage; these are called "induced" defenses. To study induced defenses is to study plant behavior, for the plant is actively responding to other organisms and even communicating with both plants and animals by chemical means. Such responses are induced by environmental conditions but are also under genetic control, so using the toolkit of molecular genetics, which enabled biologists both to silence and to enhance gene expression to study these interactions, became very important. This field of research also deals with a central problem of ecology—namely, the problem of adaptation: how are plants adapted to survive in a world full of hazards? Given that they are constantly under siege, how do they cope?

Plants produce many compounds that are not directly needed for their growth; these were regarded as "secondary" substances, and their evolutionary origins and adaptive functions were not fully understood in the mid-twentieth century. In chapter 6, I considered explanations focused on allelopathy, or how plants inhibit other plants by emitting toxic chemicals. But the coevolutionary relationships between plants and animals seemed a more plausible causal mechanism for the evolution of these substances, since they could either repel or attract animals. In 1959 the entomologist Gottfried Fraenkel proposed that secondary substances in many plant families initially arose solely as a means of attracting or repelling insects. Although many other organisms were known to be sensitive to these effects, he argued that insects likely played the main causal role. In 1964, Paul Ehrlich and Peter Raven published an influential review of coevolution between butterflies and their plant hosts, which repeatedly drew attention to the role of such secondary chemical substances in the relationship between plants and the caterpillars that ate those plants, even though the authors admitted that much had yet to be learned about these chemical roles. These arguments stimulated more research on chemical ecology, a field that was starting to develop in the 1960s.[18]

Biochemists as well as ecologists became interested in how plants mounted chemical defenses in response to attacks. At Washington State University in Pullman, Clarence A. Ryan, a biochemist, started working on this subject in the 1960s when the field was quite new. In the 1970s, Erkki Haukioja, an ecologist at the University of Turku in Finland, got interested in induced defenses and their effects on herbivore populations. David F. Rhoades, while a graduate student at the University of Washington, Seattle, found that trees decreased the quality of their tissues as food in response to attacks by herbivores, thereby making themselves less palatable to their enemies; these results supported the ideas of Haukioja and others. More surprisingly, Rhoades found that trees not under attack also showed this defensive response, suggesting that they were responding to signals coming from either the attacked trees or the attacking insects. Those signals, he thought, might be due to airborne pheromonal substances. When he presented these ideas at a meeting of the American Chemical Society in 1982, he was cautious, arguing that "the burden of proof for such an unprecedented effect should be high" and that his own experiments should not be taken as such proof.[19]

At the same time, Jack Schultz, a postdoctoral fellow at Dartmouth College, and Ian Baldwin, an undergraduate working with Schultz, were exploring similar problems, and in 1983 they published experimental evidence in favor of this hypothesis.[20] They designed experiments on potted plants in growth chambers and concluded that there was evidence that plants were able to communicate chemically. They simulated an attack by tearing the leaves of plants and found that the plants produced phenolic compounds that induced defensive responses in nearby plants not under attack. They tentatively put forward the supposition that this type of communication might also explain events such as mast fruiting (synchronous heavy fruiting every few years) and the decline of herbivore outbreaks, which if correct would suggest that this type of chemical communication had ecological significance. Their conclusions were viewed with some skepticism. It did not help that the press popularized this work as evidence of "talking trees," although no scientists were arguing that trees intentionally "spoke" or deliberately sent out danger signals to warn other plants. It would be more accurate to think of trees as "listening" or "eavesdropping" by picking up the chemical signals emitted by plants around them. It would take some time to bring ecologists around to their way of thinking.

Schultz and Baldwin became strong advocates of chemical ecology, and each has contributed in important ways to the growth of this field and to ecological and evolutionary studies. When ecologists suggested in 1988 that chemical explanations had been overemphasized in studies of herbivore diet, range, and fitness, Schultz countered with a strong argument for a more pluralistic and less exclusionary attitude. "Ecologists must accept chemistry as a part of natural history and integrate new kinds of information creatively," he contended. "Ecology should be a synthetic science, based on complete understanding of the parts but with a full appreciation of their interactions. Arguments that plant chemistry is not 'the predominant' influence on the evolution of herbivore diets are not compelling, interesting, or useful."[21] More research on chemical ecology, not less, was needed in view of how little was known about these relationships.

Baldwin completed his Ph.D. in 1989 at Cornell University, where his adviser was Thomas Eisner, an arthropod specialist and one of the founders of chemical ecology in the 1970s.[22] Baldwin then joined the faculty of the State University of New York, Buffalo, where his research explored how plants responded chemically to damage caused by herbivores. In 1997, he and Richard Karban, an entomologist at the University of California, Davis, published an extensive review of the literature on induced herbivory. Their aim was to integrate work from the schools of research that had developed by then and to demonstrate the ecological significance of the subject.[23]

Both Schultz and Baldwin recognized the importance of encouraging multidisciplinary approaches to problems of this kind, and both also emphasized the essential practical applications that could come out of this research, leading to improved crops and better methods of controlling pests that did not waste pesticides on plants that did not need them. Their ideas about multidisciplinary research extended also to the fields of agronomy and engineering and to the need for more sustainable agricultural methods. At Pennsylvania State University, Schultz founded the interdisciplinary Center for Chemical Ecology and helped to design the Huck Institutes of the Life Sciences, which were meant to promote research that might appear risky but had potentially high impact. As director of the Christopher S. Bond Life Sciences Center at the University of Missouri in Columbia, Schultz promoted interdisciplinary research ranging from plant breeding to gene therapy to electrical engineering. In collaboration with his wife, Heidi Appel, he continued to develop the field of chemical ecology, using genomic tools to study plant behaviors, and now in retirement he continues this research at the University of Toledo in Ohio, including recent research on *Arabidopsis*.

Baldwin was promoted to professor at SUNY Buffalo in 1996 and in the same year became the founding director of the new Max Planck Institute for Chemical Ecology in Jena, Germany, where he also served as director of the department of molecular ecology. Created after Germany's unification, the Max Planck Institute began its work in 1997 in rented quarters and moved into a new building in 2001. The conditions and patronage that facilitated these kinds of institutional changes—bringing disciplines together and allowing for creation of the multidisciplinary teams needed to tackle complicated questions—would be a fitting subject for further historical analysis.

The transition from chemical ecology into molecular ecology following the development of genetic engineering technologies and genome sequencing is another important theme. As work on genome sequencing in *Arabidopsis* ramped up, molecular biologists and ecologists started communicating more. A symposium at Penn State in 2000 was unusual in bringing these groups together, and it yielded papers ranging from the study of signaling pathways in *Arabidopsis* and other model plants to field studies of plants responding to insect damage.[24] Baldwin, at the meeting to present the ecological aspects of this research, discussed the work of his group at the Max Planck Institute, where they were studying a North American wild tobacco species, *Nicotiana attenuata*. Nicotine produced by the plant was a powerful poison against herbivores, but when the plant was eaten by a caterpillar resistant to nicotine, it responded by shutting off its nicotine production. The group was trying to figure out which party benefited more from this response, the plant or the insect.

In addition to laboratory studies based at the Max Planck Institute for Chemical Ecology, Baldwin established a site for field experiments on a preserve in Utah owned by Brigham Young University. Since *N. attenuata* grew naturally in this location, the Jena group could grow genetically modified plants and test their hypotheses about chemical signaling in realistic ecological settings. The team developed genetically altered strains, some with elevated emissions ("screamers") and others with suppressed emissions ("mute" plants). Similarly, plants could be engineered to be "deaf" to signals from other plants. By growing these mutants in the field biologists could test whether the signals were really doing what they thought. Although they took precautions to ensure that transgenes did not escape into wild plants, in Germany such experiments, even with precautionary measures, could not be done because of stronger popular opposition to genetically modified organisms.[25]

In 2002, Schultz drew attention to the conceptual shift that better understanding of plants' phenotypic plasticity was bringing about. Plants truly exhibited dynamic "behaviors" in their biochemical responses, and Schultz developed the argument that plants and animals had analogous signaling mechanisms. Taking this idea further, he suggested that plants and animals had the means to manipulate the other's signaling system, an idea that implied there could be an intricate type of "phylogenetic espionage" going on. Both plants and animals, for instance, might be using a form of "biowarfare" to interfere with the immune responses to pathogens in the other.[26] The way the two kingdoms were interacting, in other words, could be far more complex than was realized. These ideas have led to startling new discoveries. In 2021 an international team of scientists announced their discovery that a whitefly, a serious agricultural pest, uses a gene "hijacked" from a plant in the distant past to neutralize a toxin produced by the host plant as a defense.[27] This was the first known instance of a natural gene transfer from a plant to an insect.

In his 2002 article Schultz pointed out that in order to develop and test hypotheses about such complicated interactions, it was essential to adopt an "integrative and comparative view that encompasses plants and animals (and microbes)." Such a view would be "especially useful in grasping how plants and others interact." That kind of approach, he argued, had to "embrace multiple levels, from molecular to population, to be successful." In addition, it did not make sense to divide plants from animals when thinking about how organisms respond to their environments. Looking at convergences and commonalities between signaling mechanisms in plants and animals could, he asserted, "teach us much about adaptation as a long-term process and especially about coevolution between plants and their exploiters." That would in turn require a focus not on botany or zoology as separate spheres but on the "general phenomena of biology."[28] These comments invoke another form of synthetic thinking, in this case aiming to blur the boundary between two of the basic divisions in the life sciences, that between plants and animals. What should interest historians as well as philosophers of science are the major conceptual shifts entailed by this line of research and in particular the way these new perspectives propelled the synthesis of botany and zoology.

Schultz's suggestions about the need for an integrated "biological" perspective, especially in the study of coevolutionary processes, should not be confused with the claim that plants are like animals anatomically or behaviorally. Such claims have emerged forcefully from a small group of plant scientists in the last two decades and have given rise to a field devoted to the study of plant "neurobiology," which emerged in 2006. Proponents of this field argue that plants have something like a nervous system and possess "intelligence," "memory," and the ability to "learn" in a manner comparable to animals. The new journal *Plant Signaling and Behavior* was launched as a platform for scientific discussion of the complex mechanisms of "communication" among plants, animals, and microbes and of the notion that plants are "neuronal" individuals. Here too we find advocacy of the need to integrate disciplines and levels of analysis. Some admit that references to such things as plant "intelligence" are metaphors but defend the practice on the grounds that all metaphors have value. Others, however, have adopted the use of anthropomorphic language to such an extreme degree that it has generated strong criticisms for being imprecise, misleading, and even mystical.[29]

On the purely popular level, such language sparks the public's imagination, sells books, and turns authors into celebrities. It is easy to understand why anthropomorphic language would be a fixture in popular science writing or film. But when scientists not only defend but actually insist upon the use of such language in scientific discourse, it invites a serious examination of the underlying causes and goals as well as the consequences of this way of speaking. Is it the start of an important paradigm shift, as its proponents claim? Or does it signal a resurgence of romanticism in modern science, an echo of the "Deep Ecology" philosophical movement of the 1970s, perhaps a counter-cultural reaction against molecular reductionism or a response to other cultural forces? This subject is certainly worth scholarly attention, but it should not be confused with efforts to look deeply into the molecular mechanisms that underlie coevolutionary relationships in studies that do not employ anthropomorphic analogies.

Baldwin, despite occasional references to "talking" trees, in fact cautioned against anthropomorphism and advocated the need to acquire a firmly plant-centered frame of mind, which he termed "phytopomorphism."

He drew attention to changes in the way graduate students should be trained to make them adept at using the "-omic toolbox" while also retaining the "art of natural history discovery" that would give them a "feeling for the organism." Writing in 2012, Baldwin thought that graduate training programs still reflected the academic schism that had occurred earlier when biology departments split between those working at the cellular-molecular-developmental levels and those at the ecological and evolutionary levels. He envisioned a new breed of "genome-enabled field biologists" who would be comfortable in both worlds and who would be studying genetically manipulated organisms in the "natural laboratories" of the field.[30]

The program at Jena in Baldwin's department of molecular ecology was meant to offer this kind of hybrid training by exposing students to different ways of solving organismic-level problems. That program created three platforms for study: a molecular biology platform (for genetic modification of organisms), an analytical platform (for training in the use of diverse analytical instruments), and an ecological platform (for field research). The goal of this training was to foster a conceptual shift: students had to learn to be "phytopomorphic" in their thinking instead of anthropomorphic. They had to be tuned in to the natural history of plants and the organisms they interacted with, a skill that could only be acquired through intensive study in natural surroundings and long and close observation of organisms in their ecological contexts. The primacy of natural history, of starting in the field and ending in the field, was as relevant to the molecular ecologists of the twenty-first century as it had been for the botanists in the nineteenth century who first sought to synthesize physiology with ecology and answer the central Darwinian question of the meaning and origin of adaptive traits.

The study of plant-herbivore relationships provides an excellent lens through which to examine the development of chemical ecology and its transformation into molecular ecology. Its history, as I have suggested in this brief sketch, offers ample scope for exploring many problems of synthesis, including synthesis between disciplines, across the major boundary separating botany and zoology, and between laboratory and field sciences. Here too there are conceptual, institutional, disciplinary, and social dimensions that would all need to be examined closely to help us understand the

profound changes that have occurred in modern ecological and evolutionary biology over the past four decades. There is also the culturally intriguing phenomenon of plant "neurobiology" as a field that claims to be
advancing a new paradigm, which intersects with the literature on plant
"communication." I believe there is much to be gained by collaboration
with biologists who are engaged in these fields of research, to help us understand underlying motives, perspectives, challenges, and strategies that
may not be obvious from the published records.

Notes

Abbreviations

CIT Archives   California Institute of Technology Archives and Special Collections, Pasadena, California.

Went diary   Frits Went Journal, collection No. 79, Hunt Institute for Botanical Documentation, Carnegie Mellon University, Pittsburgh, Pennsylvania.

RAC   Rockefeller Archive Center, Sleepy Hollow, New York.

Chapter 1. The Dream of Synthesis and the Laboratory Vision

1. Christopher B. Field, "Plant Physiology of the 'Missing' Carbon Sink," *Plant Physiology* 125 (2001): 25–28, on 25.
2. Hans Lambers, F. Stuart Chapin III, and Thijs L. Pons, *Plant Physiological Ecology,* 2nd ed. (New York: Springer, 2008).
3. Patrick Manning and Mat Savelli, eds., *Global Transformations in the Life Sciences, 1945–1980* (Pittsburgh: University of Pittsburgh Press, 2018).
4. Focus section on "Fields," *Isis* 113, no. 1 (2022). There is no mention of plant sciences in this special section.
5. G. Ledyard Stebbins, "International Horizons in the Life Sciences," *AIBS Bulletin* 12, no. 6 (1962): 13–19, on 13.
6. Eugene Cittadino, *Nature as the Laboratory: Darwinian Plant Ecology in the German Empire, 1880–1900* (Cambridge: Cambridge University Press, 1990).
7. Charles F. Hottes, "The Contributions to Botany of Julius von Sachs," *Annals of the Missouri Botanical Garden* 19, no. 1 (1932): 15–30, on 20, 21.
8. Cittadino, *Nature as the Laboratory,* 2.
9. Soraya de Chadarevian, "Laboratory Science Versus Country-house Experiments: The Controversy Between Julius Sachs and Charles Darwin," *British Journal for the History of Science* 29 (1996): 17–41.
10. F. Noll, "Julius von Sachs: A Biographical Sketch with Portrait," *Botanical Gazette* 25, no. 1 (1898): 1–12, on 7–8.

11. Noll, "Julius von Sachs," 5.

12. Noll, "Julius von Sachs," 7.

13. Cittadino, *Nature as the Laboratory,* 26.

14. David Kohn, *Darwin's Garden: An Evolutionary Adventure* (New York: New York Botanical Garden, 2008), 40.

15. Malcolm Nicolson, "Humboldtian Plant Geography After Humboldt: The Link to Ecology," *British Journal for the History of Science* 29, no. 3 (1996): 289–310.

16. Andreas F. W. Schimper, *Plant-Geography upon a Physiological Basis,* trans. William R. Fisher, v. 1 (Oxford: Clarendon Press, 1903), vi–vii.

17. Cittadino, *Nature as the Laboratory,* 134–38.

18. Sharon E. Kingsland, *The Evolution of American Ecology, 1890–2000* (Baltimore: Johns Hopkins University Press, 2005), 77.

19. Robert E. Kohler, *Landscapes and Labscapes: Exploring the Lab-Field Border in Biology* (Chicago: University of Chicago Press, 2002).

20. His address is quoted in Cleveland Abbe, *A First Report on the Relations Between Climates and Crops* (Washington, D.C.: Government Printing Office, 1905), 23–27.

21. Quoted in Abbe, *Relations Between Climates and Crops,* 26.

22. Quoted in Abbe, *Relations Between Climates and Crops,* 26.

23. Quoted in Abbe, *Relations Between Climates and Crops,* 26–27.

24. Abbe, *Relations Between Climates and Crops,* 7.

25. Abbe, *Relations Between Climates and Crops,* 27.

26. Biographical information on Thompson is from Hermann Hagedorn, *The Magnate: William Boyce Thompson and His Time (1869–1930)* (New York: Reynal and Hitchcock, 1935), and William Crocker, *Growth of Plants: Twenty Years' Research at Boyce Thompson Institute* (New York: Reinhold, 1948).

27. Thompson engaged in various unofficial political intrigues trying to build support for the provisional Kerensky government, with no success. His activities are described in Claude E. Fike, "The Influence of the Creel Committee and the American Red Cross on Russian-American Relations, 1917–1919," *Journal of Modern History* 31 (1959): 93–109.

28. "Organization—Equipment—Dedication," *Contributions from Boyce Thompson Institute for Plant Research* 1 (1925): 1–58, on 30; Leonard H. Weinstein and Richard C. Staples, "Personal Views of Boyce Thompson Institute, 1974–2000," March 2005, btiscience.org/wp-content/uploads/2014/05/Personal_Views_of_BTI.pdf, accessed December 28, 2021. There is said to be a history of the early years of the institute by S. E. A. McCallan, *A Personalized History of Boyce Thompson Institute,* published in 1975, but I have not been able to locate it.

29. "W. B. Thompson Opens Plant Institute Unit," *New York Times,* September 25, 1924, 12.

30. On its later research see George L. McNew, "Profile: Boyce Thompson Institute; The Concepts and Management of Research," *BioScience* 21, no. 2 (1971): 81–84.

31. Gail Cooper, *Air-Conditioning America: Engineers and the Controlled Environment, 1900–1960* (Baltimore: Johns Hopkins University Press, 1998).

32. Crocker, *Growth of Plants;* William Crocker, "Botany of the Future," *Science* 88 (1938): 387–94.

33. Crocker, "Botany of the Future," 390.

34. The idea of "purposeful basic knowledge" remained at the core of the institute's philosophy. See the later statement by its director George L. McNew, "Basic Concepts Behind the Boyce Thompson Institute for Plant Research," *AIBS Bulletin* 6, no. 3 (1956): 6–7.

35. Kohler, *Landscapes and Labscapes,* 163.

36. Victor E. Shelford, *Laboratory and Field Ecology: The Responses of Animals as Indicators of Correct Working Methods* (Baltimore: Williams and Wilkins, 1929), 2; see also Robert A. Croker, *Pioneer Ecologist: The Life and Work of Victor Ernest Shelford, 1877–1968* (Washington, D.C.: Smithsonian Institution Press, 1991).

37. Shelford, *Laboratory and Field Ecology,* 392.

38. Shelford, *Laboratory and Field Ecology,* 424–25.

39. Charles A. Kofoid, review of *Laboratory and Field Ecology: The Responses of Animals as Indicators of Correct Working Methods,* by Victor E. Shelford, *Ecology* 11, no. 3 (1930): 609–11, on 610.

40. William C. Cook, review of *Laboratory and Field Ecology: The Responses of Animals as Indicators of Correct Working Methods,* by Victor E. Shelford, *Ecology* 11, no. 3 (1930): 611–14, on 614.

41. Charles Elton, review of *Laboratory and Field Ecology: The Responses of Animals as Indicators of Correct Working Methods,* by Victor E. Shelford, *Journal of Ecology* 19, no. 1 (1931): 216–17.

42. Victor E. Shelford, "An Experimental Approach to the Study of Plant and Animal Reproductivity and Population with a Life Science Building Plan," *Ecology* 34, no. 2 (1953): 422–26, on 425. See also V. E. Shelford, "A Ground Plan for a Biological Research Plant," *Ecology* 32, no. 4 (1951): 760–63.

43. W. W. Garner and H. A. Allard, "Effect of the Relative Length of Day and Night and Other Factors of the Environment on Growth and Reproduction in Plants," *Journal of Agricultural Research* 18, no. 11 (1920): 553–606; W. W. Garner and H. A. Allard, "Photoperiodism: The Response of the Plant to Relative Day and

Night," *Science* 55, no. 1431 (1920): 582–83. The term "photoperiodism" was suggested by O. F. Cook of the Bureau of Plant Industry. A contemporary review of these discoveries is in K. F. Kellerman, "A Review of the Discovery of Photoperiodism: The Influence of the Length of Daily Light Periods upon the Growth of Plants," *Quarterly Review of Biology* 1, no. 1 (1926): 87–94, on 90. For a full history of photoperiodism and the mechanisms underlying the response, see Linda C. Sage, *Pigment of the Imagination: A History of Phytochrome Research* (San Diego: Academic Press, 1992).

44. Barrington Moore, "The Relative Length of Day and Night," *Ecology* 1, no. 2 (1920): 234–37.

45. Shelford, *Laboratory and Field Ecology*, 309, 321.

46. H. A. Allard, "Length of Day in Relation to the Natural and Artificial Distribution of Plants," *Ecology* 13, no. 3 (1932): 221–34.

47. Shelford, "Experimental Approach," 424.

Chapter 2. An Atomic Age Laboratory

1. Frits W. Went, "Reflections and Speculations," *Annual Review of Plant Physiology* 25 (1974): 1–26; Frits W. Went, "Orchids in My Life," in *Orchid Biology, Reviews and Perspectives*, v. 5, ed. Joseph Arditti (Portland: Timber Press, 1990), 21–36.

2. Went, "Reflections and Speculations," 3.

3. Arthur W. Galston and Thomas D. Sharkey, "Frits Warmolt Went, 1903–1990," *Biographical Memoirs of the National Academy of Sciences of the United States of America* 74 (1990): 348–63; F. W. Went, "Reflections and Speculations."

4. The test is known as the "Avena test" because it is done with oat seedlings, genus *Avena*. It is described in Frits W. Went and Kenneth V. Thimann, *Phytohormones* (New York: Macmillan, 1937), 27–34.

5. F. A. F. C. Went and F. W. Went, "A Short History of General Botany in the Netherlands Indies," in *Science and Scientists in the Netherlands Indies*, ed. Pieter Honig and Frans Verdoorn (New York: Board for the Netherlands Indies, Surinam, and Curacao, 1945), 390–402; F. W. Went, "The Tjibodas Biological Station and Forest Reserve. I. A Naturalist's Paradise," in *Science and Scientists in the Netherlands Indies*, 403.

6. On the history of agricultural research in the Netherlands and the tension between universities and agricultural colleges at this time, see Harro Maat, *Science Cultivating Practice: A History of Agricultural Science in the Netherlands and Its Colonies, 1863–1986* (Dordrecht: Kluwer, 2001), chap. 2, 3.

7. Went, "Reflections and Speculations," 18.

8. Andreas F. W. Schimper, *Plant-Geography upon a Physiological Basis,* trans. William R. Fisher, v. 1 (Oxford: Clarendon Press, 1903), vi.

9. T. H. Morgan, A. H. Sturtevant, H. J. Muller, and C. B. Bridges, *The Mechanism of Mendelian Heredity* (New York: Henry Holt, 1915); T. H. Morgan, *The Theory of the Gene* (New Haven: Yale University Press, 1926).

10. Frank Lillie, a leading American embryologist, identified Morgan as one of the principal advocates of the union of genetics and physiology. Frank R. Lillie, "The Gene and the Ontogenetic Process," *Science* 66 (1927): 361–68.

11. "Biology at the California Institute of Technology," *Science* 66 (1927): 276–77, on 277. Morgan likely was consulted for this announcement.

12. Research proposal probably written by Morgan about 1930–1931, no date, in T. H. Morgan Papers, box 2, file 2.23, "Biology Research Proposal, 1930s," CIT Archives. The proposal reported on expenditures to date on experimental biology and future funding needs and was probably part of a grant proposal.

13. Nicolas Rasmussen, "Plant Hormones in War and Peace: Science, Industry, and Government in the Development of Herbicides in 1940s America," *Isis* 92 (2001): 291–316.

14. Frits Went to T. H. Morgan, July 28, 1932, in Thomas Hunt Morgan Papers, box 2, file 2.2, "Went, FAFC and FW," CIT Archives.

15. Frank B. Salisbury, "James Frederick Bonner, 1910–1996," *Biographical Memoirs of the National Academy of Sciences of the United States of America* (Washington, D.C.: National Academies Press, 1998), www.nasonline.org/publications/biographical-memoirs/memoir-pdfs/bonner-james-f.pdf.

16. Frits W. Went and Kenneth Thimann, *Phytohormones* (New York: Macmillan, 1937).

17. "Dr. Henry Eversole," *Engineering and Science,* June 1963, 18. Eversole supported the Red Cross's view that humanitarian aid aimed at saving children was crucial to maintaining economic and political stability in Europe after World War I. See Julia F. Irwin, "Sauvons les Bébés: Child Health and U.S. Humanitarian Aid in the First World War Era," *Bulletin of the History of Medicine* 86, no. 1 (2012): 337–65, Susan G. Solomon, "Knowing the 'Local': Rockefeller Foundation Officers' Site Visits to Russia in the 1920s," *Slavic Review* 62, no. 4 (2003): 710–32.

18. Edward A. White, *American Orchid Culture* (New York: A. T. de la Mare Co., 1942), 30, 72–75.

19. Went, "Reflections and Speculations," 11; Frits W. Went, "Plant Growth Under Controlled Conditions. II. Thermoperiodicity in Growth and Fruiting in the Tomato," *American Journal of Botany* 31 (1944): 135–50.

20. Frits W. Went, "Plant Growth Under Controlled Conditions. I. The Air-Conditioned Greenhouses at the California Institute of Technology," *American Journal of Botany* 30 (1943): 157–63.

21. Hiesey, "Environmental Influence and Transplant Experiments," *Botanical Review* 6 (1940): 181–203.

22. H. A. Allard, "Length of Day in Relation to the Natural and Artificial Distribution of Plants," *Ecology* 13 (1932): 221–34; W. W. Garner, "Recent Work on Photoperiodism," *Botanical Review* 3 (1937): 259–74, see discussion of work of Evans and Allard on 261; Hiesey, "Environmental Influence and Transplant Experiments," 191 on Evans's work.

23. Karl C. Hamner and James Bonner, "Photoperiodism in Relation to Hormones as Factors in Floral Initiation and Development," *Botanical Gazette* 100 (1938): 388–431. See also James Bonner, "Chapters from My Life," *Annual Review of Plant Physiology and Plant Molecular Biology* 45 (1994): 1–23.

24. For the full history of this discovery, see Linda C. Sage, *Pigment of the Imagination: A History of Phytochrome Research* (San Diego: Academic Press, 1992).

25. F. W. Went, "Effects of Light on Stem and Leaf Growth," *American Journal of Botany* 28 (1941): 83–95, on 83.

26. R. J. Downs, "Phytotrons," *Botanical Review* 46 (1980): 447–89, on 452; D. Von Wettstein and K. Perschle, "Klimakammern mit konstanten Bedingungen für die Kultur hoherer Pflanzen," *Naturwissenschaften* 28 (1940): 537–43.

27. F. W. Went, "The Earhart Plant Research Laboratory," *Chronica Botanica* 12, no. 3 (1950): 89–108, on 94.

28. Robert Millikan to Earhart Foundation, May 27, 1948, in Robert Millikan Papers, box 24, file 24.7, "Earhart Laboratories for Plant Research, 1945–1948, 1950," CIT Archives. Earhart's company, White Star Refining, was a major oil refining and distributing company that spurred Detroit's industrial growth. It later became part of Exxon Mobil Corporation.

29. Millikan to Harry Earhart, May 16, 1945, in Millikan Papers, box 24, file 24.4, "Earhart Foundation 1944–1945," CIT Archives.

30. George Beadle to Robert Millikan, December 3, 1945, in Millikan Papers, box 24, file 24.7, "Earhart Laboratories for Plant Research, 1945–1948, 1950," CIT Archives.

31. Millikan to Earhart, December 22, 1945, in Millikan Papers, box 24, file 24.7, "Earhart Laboratories for Plant Research, 1945–1948, 1950," CIT Archives.

32. Earhart to Millikan, December 7, 1945; Millikan to Earhart, December 14, 1945, and January 7, 1946, in Millikan Papers, box 24, file 24.7, "Earhart Laboratories for Plant Research, 1945–1948, 1950," CIT Archives.

33. F. W. Went, "Report on a Survey of Botanical Facilities in the United States," n.d., reporting on a trip from March 10 to April 26, 1946, in Millikan Papers, box 24, file 24.7, "Earhart Laboratories for Plant Research, 1945–1948, 1950," CIT Archives.

34. Went, "Report on a Survey of Botanical Facilities."

35. F. L. McDougall, "International Aspects of Postwar Food and Agriculture," *Annals of American Academy of Political and Social Science* 225 (1943): 122–27; Frank G. Boudreau, "The International Food Movement in Retrospect," in *Food for the World*, ed. Theodore W. Shultz (Chicago: University of Chicago Press, 1945), 1–15; League of Nations, *Nutrition: Final Report of the Mixed Committee of the League of Nations on the Relation of Nutrition to Health, Agriculture, and Economic Policy* (Geneva: League of Nations, 1937).

36. Karl Brandt, *The Reconstruction of World Agriculture* (New York: W. W. Norton, 1945), 314, 330–33.

37. John D. Black, "Food: War and Postwar," *Annals American Academy of Political and Social Science* 225 (1943): 1–5, on 3.

38. Theodore W. Schultz, ed., *Food for the World* (Chicago: University of Chicago Press, 1945); Karl Brandt, "Basic Elements of an International Food Policy," in *Food for the World*, 321–34; Brandt, *The Reconstruction of World Agriculture*.

39. Fairfield Osborn, *Our Plundered Planet* (Boston: Little, Brown, 1948). The Malthusian side of this debate is discussed in greater depth by Thomas Robertson, *The Malthusian Moment: Global Population Growth and the Birth of American Environmentalism* (New Brunswick, N.J.: Rutgers University Press, 2012), chap. 2 on Vogt and Osborn.

40. William Vogt, *Road to Survival* (New York: William Sloane, 1948). For a discussion of reviewers' responses to Vogt's and Osborn's books, see also Pierre Desrochers and Christine Hoffbauer, "The Post War Intellectual Roots of the Population Bomb. Fairfield Osborn's 'Our Plundered Planet' and William Vogt's 'Road to Survival' in Retrospect," *Electronic Journal of Sustainable Development* 1 (2009): 37–61. See citations in their article for other secondary sources dealing with Osborn and Vogt, including unpublished doctoral dissertations.

41. William Vogt, review of *Our Plundered Planet*, by Fairfield Osborn, *Science* 107 (1948): 510.

42. Guy I. Burch and Elmer Pendell, *Human Breeding and Survival: Population Roads to Peace and War* (New York: Penguin, 1947), 100.

43. Vogt, *Road to Survival*, 211.

44. For the Malthusian argument see Warren S. Thompson, "Population," *Scientific American* 182 (February 1950): 11–15, and the anti-Malthusian counter-argument by Lord John Boyd-Orr, "The Food Problem," *Scientific American* 183 (August 1950): 11–15. See also reviews of Osborn's and Vogt's books: Joseph L. Fisher, "Land Economics; Agricultural Economics; Economic Geography," *American Economic Review* 39 (June 1949): 822–25; and Jacob J. Kaplan,

"Resource Conservation and Foreign Policy," *World Politics* 1 (January 1949): 257–65.

45. Schultz, *Food for the World.*

46. Lord John Boyd-Orr, "The Food Problem," 11–15.

47. Karl Brandt, review of *Road to Survival,* by William Vogt, *Land Economics* 26 (1950): 88–90, on 88.

48. One such authority was Robert M. Salter, chief of the Bureau of Plant Industry, Soils, and Agricultural Engineering at the U.S. Department of Agriculture, Beltsville, Maryland; Salter, "World Soil and Fertilizer Resources in Relation to Food Needs," *Science* 105 (1947): 533–38.

49. F. W. Went, typescript note appended to letter from Robert Millikan to Earhart Foundation, May 27, 1948, in Millikan Papers, box 24, file 24.7, "Earhart Laboratories," CIT Archives.

50. Went to Millikan, May 27, 1948, in Millikan Papers, box 24, file 24.7, "Earhart Laboratories," CIT Archives.

51. Millikan to Earhart, June 4, 1948, in Millikan Papers, box 24, file 24.6, "Earhart Foundation, 1948–1949," CIT Archives.

52. Robert Kargon, *The Rise of Robert Millikan: Portrait of a Life in American Science* (Ithaca: Cornell University Press, 1982), 162–66.

53. Friedrich A. Hayek, *The Road to Serfdom* (University of Chicago Press, 1944), reprint, *The Road to Serfdom, Text and Documents: The Definitive Edition,* ed. Bruce Caldwell (Chicago: University of Chicago Press, 2007).

54. Earhart to Millikan, June 2, 1948, and Millikan to Earhart, June 4, 1948, Robert Millikan Papers, box 24, file 24.6, "Earhart Foundation, 1948–1949," CIT Archives. On Earhart's support of Hayek, see letters on economic education at Caltech in Robert Millikan Papers, box 24, file 24.5, "Earhart Foundation, 1946–7." The Earhart Foundation was one of the early supporters of Hayek.

55. William A. Jensen and Frank B. Salisbury, *Botany: An Ecological Approach* (Belmont, Calif.: Wadsworth Publishing, 1972), p. 719.

56. James Bonner, interview by Graham Berry, Pasadena, California, March 13–14, 1980, 16, Oral History Project, CIT Archives, retrieved January 31, 2022, resolver.caltech.edu/CaltechOH:OH_Bonner_J.

57. "Phytotron: New Experimental Greenhouse Opened at California Institute of Technology," *Plant Physiology* 24 (July 1949): 553.

Chapter 3. Big Science in a Small Pond

1. Went, "The Earhart Plant Research Laboratory," *Chronica Botanica* 12, no. 3 (1950): 89–108, on 94.

2. Frits W. Went, *The Experimental Control of Plant Growth* (Waltham, Mass.: Chronica Botanica, 1957), 23.

3. Victor Boesen, "Atomic Age Greenhouse," *Nation's Business: A General Magazine for Businessmen,* August 1952, 26–29, 66–67, on 28, 29.

4. Boesen, "Atomic Age Greenhouse," 29.

5. F. W. Went, "Climate and Agriculture," *Scientific American* 196 (July 1957): 82–98, on 86. This article includes several photos of the laboratory interior.

6. Went, *Experimental Control of Plant Growth,* 318.

7. Went, "Earhart Plant Research Laboratory," 94.

8. Went, "Earhart Plant Research Laboratory." In a later description Went had fourteen artificially lighted rooms and ten darkrooms, which suggests that the original design was slightly altered.

9. F. W. Went, "Phytotronics," in *Proceedings of Plant Science Symposium* (Camden, N.J.: Campbell Soup Company, 1962), 149–61, on 155.

10. Went, "Phytotronics," 153.

11. Went, *Experimental Control of Plant Growth,* 48.

12. Went, *Experimental Control of Plant Growth,* 319.

13. Went, *Experimental Control of Plant Growth,* 321.

14. The International Basic Economy Corporation (IBEC) was founded by Nelson Rockefeller after the war, and the IBEC Research Institute was created in November 1950. The institute's main purpose was to advance agriculture in tropical and semi-tropical areas, and in the early 1950s its work focused on agriculture in Venezuela and Brazil.

15. Biology Division papers, files 27.10, "Earhart—Budget 1954–1957" and 28.1, "Earhart Laboratory, 1951–1955," CIT Archives.

16. "Develop Tomato Plant for Hot Climates," *Science News-Letter* 72, no. 26 (1957), 404.

17. Went, "Phytotronics."

18. Henry Hellmers, "Chaparral Plants," in F. W. Went, *Experimental Control of Plant Growth,* 184–91.

19. Decades later this approach would be called into question after research revealed that the introduced exotics had unintended consequences because they could alter the fire regimes by increasing the frequency of fires in California's ecosystems. Adam M. Lambert, Carla M. D'Antonio, and Tom L. Dudley, "Invasive Species and Fire in California Ecosystems," *Fremontia* 38 (April/July 2010): 29–36.

20. F. W. Went, "Climate and Agriculture," *Scientific American* 196 (June 1957): 82–94.

21. Went, *Experimental Control of Plant Growth,* 323.

22. Went, *Experimental Control of Plant Growth,* 328.

23. Geert J. Somsen, "Committing to Internationalism: Mediating Activities of Dutch Scientists, 1900–1950," in *The Global and the Local: The History of Science and the Cultural Integration of Europe,* ed. M. Kokowski, Proceedings of the 2nd International Conference of the European Society for the History of Science (Cracow: Polish Academy of Arts and Sciences, 2007), 757–60.

24. F. A. F. C. Went, "The International Union of Biological Sciences," *Science* 68 (1928): 545–47.

25. Went, *Experimental Control of Plant Growth,* 96–97.

26. Went, *Experimental Control of Plant Growth,* 328.

27. A reference to "botanical *Drosophilas*" is in Robert O. Whyte, *Crop Production and Environment* (London: Faber and Faber, 1949), 55; Went, *Experimental Control of Plant Growth,* 97.

28. Ben Patrusky, "Drosophila Botanica," *Mosaic* 22, no. 2 (1991): 32–43; Sabina Leonelli, "Arabidopsis, the Botanical Drosophila: From Mouse Cress to Model Organism," *Endeavour* 31, no.1 (2007): 34–38.

29. Arnold Beckman Oral History Project interview with Mary Terrall, 2016-02-17-00003, on 60, CIT Archives. collections.archives.caltech.edu/repositories/2/digital_objects/228.

30. Ralph Nader, *Unsafe at Any Speed: The Designed-in Dangers of the American Automobile* (New York: Grossman Publishers, 1972), 149; Nader, "Real Junk Science: The Corruption of Science by Corporate Money," *New Solutions* 8 (1998): 33–44. On the history of photochemical smog, see James E. Krier and Edmund Ursen, *Pollution and Policy: A Case Essay on California and Federal Experience with Motor Vehicle Air Pollution, 1940–1975* (Berkeley: University of California Press, 1977); Chip Jacobs and William J. Kelly, *Smogtown: The Lung-Burning History of Pollution in Los Angeles* (Woodstock and New York: Overlook, 2008). In addition to Arnold Beckman's oral history, see the anecdotal account in Zus Haagen-Smit, Oral History Project interview with Shirley K. Cohen, 2016-02-17-000086, CIT Archives. collections.archives.caltech.edu/repositories/2/digital_objects/283.

31. A. J. Haagen-Smit, "Theory and Practice of Air Conservation," in *Science, Scientists, and Society,* ed. William Beranek, Jr. (Tarrytown-on-Hudson, N.Y., and Belmont, Calif.: Bogden and Quigley, 1972), 28–54, on 30.

32. James N. Pitts, Jr., and Edgar R. Stephens, "Arie Haagen-Smit, 1900–1977," *Journal of the Air Pollution Control Association* 28 (1978): 516–17.

33. E. R. Stephens, P. L. Hanst, R. C. Doerr, and W. E. Scott, "Reactions of Nitrogen Dioxide and Organic Compounds in Air," *Industrial and Engineering Chemistry* 48, no. 9 (1948): 1498–1504.

34. A. J. Haagen-Smit, "Abatement Strategy for Photochemical Smog," *Advances in Chemistry* 113 (1972): 169–86.

35. Frits W. Went, "The General Problem of Air Pollution and Plants," *Proceedings of the First National Air Pollution Symposium, November 10–11, 1949* (Pasadena: Stanford Research Institute, 1951), 148–49.

36. Ed Ainsworth, "Living Plants May Give Key to Smog," *Los Angeles Times,* April 17, 1950, A1.

37. Robert Millikan to Harry B. Earhart, May 11, 1951, in Robert Millikan Papers, box 39, file 39.1, "Earhart, Harry B., 1934, 1939–1940, 1950–1953," CIT Archives.

38. Ainsworth, "Living Plants May Give Key to Smog."

39. Herbert M. Hull and Frits W. Went, "Life Processes of Plants as Affected by Air Pollution," *Proceedings of the Second National Air Pollution Symposium* (Pasadena: Stanford Research Institute, 1952), 122–28; Ruth Ann Bobrov, "The Anatomical Effects of Air Pollution on Plants," *Proceedings of the Second National Air Pollution Symposium* (Pasadena: Stanford Research Institute, 1952), 129–34.

40. A. J. Haagen-Smit, "The Air Pollution Problem in Los Angeles," *Engineering and Science* 14 (December 1950): 7–13.

41. A. J. Haagen-Smit, Ellis F. Darley, Milton Zaitlin, Herbert Hull, and Wilfred Noble, "Investigation on Injury to Plants from Air Pollution in the Los Angeles Area," *Plant Physiology* 27 (January 1952): 18–34, on 34.

42. Edgar R. Stephens, "Smog Studies of the 1950s," *Eos* 68, no. 7 (1987): 89–95.

43. Jerry McAfee, "Accomplishments in Air Pollution Control by the Petroleum Industry," *Proceedings of National Conference on Air Pollution* (Washington, D.C.: Government Printing Office, 1959), 64–69, on 67.

44. Stephens et al., "Reactions of Nitrogen Dioxide."

45. Stephens, "Smog Studies of the 1950s," 92.

46. H. F. Johnstone, L. C. McCabe, and M. D. Thomas, "Chemistry of Pollutants in the Atmosphere," *Industrial and Engineering Chemistry* 48, no. 9 (1956): 1483.

47. Reported in A. J. Haagen-Smit, "A Lesson from the Smog Capital of the World," *Proceedings of the National Academy of Sciences of the United States of America* 67, no. 2 (1970): 887–97, on 890.

48. Air Pollution Foundation, *Final Report* (San Marino, Calif., 1961), 13. On this report and the Air Pollution Foundation's role in the debate, see George A. Gonzalez, "Urban Growth and the Politics of Air Pollution: The Establishment of California's Automobile Emission Standards," *Polity* 35, no. 2 (2002): 213–36.

49. Haagen-Smit, "A Lesson from the Smog Capital of the World," 890.

50. Stephens, "Smog Studies of the 1950s."

51. Alberto Cambrosio and Peter Keating, "The Disciplinary Stake: The Case of Chronobiology," *Social Studies of Science* 13 (1983): 323–52; see also Jole Shackelford, *The Search for Biological Clocks: Metaphors, Models, and Mechanisms* (Pittsburgh: University of Pittsburgh Press, 2022).

52. K. C. Hamner, "Photoperiodism and Circadian Rhythms," in *Biological Clocks: Cold Spring Harbor Symposia on Quantitative Biology* 25 (1960): 269–77.

53. Frits W. Went, "Ecological Implications of the Autonomous 24-hour Rhythm in Plants," *Annals of New York Academy of Sciences* 98 (1962): 866–75.

54. J. Woodland Hastings and Beatrice Sweeney, "On the Mechanism of Temperature Independence in a Biological Clock," *Proceedings of the National Academy of Sciences of the United States of America* 43, no. 9 (1957): 804–11.

55. Colin Pittendrigh, "On Temperature Independence in the Clock System Controlling Emergence Time in Drosophila," *Proceedings of the National Academy of Sciences of the United States of America* 40, no. 10 (1954): 1018–29, on 1027.

56. Harry R. Highkin, "The Effect of Constant Temperature Environments and of Continuous Light on the Growth and Development of Pea Plants," in *Biological Clocks: Cold Spring Harbor Symposia in Quantitative Biology* 25 (1960): 231–38.

57. *Biological Clocks: Cold Spring Harbor Symposia on Quantitative Biology* 25 (1960). The volume is available on the Cold Spring Harbor Symposia website: symposium.cshlp.org/content/25.

58. James Bonner, interview by Graham Berry, Pasadena, California, March 13–14, 1980, 16, Oral History Project, CIT Archives, retrieved January 31, 2022, resolver.caltech.edu/CaltechOH:OH_Bonner_J.

59. "Chemical Biology," *Engineering and Science* 17 (February 1954): 9–13.

60. Earnest C. Watson to Warren Weaver, September 25, 1951, in Rockefeller Foundation Archives, R.G. 1.2, Series 200D, United States, Natural Sciences and Agriculture, box 177, folder 1632, "California Institute of Technology—Plant Physiology," RAC.

61. J. G. Harrar, memo dated November 15, 1952, in Rockefeller Foundation Archives, R. G. 1.2, series 200D, United States: Natural Sciences and Agriculture, box 177, folder 1632, "California Institute of Technology—Plant Physiology," RAC.

62. J. G. Harrar, memo dated November 15, 1952, in Rockefeller Foundation Archives, R. G. 1.2, series 200D, United States: Natural Sciences and Agriculture, box 177, folder 1632, "California Institute of Technology—Plant Physiology," RAC.

63. J. G. Harrar to George Beadle, September 21, 1956, in Rockefeller Foundation Archives, R. G. 1.2, Series 200D, United States: Natural Sciences and Agriculture,

box 177, folder 1633, "California Institute of Technology—Plant Physiology," RAC. In the same folder see also Bonner to J. G. Harrar, June 1, 1956; Harrar to G. Beadle, September 21, 1956; Lee Dubridge to J. G. Harrar, October 16, 1956, RAC.

64. Went to Harrar, June 27, 1956, in Rockefeller Foundation Archives, R. G. 1.2, series 200D, United States: Natural Sciences and Agriculture, box 177, folder 1633, "California Institute of Technology—Plant Physiology," RAC.

65. Anton Lang, "Some Recollections and Reflections," *Annual Review of Plant Physiology* 31 (1980): 1–28, on 20.

66. Lang, "Some Recollections and Reflections."

67. R. J. Downs, "Phytotrons," *Botanical Review* 46 (1980): 447–89, on 477.

68. James Bonner, "Chapters from My Life," *Annual Review of Plant Physiology and Plant Molecular Biology* 45 (1994): 1–23.

69. Went, "Phytotronics," 151.

70. Went, "Phytotronics."

71. Anton Lang, "Some Recollections and Reflections," 20.

Chapter 4. Crossing Borders

1. J. P. Hudson, ed., *Control of the Environment* (London: Butterworths Scientific, 1957), ix.

2. Frits W. Went, *The Experimental Control of Plant Growth* (Waltham, Mass.: Chronica Botanica, 1957), 73.

3. Went to Harry B. Earhart, July 18, 1950, in Robert Millikan Papers, box 24, file 24.7, "Earhart Laboratories," CIT Archives.

4. His ecological paper was F. W. Went, "The Effects of Rain and Temperature on Plant Distribution in the Desert," in *Desert Research: Proceedings, International Symposium Held in Jerusalem, May 7–14, 1952* (Jerusalem: Research Council of Israel, 1953), 230–37.

5. Went diary, May 10, 1952.

6. Went diary, November 7, 1952.

7. R. Bouillenne and M. Bouillenne-Walrand, "Le Phytotron de l'Institut Botanique de l'Université de Liège," *Archives de l'Institut de Botanique de l'Université de Liège* 20 (1950): 1–61.

8. Harro Maat, *Science Cultivating Practice: A History of Agricultural Science in The Netherlands and Its Colonies, 1863–1986* (Dordrecht: Kluwer Academic, 2001); O. Banga, *The Institute of Horticultural Plant Breeding* (brochure) (Wageningen, The Netherlands: no publisher., n.d.); J. P. Braak and L. Smeets, "The Phytotron of the Institute of Horticultural Plant Breeding at Wageningen,

Netherlands," *Euphytica* 5 (1956): 205–17; L. Smeets, "The Phytotron of the Institute for Horticultural Plant Breeding, Wageningen, the Netherlands: A Revision of Previous Descriptions," *Netherlands Journal of Agricultural Science* 26 (1978): 8–12.

9. Diter von Wettstein, "The Phytotron in Stockholm," *Studia Forestalia Suicica* 44 (1967): 3–23. The Stockholm phytotron, constructed between 1958 and 1964 and opened in 1965, was designed for forestry research and received support from the Rockefeller Foundation. On the Austrian phytotron, see L. Roussel, "Le Phytotron du Patscherkofel," *Revue Forestière Française,* no. 12 (1960): 769–74.

10. "Existing Phytotrons and Research Institute[s] with Environment Controlled Installations," *Phytotronic Newsletter,* no. 21, July 1980, 23–30; Robert J. Downs, "Phytotrons," *Botanical Review* 46 (1980): 447–89.

11. Michio Konishi, "Phytotrons in Japan and the Japanese Society of Environment Control in Biology and Its Activities, Including the Plan of the 'National Biotron Center,'" *Environmental Control in Biology* 10, no. 3 (1972): 91–100.

12. "Phytotron in Shanghai Institute of Plant Physiology, Academia Sinica," *Phytotronic Newsletter,* no. 21, July 1980, 31.

13. As reported in S. B. Hendricks and F. W. Went, "Controlled-Climate Facilities for Biologists," *Science* 128 (1958): 510–12, on 511.

14. Gladys L. Baker, Wayne D. Rasmussen, Vivian Wiser, and Jane M. Porter, *Century of Service: The First 100 Years of the United States Department of Agriculture* (Washington, D.C.: Government Printing Office, 1963).

15. Byron T. Shaw, "Biology Unlimited," *AIBS Bulletin,* 6, no. 5 (1956): 7–9; Carl P. Heisig, "New Arrangements for Doing Pioneering Research in USDA," *Journal of Farm Economics* 41 (1959): 1467–74; Margarette M. Hedge, "Beltsville," *USDA Yearbook of Agriculture* (Washington, D.C.: Government Printing Office, 1962), 38–44.

16. Hendricks and Went, "Controlled-Climate Facilities," 510.

17. Dwight Eisenhower, "Science: Handmaiden of Freedom," in *Symposium on Basic Research,* ed. Dael Wolfle (Washington, D.C.: American Association for the Advancement of Science, 1959), 133–42, on 138.

18. Toby Appel, *Shaping Biology: The National Science Foundation and American Biological Research, 1945–1975* (Baltimore: Johns Hopkins University Press, 2000), 183–86.

19. John T. Wilson, "Support of Research by NSF," *AIBS Bulletin* 11, no. 6 (1961): 21–24; 31.

20. Jim Feldman, *The Buildings of the University of Wisconsin* [no city, no publisher], 1997, 351–53, www.williamcronon.net/uw-campus-atlas/feldman-

buildings-of-university-of-wisconsin-1997-alphabetical.pdf, accessed January 31, 2022. See also the website of the biotron at the University of Wisconsin: biotron.wisc.edu/. It ended its run as a research center in spring 2021, after fifty years of operation. See also David P. D. Munns, *Engineering the Environment: Phytotrons and the Quest for Climate Control in the Cold War* (Pittsburgh: University of Pittsburgh Press, 2017), chap. 6.

21. Paul J. Kramer, H. Hellmers, and R. J. Downs, "SEPEL: New Phytotrons for Environmental Research," *BioScience* 20, no. 22 (1970): 1201–8.

22. Paul J. Kramer, "Some Reflections After 40 Years in Plant Physiology," *Annual Review of Plant Physiology* 24 (1973): 1–24, on 22.

23. Downs, "Phytotrons."

24. T. T. Kowlowski, "Resolution of Respect: Paul Jackson Kramer, 1904–1995," *Bulletin of the Ecological Society of America* 76, no. 4 (1995): 181–82.

25. Harold A. Mooney, "Paul Jackson Kramer (8 May 1904–24 May 1995)," *Proceedings of the American Philosophical Society* 143, no. 2 (1999): 341–43, on 341.

26. Mooney, "Paul Jackson Kramer," 342.

27. Kramer, "Some Reflections," 8.

28. Kramer, "Some Reflections," 10.

29. Kramer, Hellmers, and Downs, "SEPEL."

30. Appel, *Shaping Biology,* 183–86.

31. Downs, "Phytotrons," 485.

32. Roger Jacques, *Etudes de Biologie Végétale: Hommage au Professeur Pierre Chouard* (Paris: Centre National de la Recherche Scientifique, 1976). See especially the biographical essays by G. Drouineau, J. Lavollay, C. Bresson, and H. Gaussen. Several leading phytotronists contributed research papers to this festschrift.

33. P. Chouard, R. Jacques, and N. de Bilderling, "Phytotrons and Phytotronics," *Endeavour* 31, no. 112 (January 1972): 41–45.

34. "Dr. Jean Nitsch," *Nature* 234 (December 24, 1971): 494.

35. Chouard, Jacques, and de Bilderling, "Phytotrons and Phytotronics," 42.

36. Gildas Beauchesne, "Le Phytotron de Gif-sur-Yvette," *Etudes* 317 (April 1963): 65–69.

37. Chouard, Jacques, and de Bilderling, "Phytotrons and Phytotronics," 42.

38. Pierre Chouard, "Phytotronique," *Encyclopedia Universalis,* v. 18 (Paris: 1990), 291–95.

39. Chouard, Jacques, and de Bilderling, "Phytotrons and Phytotronics," 42.

40. Ronald Fraser, "International Frontiers of Biological Research," *AIBS Bulletin* 12, no. 1 (February 1962): 19–20.

41. Chouard, Jacques, and de Bilderling, "Phytotrons and Phytotronics," 42, 45.

42. Pierre Chouard (Director), *Le Phytotron*, Sciencefilm, Centre National de la Recherche Scientifique, 1969, videotheque.cnrs.fr/doc=1342?langue=FR, accessed January 31, 2022.

43. G. Drouineau, "Pierre Chouard et l'Agronomie," in *Etudes de Biologie Vegetale, Hommage au Professeur Pierre Chouard*, ed. Roger Jacques (Paris: Centre National de la Recherche Scientifique, 1976), 13–16.

44. "Cooperation Internationale: Aide de l'UNESCO," in *Phytotronique: Science, Technique et Recherches sur les Rapports entre l'Environment et la Biologie des Végétaux, Compte rendu de la table ronde tenue avec l'aide de l'Unesco, London, 1964* (Paris: Centre National de la Recherche Scientifique, 1969), 100.

45. Denis Guthleben, "Toutes les saisons du monde . . ." *La Revue pour l'Histoire du CNRS*, 19 (Winter 2007), journals.openedition.org/histoire-cnrs/5332 placed online December 31, 2009, accessed January 31, 2022; DOI: doi. org/10.4000/histoire-cnrs.5332. This is a transcribed interview with a group of technicians who had worked at the phytotron, which presents their views of the final years and closing of the laboratory.

46. A series of reports in *Nature* tracked the changing policies toward French science in this period. See Robert Walgate, "Chirac Takes Knife to Fatted Calf," *Nature* 321 (May 1, 1986): 8; Walgate, "Old Guard Takes High Jump," *Nature* 321 (June 12, 1986): 642.

47. F. R. Fosberg, "The Study of Vegetation in Europe," *AIBS Bulletin* 11, no. 3 (1961): 17–19, on 19.

48. Daniel Behrman, "La Botanique; Science Presse-bouton," *Le Courier*, September 1960, 14–15.

49. "Ecotrons Phylogenies," on the CNRS Ecotron website under Governance and Ecotron Origins: ecotron.cnrs.fr/en/governance, accessed January 31, 2022.

50. F. E. Eckardt, ed., *Methodology of Plant Eco-Physiology: Proceedings of the Montpellier Symposium* (Paris: UNESCO, 1965).

51. Jacques Roy, Bernard Saugier, and Harold A. Mooney, *Terrestrial Global Productivity* (San Diego: Academic Press, 2001), xix.

52. The ecotron's website is ecotron.cnrs.fr/en/ (accessed January 31, 2022). For discussion of ecotrons and artificial biospheres, see Céline Grandjou, *Environmental Changes: The Futures of Nature* (London: ISTE Press, 2016).

53. Lloyd Evans, "Conjectures, Refutations, and Extrapolations," *Annual Review of Plant Biology* 54 (2003): 1–21, on 5.

54. Evans, "Conjectures, Refutations, and Extrapolations," 5; "Dr. Lloyd Evans (1927–2015), Plant Scientist," interview by Bob Crompton, Australian Academy of Science, 2003, science.org.au/learning/general-audience/history/

interviews-australian-scientists/dr-lloyd-evans-1927-2015-plant, accessed January 31, 2022. There is discrepancy in the dates mentioned in these two sources, so it is unclear exactly when Evans was at Caltech and when Frankel visited, but my conjecture is that Evans was there in 1954–1955 and Frankel visited in 1955.

55. Frits W. Went, *Some Aspects of Plant Research in Australia: A Report on a Visit to Australia,* July–October 1955 (Melbourne: Commonwealth Scientific and Industrial Research Organisation, 1956).

56. Went, *Some Aspects of Plant Research in Australia,* 8.

57. Evans, "Conjectures, Refutations, and Extrapolations."

58. Otto Frankel, "The IRRI Phytotron: Science in the Service of Human Welfare," in *Proceedings of the Symposium on Climate and Rice* (Los Baños: International Rice Research Institute, 1976), 3–9. See also Lloyd T. Evans, "Otto Herzberg Frankel, 4 November 1900–21 November 1998," *Biographical Memoirs of Fellows of the Royal Society* 45 (1999): 166–81; Roderick W. King, "Lloyd Thomas Evans AO FAA, 6 August 1927–23 March 2015," *Biographical Memoirs of Fellows of the Royal Society* 62 (2016): 125–46.

59. "Dr. Lloyd Evans, Plant Scientist," interview by Crompton.

60. Phytotron page on the CSIRO website, csiro.au/en/about/locations/heritage-management/land-and-buildings/phytotron-building, accessed January 31, 2022.

61. Lloyd T. Evans, ed., *Environmental Control of Plant Growth: Proceedings of a Symposium Held at Canberra, Australia, August 1962* (New York: Academic Press, 1963), ix.

62. Frits W. Went, "The Concept of a Phytotron," in *Environmental Control of Plant Growth,* ed. L. T. Evans (New York: Academic Press, 1963), 1–4.

63. Evans, *Environmental Control,* 421.

64. Evans, *Environmental Control,* 435.

65. "Lloyd T. Evans, Plant Scientist," interview by Crompton.

66. Lloyd T. Evans, "The Plant Physiologist as Midwife," *Search* 8 (August 1977): 262–68.

67. Lloyd Evans, "The Role of Phytotrons in Agricultural Research," in *Proceedings of the Symposium on Climate and Rice* (Los Baños: International Rice Research Institute, 1976), 11–27.

68. L. T. Evans, "Plants and Environment: Two Decades of Research at the Canberra Phytotron," *Botanical Review* 51, no. 2 (1985): 203–72.

69. Bruce Griffing and Randall L. Scholl, "Qualitative and Quantitative Genetic Studies of Arabidopsis thaliana," *Genetics* 129 (1991): 605–9, on 606; John Langridge, "The Genetic Basis of Climatic Response," in *Environmental Control of Plant Growth,* ed. L. T. Evans (New York: Academic Press, 1963), 367–79.

70. Evans, "Conjectures, Refutations, and Extrapolations," 6. See also R. N. Morse and L. T. Evans, "Design and Development of CERES—An Australian Phyto-tron," *Journal of Agricultural and Engineering Research* 7 (1962): 128–40.

71. David P. D. Munns, "Controlling the Environment: The Australian Phytotron, the Colombo Plan, and Postcolonial Science," *British Scholar* 2, no. 2 (March 2010): 197–226; David P. D. Munns, *Engineering the Environment: Phytotrons and the Quest for Climate Control in the Cold War,* chap. 4.

72. The global linkup is remembered more for its British segment, during which a group of musicians joined the Beatles to sing their song written for the occa-sion, "All You Need is Love."

73. Lloyd Evans, "The Phytotron Design," in *Phytotronique: Science, Technique et Recherches sur les Rapports entre l'Environment et la Biologie des Végétaux, Compte rendu de la table ronde tenue avec l'aide de l'Unesco, London, 1964* (Paris: Centre National de la Recherche Scientifique, 1969), 44. I have cor-rected obvious errors in the transcript of this meeting.

74. E. O. Wilson, *Naturalist* (Washington, D.C.: Island Press, 1994), 219.

75. Hans Mohr, "Is the Program of Molecular Biology Reductionist?" in *Reductionism and Systems Theory in the Life Sciences,* ed. Paul Hoyningen-Huene and Franz M. Wuketits (Dordrecht: Kluwer Academic, 1989), 137–53, on 141.

76. F. C. Steward, "Effects of Environment on Metabolic Patterns," in *Environ-mental Control of Plant Growth,* ed. L. T. Evans (New York: Academic Press, 1963), 195–214, on 210–11.

77. On the history of the term "epigenetics" see Jan Sapp, "Epigenetics and Be-yond," in *Visions of Cell Biology: Reflections Inspired by Cowdry's "General Cy-tology,"* ed. Karl S. Matlin, Jane Maienschein, and Manfred D. Laubichler (Chicago: University of Chicago Press, 2018), 183–208.

78. O. H. Frankel, "Concluding Remarks: The Next Decade," in *Environmental Control of Plant Growth,* ed. L. T. Evans (New York: Academic Press, 1963), 439–41, on 440.

79. The First International Plant Phenomics Symposium was held at Canberra in 2009, launching a collaborative international approach to the subject. See Robert T. Furbank, "Plant Phenomics: From Gene to Form and Function," *Functional Plant Biology* 36 (2009): v–vi; D. Houle, D. R. Govindaraju, and S. Omholt, "Phenomics: The Next Challenge," *Nature Reviews Genetics* 11 (2010): 855–66.

80. Frankel, "The IRRI Phytotron."

81. William S. Gaud, "Foreign Aid Today: Facts and Fancies," *Educational Hori-zons* 47, no. 1 (1968): 41–45.

82. Jonathan Harwood, "Global Visions vs. Local Complexity: Experts Wrestle with the Problem of Development," in *Rice: Global Networks and New Histories*, ed. Francesca Bray, Peter A. Coclanis, Edda L. Fields-Black, and Dagmar Schaefer (New York: Cambridge University Press, 2015), 41–55.

83. Nicholas Wade, "Green Revolution (I): A Just Technology, Often Unjust in Use," *Science* 186 (1974): 1093–96; Wade, "Green Revolution (II): Problems of Adapting a Western Technology," *Science* 186 (1974): 1186–92.

84. A political ecology perspective on the history of the green revolution and its antecedents is provided by John H. Perkins, *Geopolitics and the Green Revolution: Wheat, Genes, and the Cold War* (New York: Oxford University Press, 1997). On the decision not to adopt a peasant-centered approach see Jonathan Harwood, "Peasant Friendly Plant Breeding and the Early Years of the Green Revolution in Mexico," *Agricultural History* 83, no. 3 (2009): 384–410. On the sidelining of women farmers during the green revolution, see Patrick Kilby, *The Green Revolution: Narratives of Politics, Technology, and Gender* (New York: Routledge, 2019).

85. Marci Baranski, "The Wide Adaptation of Green Revolution Wheat" (Ph.D. diss., Arizona State University, 2015); Harwood, "Global Visions."

86. Frankel, "The IRRI Phytotron," 4, 8.

87. R. Huke, "Geography and Climate of Rice," in *Proceedings of the Symposium on Climate and Rice* (Los Baños: International Rice Research Institute, 1976), 31–50.

88. B. S. Vergara, "Physiological and Morphological Adaptability of Rice Varieties to Climate," in *Proceedings of the Symposium on Climate and Rice* (Los Baños: International Rice Research Institute, 1976), 67–86; T. T. Chang and H. I. Oka, "Genetic Variousness in the Genetic Adaptation of Rice Cultivars," in *Symposium on Climate and Rice*, 87–111, on 110.

89. Lloyd Evans, "The Role of Phytotrons in Agricultural Research," in *Proceedings of the Symposium on Climate and Rice* (Los Baños: International Rice Research Institute, 1976), 25

90. Kaori Iida, "Postwar Reconstruction of Japanese Genetics: Kihara Hitoshi and the Rockefeller Foundation Rice Project in Cold War Asia," *Historia Scientiarum* 30 (2021): 176–94.

91. Chang and Oka, "Genetic Variousness," 103.

92. William A. Bailey, Donald T. Krizek, and Herschel H. Kleuter, "Controlled Environment and the Genies of Growth," in *Science for Better Living*, USDA Yearbook for 1968 (Washington, D.C.: Government Printing Office, 1968), 2–12.

93. Dana G. Dalrymple, *Controlled Environment Agriculture: A Global View of Greenhouse Food Production* (Washington, D.C.: U.S. Department of Agriculture, 1973).

94. W. D. Billings, "Two Symposia on Desert Ecology," *Ecology* 39 (1958): 563–64, on 564. Billings was reviewing proceedings from symposia held in 1952 and 1955.

95. Reports of the Arid Zone Study Group and Arid Zone Research Council from 1949 and 1950, available from the UNESCO website: unesco.org/new/en/unesco/resources/publications/unesdoc-database/. Search under "Arid Zone Study Group" and "Arid Zone Research Council" for documents online. See also the report on a UNESCO-sponsored survey of arid zones by Ritchie Calder, *Men Against the Desert* (London: G. Allen and Unwin, 1951).

96. Walter C. Lowdermilk, *Palestine, Land of Promise* (New York: Harper, 1944).

97. W. C. Lowdermilk, "Reconquest of the Desert," in *Desert Research: Proceedings of the International Symposium Held in Jerusalem, May 7–14, 1952* (Jerusalem: Israel Research Council, 1953), no page numbers.

98. Sir Ben Lockspeiser, "Closing Address to the International Symposium on Desert Research," in *Desert Research: Proceedings of the International Symposium Held in Jerusalem, May 7–14, 1952* (Jerusalem: Israel Research Council, 1953), 618–23.

99. For a review of these initiatives, see Gilbert F. White, "International Cooperation in Arid Zone Research," *Science* 123 (1956): 537–38.

100. United Nations Educational, Scientific, and Cultural Organization, Major Project on Scientific Research on Arid Lands, UNESCO and *Arid Zone Research* 24 (January 24, 1958), unesdoc.unesco.org/images/0014/001488/148896eb.pdf, accessed February 2, 2022.

101. Charles F. Hutchinson and Stefanie M. Herrmann, *The Future of Arid Lands—Revisited: A Review of 50 Years of Drylands Research* (Paris: UNESCO, and Dordrecht: Springer, 2008); Diana K. Davis, *The Arid Lands: History, Power, Knowledge* (Cambridge, Mass.: MIT Press, 2016).

102. Michael Evenari and Dov Koller, "Desert Agriculture: Problems and Results in Israel," in *The Future of Arid Lands: Papers and Recommendations from the International Arid Lands Meetings,* ed. Gilbert F. White (Washington, D.C.: American Association for the Advancement of Science, 1956), 390–413.

103. Michael Evenari, *The Awakening Desert: The Autobiography of an Israeli Scientist* (Berlin: Springer, 1989), 106–9. The original edition was published in German in 1987. For biographical information see also Otto L. Lange and Ernst-Detlef Schulze, "In Memoriam: Michael Evenari (formerly Walter Schwarz), 1904–1989," *Oecologia* 81 (1989): 433–36.

104. Evenari, *Awakening Desert,* chap. 6 and 7 on postwar Palestine and his trip to the United States and South America.

105. Evenari, *Awakening Desert,* chap. 8.

106. Evenari, *Awakening Desert,* 106.

107. Evenari, *Awakening Desert,* 111.

108. Evenari and Koller, "Desert Agriculture."

109. Evenari, *Awakening Desert,* chap. 11 and 12; Michael Evenari and Dov Koller, "Ancient Masters of the Desert," *Scientific American* 194 (April 1956): 39–45.

110. Nelson Glueck, *The Other Side of the Jordan* (New Haven: American Schools of Oriental Research, 1940).

111. Evenari and Koller, "Ancient Masters of the Desert."

112. Michael Evenari, L. Shanan, N. Tadmor, and Y. Aharoni, "Ancient Agriculture in the Desert," *Science* 133 (March 31, 1961): 979–96; Anil Agarwal, "Coaxing the Barren Deserts Back to Life," *New Scientist* 75 (September 15, 1977): 674–75; Michael Evenari, Leslie Shanan, and Naphtali Tadmor, *The Negev: The Challenge of a Desert,* 2nd ed. (Cambridge, Mass.: Harvard University Press, 1982).

113. Michael Evenari, "Plant Physiology and Arid Zone Research," in *The Problems of the Arid Zone: Proceedings of the Paris Symposium* (Paris: UNESCO, 1962), 175–91.

114. F. Dixey, "Arid Zone Problems," *Nature* 187, no. 4734 (July 23, 1964): 295–96.

115. Evenari, "Plant Physiology and Arid Zone Research," 190.

116. Evenari, "Plant Physiology and Arid Zone Research," 190.

117. Pierre Chouard, "Report to the Advisory Committee on Arid Zone Research," in *The Problems of the Arid Zone* (Paris: UNESCO, 1962), 192–93.

118. R. O. Slatyer, ed., *Plant Response to Climatic Factors: Proceedings of the Uppsala Symposium* (Paris: UNESCO, 1973); see the volume's foreword, no page numbers.

119. Ditza Koller, email to author, February 2016.

120. Dov Koller, "Germination," *Scientific American* 200 (April 1959): 75–84.

121. Dov Koller and M. Negbi, "The Regulation of Germination in *Oryzopsis Miliacea,*" *Ecology* 40, no. 1 (1959): 20–36. Koller worked on this problem at Caltech. See also Dov Koller, "Preconditioning of Germination in Lettuce at Time of Fruit Ripening," *American Journal of Botany* 49 (1962): 841–44. These experiments were done at the Earhart laboratory.

122. "A Phytotron in Israel," Report for the National Council for Research and Development, Jerusalem, 1961 (unpublished).

123. "Proposal to Fund Construction of a New Phytotron at the Hebrew University's Faculty of Agriculture in Rehovot," Hebrew University of Jerusalem, Division for Development and Public Relations, June 1994. I am grateful to Jaime Kigel, the phytotron's director at the Hebrew University, for providing me with a copy of this proposal.

124. Evenari, *Awakening Desert,* 173.
125. Evenari, *Awakening Desert,* 173.
126. Hutchinson and Herrmann, *Future of Arid Lands—Revisited,* 19.
127. Hutchinson and Herrmann, *Future of Arid Lands—Revisited,* 20.

Chapter 5. Phytotrons Under Communism

1. "Yarovization Process Turns Biennial Plants into Annuals," *Science News Letter* 24 (November 11, 1933): 309–10.
2. The mechanisms behind the ability of plants to "remember" their exposure to cold are still being studied. See Jean Marx, "Remembrance of Winter Past," *Science* 303 (2004): 607; Franziska Turck and George Coupland, "When Vernalization Makes Sense," *Science* 331 (2011): 36–37.
3. This is mentioned in "Vernalization and Phasic Development of Plants," *Imperial Bureau of Plant Genetics: Herbage Plants, Bulletin* 17, 1935. This bulletin appears to be a synopsis of views taken from different sources and has no author.
4. N. A. Maximov, "The Theoretical Significance of Vernalization," *Imperial Bureau of Plant Genetics: Herbage Plants, Bulletin* 16, 1934.
5. Zhores Medvedev, *The Rise and Fall of T. D. Lysenko,* trans. I. Michael Lerner (New York: Columbia University Press, 1969), 15.
6. Nils Roll-Hansen, *The Lysenko Effect: The Politics of Science* (Amherst, N.Y.: Humanity Books, 2005), 58.
7. Nils Roll-Hansen, "A New Perspective on Lysenko," *Annals of Science* 42 (1985): 261–78; Roll-Hansen, *Lysenko Effect.*
8. On Russian Darwinism, see Daniel P. Todes, *Darwin Without Malthus: The Struggle for Existence in Russian Evolutionary Thought* (New York: Oxford University Press, 1989).
9. A. J. Bruman, "The Place of Iarovization in Plant Breeding," *Journal of Heredity* 28 (1937): 31–33, on 31.
10. Roll-Hansen, *Lysenko Effect.*
11. Medvedev, *Rise and Fall of T. D. Lysenko.*
12. Nikolai Krementsov, *Stalinist Science* (Princeton: Princeton University Press, 1997), 174–83.
13. David Joravsky, *The Lysenko Affair* (Chicago: University of Chicago Press, 1970), 140.
14. Medvedev, *Rise and Fall of T. D. Lysenko,* 14.
15. E. M. Senchenkova, "Maksimov, Nikolay Aleksandrovich," Encyclopedia.com, encyclopedia.com/science/dictionaries-thesauruses-pictures-and-press-

releases/maksimov-nikolay-aleksandrovich, accessed February 4, 2022. The dates in this biographical entry seem inconsistent in placing him both at Saratov and in Moscow at the same time in the late 1930s.

16. Zhores Medvedev, "A Dangerous Occupation," in *Mill Hill Essays, 2010* (London: National Institute for Medical Research, 2010), 12–23. Medvedev was referring to 1950, when he was completing his dissertation in botany and had to present it to the institute's faculty for approval instead of to his own faculty at the Agricultural Academy, which had succumbed to Lysenkoism.

17. Joravsky, *Lysenko Affair,* 40.

18. Joravsky, *Lysenko Affair,* 201.

19. Joravsky, *Lysenko Affair,* 202.

20. Krementsov, *Stalinist Science,* 253.

21. Mark B. Adams, "Networks in Action: The Khrushchev Era, the Cold War, and the Transformation of Soviet Science," in *Science, History, and Social Activism: A Tribute to Everett Mendelsohn,* ed. Garland E. Allen and Roy M. MacLeod (Dordrecht: Kluwer, 2001), 255–76, on 269.

22. M. N. Zaprometov, "Andrei L'vovich Kursanov (1902–1999)," *Russian Journal of Plant Physiology* [*Fiziologiya Rastenii*] 49 (2002): 3–7. This journal was at different times titled *Soviet Plant Physiology* and *Russian Journal of Plant Physiology* in its English translation. I refer to it consistently as *Russian Journal of Plant Physiology,* which conforms to the modern title.

23. V. N. Zholkevich, "On the Centenary of Pavel Aleksandrovich Henkel," *Russian Journal of Plant Physiology* 49 (2002): 845–47.

24. "Fitotron" [in Russian], *Great Soviet Encyclopedia,* bse.slovaronline.com/47459-FITOTRON, accessed February 4, 2022.

25. The photograph of the institute's front is at https://commons.wikimedia.org/wiki/File:Timiryazev_Institute_of_Plant_Physiology_-_Phytotron_lab_[30795237604].jpg; interior photographs appear in I. I. Tumanov, "The Soviet Phytotron" [in Russian], *Priroda* [no volume number], no. 1 (1959): 112–17.

26. K. I. Sergeev and G. I. Kuleshova, "Seven Decades of Service to Domestic Science," *Herald of the Russian Academy of Sciences* 78 (2008): 537–45.

27. Went diary, May 3, 1965.

28. Krementsov, *Stalinist Science,* 244–45.

29. Tumanov, "The Soviet Phytotron," 112, 117.

30. M. N. Zaprometov, "Andrei L'vovich Kursanov (1902–1999)," *Russian Journal of Plant Physiology* 49 (2002): 3–7. On Kursanov's school of research, see M. S. Krasavina and S. V. Sokolova, "A. L. Kursanov: His Scientific School and the Development of Russian Plant Physiology," *Russian Journal of Plant Physiology* 44 (1997): 697–705.

31. "Plant Physiology in the Soviet Union During the Last Forty Years (1917–1957)," *Russian Journal of Plant Physiology* 4 (1957): 369–75.

32. "Plant Physiology in the Soviet Union," 370.

33. "Plant Physiology in the Soviet Union," 371.

34. "Plant Physiology in the Soviet Union," 372.

35. "Plant Physiology in the Soviet Union," 375.

36. "Plant Physiology in the Soviet Union," 375.

37. Resolution of the conference on the Physiology of Plant Hardiness, adopted March 7, 1959, reported in *Russian Journal of Plant Physiology* 6 (1959): 652–58.

38. P. Chouard, R. Jacques, and N. de Bilderling, "Phytotrons and Phytotronics," *Endeavour* 31, no. 12 (January 1972): 41–45, on 41.

39. Charles G. Whiteford, "Beltsville's Superintendent Can Sleep at Night Again," *Baltimore Sun,* September 18, 1959, 1, ProQuest.

40. A. L. Kursanov, "Recent Advances in Plant Physiology in the U.S.S.R.," *Annual Review of Plant Physiology* 7 (1956): 401–36. Lysenko is cited on 420.

41. B. Strogonov, "Origins and Development of Plant Physiology at the Academy of Sciences of the USSR," *Russian Journal of Plant Physiology* 21 (1974): 359–66.

42. Robert J. Downs, "Phytotrons," *Botanical Review* 46 (1980): 447–89.

43. The biotron's program of research is described on the website of the Institute of Bioorganic Chemistry, ibch.ru/en/branch/research/biotron, accessed February 4, 2022.

44. L. N. Ivanenko, "Mass Simulation Games as a Social Management Tool," *Cybernetics and Systems Analysis* [translation of *Kibernetika*] 23, no. 6 (1987): 796–806.

45. O. N. Kulaeva, "Kursanov's Ideas in Phytohormone Investigations," *Russian Journal of Plant Physiology* 49 (2002): 74–75.

46. Miklós Müller, "Lysenkoism in Hungary," lecture given at the International Workshop on Lysenkoism, Columbia University, December 4–5, 2009; available on YouTube, youtube.com/watch?v=8ZGKoGzCe_0, accessed February 4, 2022.

47. Z. Bedő, "Prof. Sándor Rajki (1921–2007)," *Acta Agronomica Hungarica* 55, no. 3 (2007): 393–96.

48. Bedő's obituary of Rajki does not mention Lysenko's influence.

49. S. Rajki, "Data on the Genetics of Converting Spring Wheat into Winter Wheat," *Symposium on Genetics and Breeding of Wheat* (Martonvásár: Hungarian Academy of Sciences, 1962), 96.

50. *Symposium on Genetics and Breeding of Wheat,* report on discussion, 179.

51. *Symposium on Genetics and Breeding of Wheat,* 180.

52. *Symposium on Genetics and Breeding of Wheat,* 182.

53. Matthew Cobb, "60 Years Ago, Francis Crick Changed the Logic of Biology," *PLoS Biology* 15, no. 9 (2017): e2003243, doi.org/10.1371/journal.pbio.2003243.

54. Sándor Rajki, "On the Situation in Genetics" (Martonvásár: Mezőgazdasági Kutató Intéte, 1966); *Autumnization and Its Genetic Interpretation* (Budapest: Akadémiai Kiadó, 1967). "On the Situation in Genetics" was published in English, Hungarian, and Russian; the English translation may not be accurate, as it contains statements that appear to be inconsistent.

55. Barry Commoner, "In Defense of Biology," *Science* 133 (1961): 1745–48; Commoner, "Roles of Deoxyribonucleic Acid in Inheritance," *Nature* 202, no. 4936 (1964): 960–68.

56. Rajki, "Situation in Genetics," 12.

57. G. Ledyard Stebbins, review of *Autumnization and Its Genetic Interpretation,* by Sándor Rajki, *Quarterly Review of Biology* 44, no. 4 (1969): 413.

58. Ralph Riley, "More Lysenkoism," *Nature* 217, no. 5125 (1968): 291–92.

59. Riley, "More Lysenkoism," 292.

60. Robert C. Gallo, "Howard M. Temin, 1934–1994," *Nature* 368, no. 6466 (March 1994): 17; Daniel J. Kevles, "Howard Temin: Rebel of Evidence and Reason," in *Rebels, Mavericks, and Heretics in Biology,* ed. Oren Harman and Michael R. Dietrich (New Haven: Yale University Press, 2008), 248–64.

61. Francis Crick, "Central Dogma of Molecular Biology," *Nature* 227, no. 5258 (1970): 561–63.

62. S. Rajki, address at the inauguration of the Hungarian phytotron on November 3, 1972, in *Phytotronic Newsletter,* nos. 4, 5, 6 (issued together), November 3, 1973, 41.

63. M. Sawatatsky, "The Cordial and Productive Relationship Between Conviron and Martonvásár," in *Cereal Adaptation to Low Temperature Stress in Controlled Environments,* ed. Z. Bedő (Martonvásár: Hungarian Academy of Sciences, 1997), 285–89.

64. Sawatatsky, "Cordial and Productive Relationship," 287.

65. S. Rajki, "Research Strategy of the Martonvásár Phytotron," *Phytotronic Newsletter,* nos. 4, 5, 6 (issued together), November 3, 1973, 45.

66. Bedő, "Prof. Sándor Rajki," 396.

67. Harry R. Highkin, "Temperature-Induced Variability in Peas," *American Journal of Botany* 45 (1958): 626–31; F. W. Went, "Phytotronics," *Proceedings of the Plant Science Symposium* (Camden, N.J.: Campbell Soup Company, 1962), 149–61.

68. Harry R. Highkin, "The Effect of Constant Temperature Environments and of Continuous Light on the Growth and Development of Pea Plants," in *Biological Clocks: Cold Spring Harbor Symposia on Quantitative Biology* 25 (1960):

231–37. On Jollos's work in the context of growing interest in cytoplasmic heredity see Jan Sapp, *Beyond the Gene: Cytoplasmic Inheritance and the Struggle for Authority in Genetics* (New York: Oxford University Press, 1987).

69. Alan Durrant, "The Environmental Induction of Heritable Change in *Linum*," *Heredity* 17 (1962): 27–61.

70. Anton Lang, "Achievements, Challenges, and Limitations of Phytotrons," in *Environmental Control of Plant Growth*, ed. Lloyd T. Evans (New York: Academic Press, 1963), 405–19, on 409–10.

71. A. Durrant, "Unstable Genotypes," *Philosophical Transactions of the Royal Society of London*, ser. B, 292 (1981): 467–74.

72. Snait B. Gissis and Eva Jablonka, eds., *Transformations of Lamarckism: From Subtle Fluids to Molecular Biology* (Cambridge, Mass.: MIT Press, 2011). On the development of epigenetics from the 1970s to 1990s, see Vincenzo E. A. Russo, Robert A. Martienssen, and Arthur D. Riggs, eds., *Epigenetic Mechanisms of Gene Regulation* (Cold Spring Harbor: Cold Spring Harbor Laboratory Press, 1996).

Chapter 6. Physiological Ecology Comes of Age

1. C. R. Tracy, J. S. Turner, G. A. Bartholomew, A. Bennett, W. D. Billings, B. F. Chabot, D. M. Gates, et al., "What Is Physiological Ecology?" *Bulletin of the Ecological Society of America* 63, no. 4 (1982): 340–47, on 342.

2. W. D. Billings, "Address of the Past President: American Deserts and Their Mountains: An Ecological Frontier," *Bulletin of the Ecological Society of America* 61, no. 4 (1980): 203–9, on 207.

3. Boyd R. Strain, "Resolution of Respect: William Dwight Billings, 1910–1997," *Bulletin of the Ecological Society of America* 78, no. 2 (1997): 115–17; Kim M. Peterson, "William Dwight Billings (1910–1997)," *Arctic* 50, no. 3 (1997): 275–76.

4. T. T. Kozlowski, "Resolution of Respect: Paul Jackson Kramer, 1904–1995," *Bulletin of the Ecological Society of America* 76, no. 4 (1995): 181–82. See also the short history by Jessica Harland-Jacobs, *Balancing Tradition and Innovation: The History of Botany at Duke University, 1849–1996* (Durham: Department of Botany, Duke University, 1996).

5. Strain, "Resolution of Respect: William Dwight Billings, 1910–1997," 115–17.

6. W. D. Billings, "Physiological Ecology," *Annual Review of Plant Physiology* 8 (1957): 375–92, on 375.

7. W. D. Billings, "The Historical Development of Physiological Plant Ecology," in *Physiological Ecology of North American Plant Communities*, ed. B. F. Chabot and H. A. Mooney (New York: Chapman and Hall, 1985), 1–15.

8. Billings, "Historical Development," 9.
9. See R. P. McIntosh, "Plant Ecology, 1947–1972," *Annals of the Missouri Botanical Garden* 61 (1974): 132–65; H. A. Mooney, "Plant Physiological Ecology—Determinants of Progress," *Functional Ecology* 5 (1991): 127–35; H. A. Mooney, "On the Road to Global Ecology," *Annual Review of Energy and the Environment* 24 (1999): 1–31; Arthur P. Galston, "Plant Biology—Retrospect and Prospect," *Current Science* 80, no. 2 (2001): 143–52.
10. Billings, "Historical Development," 1.
11. Tracy et al., "What Is Physiological Ecology?" on 342.
12. Sharon E. Kingsland, *The Evolution of American Ecology, 1890–2000* (Baltimore: Johns Hopkins University Press, 2005). See also Patricia Craig, *Centennial History of the Carnegie Institution of Washington*, v. 4, *The Department of Plant Biology* (New York: Cambridge University Press, 2005).
13. Lloyd Tevis, Jr., "Caltech's Roving Laboratory," *Engineering and Science* 20, no. 5 (1957): 19–23. It is not known when Went first started talking about a mobile field laboratory, but Tevis wrote that he had "long advocated" such a laboratory.
14. Emanuel D. Rudolph, "One Hundred Years of the Missouri Botanical Garden," *Annals of the Missouri Botanical Garden* 78, no. 1 (1991): 1–18. On Went's role in designing and building the Climatron, see David P. D. Munns, *Engineering the Environment: Phytotrons and the Quest for Climate Control in the Cold War* (Pittsburgh: University of Pittsburgh Press, 2017), chap. 3.
15. Reported in the *National Science Foundation 10th Annual Report for Fiscal Year Ending June 30, 1960* (Washington, D.C.: Government Printing Office, 1960), 239.
16. In 1964 Went received an NSF grant of $43,700 for two years for "Operation of a mobile laboratory for chromatographic air analysis." Reported in *National Science Foundation Grants and Awards for Fiscal Year Ending June 30, 1964* (Washington, D.C.: Government Printing Office, 1964), 6.
17. Philip M. Boffey, "Desert Research Institute: A Formula for Growth," *Science* 161 (1968): 866–68; Boffey, "Trouble at Nevada Research Center," *Science* 165 (1969): 880.
18. Arthur W. Galston and Thomas D. Sharkey, "Frits Warmolt Went," *Biographical Memoirs of the National Academy of Sciences of the United States of America* 74 (1998): 348–63; F. W. Went, "The Mobile Laboratories of the Desert Research Institute," *BioScience* 18, no. 4 (1968): 293–97.
19. F. W. Went, *The Experimental Control of Plant Growth* (Waltham, Mass.: Chronica Botanica, 1957), 96–97 and chap. 8.
20. F. W. Went, "Ecology of Desert Plants. I. Observations on Germination in the Joshua Tree National Monument, California," *Ecology* 29 (1948): 242–53; Went,

"Ecology of Desert Plants. II. The Effect of Rain and Temperature on Germination and Growth," *Ecology* 30 (1949): 1–13; Marcella Juhren, F. W. Went, and Edwin Phillips, "Ecology of Desert Plants. IV. Combined Field and Laboratory Work on Germination of Annuals in the Joshua Tree National Monument, California," *Ecology* 37 (1956): 318–30; Went, *Experimental Control of Plant Growth,* 248–53; Went, "The Ecology of Desert Plants," *Scientific American* 192 (April, 1955): 68–75.

21. F. W. Went, "Soziologie der Epiphyten eines Tropischen Urwaldes," *Annales du Jardin Botanique de Buitenzorg* 50 (1940): 1–98.

22. F. W. Went, "The Dependence of Certain Annual Plants on Shrubs in Southern California Deserts," *Bulletin of the Torrey Botanical Club* 69, no. 2 (1942): 100–114.

23. Went, "Dependence of Certain Annual Plants," 112.

24. Harlan Lewis and F. W. Went, "Plant Growth Under Controlled Conditions. IV. Response of California Annuals to Photoperiod and Temperature," *American Journal of Botany* 32 (1945): 1–12.

25. James Bonner, "The Role of Toxic Substances in the Interactions of Higher Plants," *Botanical Review* 16 (1950): 51–65.

26. James Bonner, "Further Investigation of Toxic Substances Which Arise from Guayule Plants: Relation of Toxic Substances to the Growth of Guayule in Soil," *Botanical Gazette* 107 (1946): 343–51.

27. Reed Gray and James Bonner, "An Inhibitor of Plant Growth from the Leaves of *Encelia farinosa*," *American Journal of Botany* 35 (1948): 52–57.

28. Bonner, "The Role of Toxic Substances in the Interactions of Higher Plants," 51–65.

29. C. H. Muller, "The Association of Desert Annuals with Shrubs," *American Journal of Botany* 40 (1953): 53–60; Walter H. Muller and Cornelius Muller, "Patterns Involving Desert Plants That Contain Toxic Products," *American Journal of Botany* 43 (1956): 354–61.

30. Harold A. Mooney, "On the Road to Global Ecology," *Annual Review of Energy and the Environment* 24 (1992): 1–31, on 4.

31. Mooney, "On the Road to Global Ecology," 4.

32. Cornelius H. Muller, "This Week's Citation Classic," *Current Contents,* no. 45, November 8, 1982, 20.

33. Muller, "This Week's Citation Classic," 20.

34. Cornelius H. Muller, Walter H. Muller, and Bruce L. Haines, "Volatile Growth Inhibitors Produced by Aromatic Shrubs," *Science* 143 (1964): 471–73.

35. Cornelius H. Muller, "The Role of Chemical Inhibition (Allelopathy) in Vegetational Composition," *Bulletin of the Torrey Botanical Club* 93, no. 5 (1966): 332–51.

36. R. H. Whittaker, "Eminent Ecologist, 1975: Cornelius Muller," *Bulletin of the Ecological Society of America* 56, no. 4 (1975): 23–24. For an overview of chemical ecology as a field at this time, see Ernest Sondheimer and John B. Simeone, eds., *Chemical Ecology* (New York: Academic Press, 1970).

37. Cornelius H. Muller, "Allelopathy as a Factor in Ecological Process," *Vegetatio* 18 (1969): 348–57.

38. Muller, "This Week's Citation Classic."

39. Muller, "Allelopathy as a Factor, 351, 356.

40. Rachel N. M. Dentinger, "The Nature of Defense: Coevolutionary Studies, Ecological Interaction, and the Evolution of 'Natural Insecticides,' 1959–1983" (Ph.D. diss., University of Minnesota, 2009), chap. 4. The meeting between Bartholomew, Mooney, and Muller's students is discussed in Richard W. Halsey, "In Search of Allelopathy: An Eco-historical View of the Investigation of Chemical Inhibition in California Coastal Sage Scrub and Chemise Chaparral," *Journal of the Torrey Botanical Society* 131, no. 4 (2004): 343–67. See also Bruce Bartholomew, "Bare Zone Between California Shrub and Grassland Communities: The Role of Animals," *Science* 170 (1970): 1210–12; rebuttal by C. H. Muller and Roger del Moral with reply by Bruce Bartholomew, "Role of Animals in Suppression of Herbs by Shrubs," *Science* 173 (1971): 462–63.

41. Halsey, "In Search of Allelopathy."

42. Muller, "This Week's Citation Classic."

43. R. H. Whittaker, "The Biochemical Ecology of Higher Plants," in *Chemical Ecology,* ed. Ernest Sondheimer and John B. Simeone (New York: Academic Press, 1970), 43–70, on 63.

44. Whittaker, "Biochemical Ecology of Higher Plants," 64.

45. R. H. Whittaker and P. P. Feeny, "Allelochemics: Chemical Interactions Between Species," *Science* 171 (1971): 757–70.

46. John L. Harper, review of *Allelopathy (Physiological Ecology),* by Elroy L. Rice, *Quarterly Review of Biology* 50 (1975): 493–95. A revised second edition of Rice's book was published in 2012.

47. Alastair Fitter, "Making Allelopathy Respectable," *Science* 301 (2003): 1337–38.

48. John L. Harper, *Population Biology of Plants* (London: Academic Press, 1977), 369.

49. Ragan M. Callaway, Nalini M. Nadkarni, and Bruce E. Mahall, "Facilitation and Interference of *Quercus douglasii* on Understory Productivity in Central California," *Ecology* 72 (1991): 1484–99.

50. Bruce E. Mahall and Ragan M. Callaway, "Root Communication Among Desert Shrubs," *Proceedings of the National Academy of Sciences of the United States of America* 88, no. 3 (1991): 874–76; B. E. Mahall and R. M. Callaway,

"Root Communication Mechanisms and Intracommunity Distributions of Two Mojave Desert Shrubs," *Ecology* 73 (1992): 2145–51.

51. Mahall and Callaway, "Root Communication Mechanisms," 2150.

52. Ragan M. Callaway and Erik T. Aschehoug, "Invasive Plants Versus Their New and Old Neighbors: A Mechanism for Exotic Invasion," *Science* 290 (2000): 521–23.

53. Harsh P. Bais, Ramarao Vepachedu, Simon Gilroy, Ragan M. Callaway, and Jorge M. Vivanco, "Allelopathy and Exotic Plant Invasion: From Molecules and Genes to Species Interactions," *Science* 301 (2003): 1377–80, on 1377.

54. Fitter, "Making Allelopathy Respectable," 1338. See also the recent review by José L. Hierro and Ragan M. Callaway, "The Ecological Importance of Allelopathy," *Annual Review of Ecology, Evolution, and Systematics* 52 (2021): 25–45.

55. Jack C. Schultz, "Shared Signals and the Potential for Phylogenetic Espionage Between Plants and Animals," *Integrative and Comparative Biology* 42, no. 3 (2002): 454–62.

56. This point was made by Herbert G. Baker, "Evolution in the Tropics," *Biotropica* 2, no. 2 (1970): 101–11.

57. Went diary, entries for February 18 and 25, 1954, quote from February 25.

58. F. W. Went, "Air Pollution," *Scientific American* 192 (May 1955): 62–72.

59. Frits W. Went, *Some Aspects of Plant Research in Australia: A Report on a Visit to Australia* (Melbourne: Commonwealth Scientific and Industrial Research Organisation, 1956). The trip to Australia was from July to October 1955. A copy of the report was obtained from the National Agricultural Library, Beltsville, Maryland.

60. Went, "Air Pollution."

61. F. W. Went, "Blue Hazes in the Atmosphere," *Nature* 187, no. 4738 (1960): 641–43; F. W. Went, "Organic Matter in the Atmosphere, and Its Possible Relation to Petroleum Formation," *Proceedings of the National Academy of Sciences of the United States of America*, no. 2 (February 15, 1960): 212–21.

62. Went, "Organic Matter in the Atmosphere."

63. Susan Warren, "Dr. Rasmussen Knew It All Along: Trees Do Help Make Smog," *Wall Street Journal*, Eastern edition, May 16, 1999, A1, ProQuest.

64. Reinhold A. Rasmussen, "Terpenes: Their Analysis and Fate in the Atmosphere" (Ph.D. diss., Washington University, 1964).

65. Eugene Warner, "What Makes the Blue Ridge Mountains Blue," *New York Times,* July 26, 1964, XX13, ProQuest.

66. Reinhold A. Rasmussen and F. W. Went, "Volatile Organic Material of Plant Origin in the Atmosphere," *Proceedings of the National Academy of Sciences of the United States of America* 53, no. 1 (January 15, 1965): 215–20.

67. "Botany: Arboreal Pollution," *Time,* September 9, 1966, content.time.com/time/subscriber/article/0,33009,836314,00.html, accessed February 4, 2022.

68. F. W. Went, "Reflections and Speculations," *Annual Review of Plant Physiology* 25 (1974): 1–26.

69. Reinhold Rasmussen, "Isoprene: Identified as a Forest-Type Emission to the Atmosphere," *Environmental Science and Technology* 4 (1970): 667–71.

70. G. A. Sanadze, "Biogenic Isoprene (A Review)," *Russian Journal of Plant Physiology* 51, no. 6 (2004): 810–24. In 1967 Sanadze received a doctoral degree from the Institute of Plant Physiology in Moscow for his dissertation on isoprene emission from leaves. See also Thomas D. Sharkey, "The Future of Isoprene Research," *Bulletin of the Georgian National Academy of Sciences* 3, no. 3 (2009): 106–13.

71. Reinhold Rasmussen, "What Do the Hydrocarbons from Trees Contribute to Air Pollution?" *Journal of the Air Pollution Control Association* 22, no. 7 (1972): 537–43.

72. Patrick R. Zimmerman, *Testing of Hydrocarbon Emissions from Vegetation, Leaf Litter, and Aquatic Surfaces, and Development of a Methodology for Compiling Biogenic Emission Inventories.* Final Report (Research Triangle Park, N.C.: U.S. Environmental Protection Agency, March 1979).

73. Carl Pope, "The Candidates and the Issues," *Sierra* 65 (1980): 15–17. The radio broadcast was in 1979.

74. Jack Nelson, "Pollution Curbed, Reagan Says; Attacks Air Cleanup," *Los Angeles Times,* October 9, 1980, B1, ProQuest.

75. Arthur Galston, letter to the editor, *New York Times,* November 30, 1980, E16, ProQuest.

76. W. L. Chameides, R. W. Lindsay, J. Richardson, and C. S. Kiang, "The Role of Biogenic Hydrocarbons in Urban Photochemical Smog: Atlanta as a Case Study," *Science* 241 (1988): 1473–75. Project description is on the SOS website, projects.ncsu.edu/sos/index.html, accessed February 4, 2022.

77. Thomas D. Sharkey and Russell K. Monson, "Isoprene Research—60 Years Later, the Biology Is Still Enigmatic," *Plant, Cell and Environment* 40 (2017): 1671–78.

78. Sharkey and Monson, "Isoprene Research."

79. Russell K. Monson, email to author, August 31, 2021. The study was published as Russell K. Monson and Ray Fall, "Isoprene Emission from Aspen Leaves," *Plant Physiology* 90 (1989): 267–74.

80. Sharkey, "Future of Isoprene Research."

81. Francesco Loreto and Thomas D. Sharkey, "A Gas Exchange Study of Photosynthesis and Isoprene Emission in *Quercus rubra* L." *Planta* 182, no. 4 (1990): 523–31.

82. Thomas D. Sharkey, Elizabeth A. Holland, and Harold A. Mooney, eds., *Trace Gas Emissions by Plants* (San Diego: Academic Press, 1991).

83. Russell K. Monson, email to author, August 31, 2021.

84. Ray Fall, "Isoprene Emission from Plants: Summary and Discussions," in *Trace Gas Emissions by Plants,* ed. T. D. Sharkey, E. A. Holland, and H. A. Mooney (San Diego: Academic Press, 1991), 209–16, on 209.

85. Fall, "Isoprene Emissions from Plants," 215.

86. Alex Brian Guenther, "Wind Tunnel, Field, and Numerical Investigations of Plume Downwash and Dispersion at an Arctic Industrial Site" (Ph.D. diss., Washington State University, 1989).

87. Alex B. Guenther, Russell K. Monson, and Ray Fall, "Isoprene and Monoterpene Emission Rate Variability: Observations with Eucalyptus and Emission Rate Algorithm Development," *Journal of Geophysical Research* 96 (1991): 10799–808; Alex B. Guenther, Patrick R. Zimmerman, Peter C. Harley, Russell K. Monson, and Ray Fall, "Isoprene and Monoterpene Emission Rate Variability: Model Evaluations and Sensitivity Analyses," *Journal of Geophysical Research* 98 (1993): 12609–617.

88. Sharkey and Monson, "Isoprene Research," 1674.

89. Thomas D. Sharkey and Eric L. Singsaas, "Why Plants Emit Isoprene," *Nature* 374, no. 6525 (1995): 769; Thomas D. Sharkey, "Isoprene Synthesis by Plants and Animals," *Endeavour* 20 (1996): 74–78.

90. Barry A. Logan and Russell K. Monson, "Thermotolerance of Leaf Disks from Four Isoprene-Emitting Species Is Not Enhanced by Exposure to Exogenous Isoprene," *Plant Physiology* 120 (1999): 821–25.

91. Thomas D. Sharkey and Manuel T. Lerdau, "Atmospheric Chemistry and Hydrocarbon Emission from Plants," *Ecological Applications* 9, no. 4 (1999): 1107–8.

92. The program of the 2000 conference on Biogenic Hydrocarbons and the Atmosphere is on the Gordon Research Conference website, grc.org/biogenic-hydrocarbons-and-the-atmosphere-conference/2000/, accessed April 6, 2022.

93. Russell K. Monson, email to author, August 31, 2021.

94. For a review of this research see Claudia E. Vickers, Mareike Bongers, Qing Liu, Thierry Delatte, and Harro Bouwmeester, "Metabolic Engineering of Volatile Isoprenoids in Plants and Microbes," *Plant, Cell and Environment* 37 (2014): 1753–75, doi.org/10.1111/pce.12316.

95. Russell Monson, Ryan T. Jones, Todd N. Rosenstiel, and Jörg-Peter Schnitzler, "Why Only Some Plants Emit Isoprene," *Plant, Cell and Environment* 36 (2013): 503–16.

96. Christopher M. Harvey and Thomas D. Sharkey, "Exogenous Isoprene Modulates Gene Expression in Unstressed Arabidopsis Thaliana Plants," *Plant,*

Cell and Environment 39 (2016): 1251–63; Zhaojiang Zuo, Sarathi M. Weraduwage, Alexandra T. Lantz, Lydia M. Sanchez, Sean E. Weise, Jie Wang, Kevin L. Childs, and Thomas D. Sharkey, "Isoprene Acts as a Signaling Molecule in Gene Networks Important for Stress Responses and Plant Growth," *Plant Physiology* 180 (2019): 124–52.

97. Russell K. Monson, Sarathi M. Weraduwage, Maaria Rosenkranz, Jörg-Peter Schnitzler, and Thomas D. Sharkey, "Leaf Isoprene Emission as a Trait That Mediates the Growth-Defense Tradeoff in the Face of Climate Stress," *Oecologia* 197 (2021): 885–902, doi.org/10.1007/s00442-020-04813-7.

98. Daniel A. Herms and William J. Mattson, "The Dilemma of Plants: To Grow or Defend," *Quarterly Review of Biology* 67, no. 3 (1992): 283–335.

99. Monson et al., "Leaf Isoprene Emission as a Trait," 886.

100. Russell K. Monson, "Volatile Organic Compound Emissions from Terrestrial Ecosystems: A Primary Biological Control over Atmospheric Chemistry," *Israel Journal of Chemistry* 42 (2002): 29–42, on 31.

101. Thomas D. Sharkey, Amy E. Wiberley, and Autumn R. Donohue, "Isoprene Emission from Plants: Why and How," *Annals of Botany* 101 (2008): 5–18, on 5.

102. Ülo Niinemets and Russell K. Monson, eds., *Biology, Controls, and Models of Tree Volatile Organic Compound Emissions* (Dordrecht: Springer, 2013); Russell Monson and Dennis Baldocchi, *Terrestrial Biosphere-Atmosphere Fluxes* (Cambridge: Cambridge University Press, 2014).

103. Todd N. Rosenstiel, Mark J. Potosnak, Kevin L. Griffin, Ray Fall, and Russell K. Monson, "Increased CO_2 Uncouples Growth from Isoprene Emission in an Agriforest Ecosystem," *Nature* 421, no. 6920 (2003): 256–59.

104. Tiffany O'Callaghan, "A Biosphere Reborn," *New Scientist* 219, no. 2927 (July 27, 2013): 41–45; John Allen, *Biosphere 2: The Human Experiment* (New York: Viking Penguin, 1991); Sabine Höhler, "The Environment as a Life Support System. The Case of Biosphere 2," *History and Technology* 26, no. 1 (2010): 39–58.

105. Russell K. Monson, Barbro Winkler, Todd N. Rosenstiel, Katja Block, Juliane Merl-Pham, Steven H. Strauss, Kori Ault, et al., "High Productivity in Hybrid-Poplar Plantations Without Isoprene Emission to the Atmosphere," *Proceedings of the National Academy of Sciences of the United States of America* 117, no. 3 (2020): 1596–1605.

106. Barry Osmond, "Experimental Ecosystem and Climate Change Research in Controlled Environments: Lessons from the Biosphere 2 Laboratory, 1996–2003," in *Plant Responses to Air Pollution and Global Change,* ed. K. Omasa, I. Nouchi, and L. J. De Kok (Tokyo: Springer, 2005), 173–84.

107. Barry Osmond, "Our Eclectic Adventures in the Slower Eras of Photosynthesis: From New England Down Under to Biosphere 2 and Beyond," *Annual Review of Plant Biology* 65 (2014): 1–32, on 15.

108. Barry Osmond, Gennady Ananyev, Joseph Berry, Chris Langdon, Zbigniew Kolber, Gunghui Lin, Russell Monson, et al., "Changing the Way We Think About Global Change Research: Scaling Up in Experimental Ecosystem Science," *Global Change Biology* 10 (2004): 393–407, on 395.

109. Osmond et al. "Changing the Way We Think," 395, 405.

110. Osmond, "Our Eclectic Adventures," 17.

111. Joseph A. Berry, "There Ought to Be an Equation for That," *Annual Review of Plant Biology* 63 (2012): 1–17, on 13.

112. J. Ortega, A. Turnipseed, A. B. Guenther, T. G. Karl, D. A. Day, D. Gochis, J. A. Huffman, et al., "Overview of the Manitou Experimental Forest Observatory: Site Description and Selected Science Results from 2008 to 2013," *Atmospheric Chemistry and Physics* 14 (2014): 6345–67. See also "Pine Trees, Terpenes, and Clouds," a film describing some of the scientific work of this project, which was produced in 2011 by Russell Monson and Ben Arfmann and filmed and edited by Greg Monson; vimeo.com/33612217, accessed February 4, 2022.

113. "Pine Trees, Terpenes, and Clouds."

Chapter 7. Exposing the Roots

1. Knut Schmidt-Nielsen, "Per Scholander, 1905–1980," *Biographical Memoirs of the National Academy of Sciences of the United States of America* 56 (1987): 387–412; Elizabeth Noble Shor, *Scripps Institution of Oceanography: Probing the Oceans, 1936–1976* (San Diego: Tofua Press, 1978), chap. 8.

2. The term "expeditionary physiology" likely originated in Scholander's wartime work at Harvard University's Fatigue Laboratory. See G. E. Folk, "The Harvard Fatigue Laboratory: Contributions to World War II," *Advances in Physiology Education* 34 (2010): 119–27; see also the study of physiological animal ecology by Joel B. Hagan, *Life Out of Balance: Homeostasis and Adaptation in a Darwinian World* (Tuscaloosa: University of Alabama Press, 2021).

3. Knut Schmidt-Nielsen, "The Alpha Helix, a Research Opportunity," *BioScience* 19, no. 1 (1969): 59.

4. "Alpha Helix (Ship) Sails for the Amazon," news release February 1, 1967, in digital collections archive of the University of California, San Diego, library. ucsd.edu/dc/object/bb5598587h, accessed February 7, 2022. For more detail on the ship's design and facilities see also Penelope H. Hardy, "Where Science

Meets the Sea: Research Vessels and the Construction of Knowledge in the Nineteenth and Twentieth Centuries" (Ph.D. diss., Johns Hopkins University, 2017), chap. 5.

5. Nellie May Beetham, "The Ecological Tolerance Range of the Seedling Stage of *Sequoia gigantea*" (Ph.D. diss., Duke University, 1962); published as N. Stark, "The Environmental Tolerance of the Seedling Stage of *Sequoiadendron giganteum*," *American Midland Naturalist* 80, no. 1 (1968): 84–95.

6. Jean H. Langenheim, "Early History and Progress of Women Ecologists: Emphasis upon Research Contributions," *Annual Review of Ecology and Systematics* 27 (1996): 1–53.

7. Nellie Beetham Stark, letter to author, March 20, 2021.

8. Per Scholander and a Brazilian colleague were studying a related problem, the impact of flooding and subsequent drying on sap pressure, as part of the *Alpha Helix* expedition. See P. F. Scholander and M. de Oliveira Perez, "Sap Tension in Flooded Trees and Bushes of the Amazon," *Plant Physiology* 43 (1968): 1870–73.

9. Nellie Beetham Stark, letter to author, March 20, 2021.

10. Went diary, August 14, 1967.

11. Nellie Beetham Stark, letter to author, March 20, 2021.

12. Arthur W. Galston and Thomas D. Sharkey, "Frits Warmolt Went, 1903–1990," *Biographical Memoirs of the National Academy of Sciences of the United States of America* 74 (1998): 348–63.

13. Nellie Beetham Stark, letter to author, March 20, 2021.

14. The technique is demonstrated in this YouTube video, where a strip of inner bark is used to make the loop: "Climbing into the Rain Forest Canopy: Life in the Amazon Jungle," youtube.com/watch?v=3Gs_Xn2p0uM, accessed February 21, 2022.

15. Went diary, September 15, 1967.

16. B. Frank, "On the Nutritional Dependence of Certain Trees on Root Symbiosis with Below-Ground Fungi (An English Translation of A. B. Frank's Classic Paper of 1885)," trans. James M. Trappe, *Mycorrhiza* 15 (2005): 267–75, on 272, 274.

17. F. W. Went and N. Stark, "Mycorrhiza," *BioScience* 18, no. 11 (1968): 1035–39; F. W. Went and N. Stark, "The Biological and Mechanical Role of Soil Fungi," *Proceedings of the National Academy of Sciences of the United States of America* 60, no. 2 (1968): 497–504.

18. P. W. Richards, *The Tropical Rain Forest: An Ecological Study* (Cambridge: Cambridge University Press, 1952). See also Philip E. Stanley, George C. G. Argent, and Harold L. K. Whitehouse, "A Botanical Biography of Professor Paul Richards, C. B. E.," *Journal of Bryology* 20 (1998): 323–70.

19. Richards, *Tropical Rain Forest,* 220.

20. Sally E. Smith and David J. Read, *Mycorrhizal Symbioses,* 3rd ed. (Amsterdam: Academic Press, 2008).

21. Later studies revealed that arbuscules, which were specialized branched structures, did not enter into the protoplast of the plant cell; rather, the plant cell wall "accommodated the entering arbuscule by forming a new and distinct periarbuscular membrane continuous with the plasma membrane of the cortical cell." See Scott W. Behie and Michael J. Bidochka, "Nutrient Transfer in Plant-Fungal Symbioses," *Trends in Plant Science* 19, no. 11 (2014): 734–40, on 735.

22. A. F. Blakeslee, "A Paradise for Plant Lovers: The Biological Station at Alto da Serra, Brazil," *Scientific Monthly* 25, no. 1 (1927): 5–18.

23. Went diary, September 15, 1967.

24. Nellie Beetham Stark, letter to author, April 2021.

25. Went and Stark, "Mycorrhiza," 1038.

26. Went and Stark, "Biological and Mechanical Role of Soil Fungi."

27. Went and Stark, "Mycorrhiza," 1038.

28. "Research to Save the Fragile Green Hell," *Science News* 95, no. 6 (1969): 134–35. This report suggests that Stark was present at the meeting in Colombia, but in a follow-up letter Went wrote that she was "unable to participate personally at the meeting," so her paper may have been delivered by another person. F. W. Went, letter to the editor of *Science News* 95, no. 10 (1969): 228. See also the report of their nutrient cycling theory in Christopher Weathersbee, "Spoiling the Jungle Yields Few Riches," *Science News* 95, no. 13 (1969): 312–13. An early analysis of the ecological consequences of this development can be found in Emilio F. Moran, *Developing the Amazon* (Bloomington: Indiana University Press, 1981).

29. Went and Stark, "Biological and Mechanical Role of Soil Fungi."

30. James M. Trappe, "A. B. Frank and Mycorrhizae: The Challenge to Evolutionary and Ecologic Theory," *Mycorrhiza* 15 (2005): 277–81, on 280.

31. F. W. Went, "The Experimental Approach to Ecological Problems," in *Structure and Functioning of Plant Populations,* ed. A. H. J. Freysen and J. W. Woldendorp (Amsterdam: North Holland Publishing, 1978), 139–51, on 140.

32. Nellie Stark, "The Nutrient Content of Plants and Soils from Brazil and Surinam," *Biotropica* 2, no. 1 (1970): 51–60; Stark, "Nutrient Cycling II. Nutrient Distribution in Amazonian Vegetation," *Tropical Ecology* 12, no. 2 (1971): 177–201; Stark, "Nutrient Cycling Pathways and Litter Fungi," *BioScience* 22, no. 6 (1972): 355–60.

33. Edward Hacskaylo, "Metabolic Exchanges in Ectomycorrhizae," in *Mycorrhizae: Proceedings of the First North American Conference on Mycorrhizae, April*

1969, ed. Edward Hacskaylo (Washington, D.C.: Government Printing Office, 1971), 175–82, on 181.

34. Edward Hacskaylo, "Mycorrhizae: The Ultimate in Reciprocal Parasitism?" *BioScience* 22, no. 10 (1972): 577–83.

35. Paul W. Richards, "The Tropical Rain Forest," *Scientific American* 229, no. 6 (1973): 58–68, on 63.

36. Stark, "Nutrient Cycling Pathways and Litter Fungi," 355–60.

37. Stark, "Nutrient Cycling Pathways," 360.

38. Nellie Stark," Nutrient Cycling I. Nutrient Distribution in Some Amazonian Soils," *Tropical Ecology* 12 (1971): 24–50, on 26.

39. Stark, "Nutrient Cycling II," 191.

40. Nalini Nadkarni, "Canopy Roots: Convergent Evolution in Rainforest Nutrient Cycles," *Science* 214 (1981): 1023–24. See also Nalini M. Nadkarni, "The Effects of Epiphytes on Nutrient Cycles Within Temperate and Tropical Rainforest Tree Canopies" (Ph.D. diss., University of Washington, 1983).

41. Nellie Stark, "Mycorrhizae and Nutrient Cycling in the Tropics," in *Mycorrhizae: Proceedings of the First North American Conference on Mycorrhizae, April 1969*, ed. Edward Hasckaylo (Washington, D.C.: Government Printing Office, 1971), 228–29.

42. Gösta Lindberg, "Elias Melin, Pioneer Leader in Mycorrhizal Research," *Annual Review of Phytopathology* 27 (1989): 49–57.

43. Edward G. Farnworth and Frank B. Golley, eds., *Fragile Ecosystems: Evaluation of Research and Applications in the Neotropics; A Report of the Institute of Ecology* (New York: Springer, 1974).

44. Rafael Herrera, email to author, September 11, 2021.

45. Carl F. Jordan and Ernesto Medina, "Ecosystem Research in the Tropics," *Annals of the Missouri Botanical Garden* 64 (1977): 737–45.

46. For discussion of the significance of this project see Eberhard F. Brünig, "The Tropical Rain Forest—A Wasted Asset or an Essential Biospheric Resource?" *Ambio* 6, no. 4 (1977): 187–91.

47. Ernesto Medina, Rafael Herrera, Carl Jordan, and Hans Klinge, "The Amazon Project of the Venezuelan Institute for Scientific Research," *Nature and Resources* 13, no. 3 (1977): 4–6; R. Herrera, C. Jordan, E. Medina, and H. Klinge, "How Human Activities Disturb the Nutrient Cycles of a Tropical Rainforest in Amazonia," *Ambio* 10 (1981): 109–14.

48. ForestPlots.net website, forestplots.net, accessed February 7, 2022.

49. Jordan and Medina, "Ecosystem Research in the Tropics," 739.

50. Peter H. Nye and Dennis J. Greenland, *The Soil Under Shifting Cultivation*, Technical communication no. 51, Commonwealth Bureau of Soil Science

(Farnham Royal: Commonwealth Agricultural Bureaux, 1960); Herrera, email to author, September 11, 2021.

51. Howard T. Odum, *Rain Forest Project Annual Report: Terrestrial Ecology Program I, 1965* (Rio Piedras: Puerto Rico Nuclear Center, 1965). For the history of the long discussion about the sea-level canal, see Christine Keiner, *Deep Cut: Science, Power, and the Unbuilt Interoceanic Canal* (Athens: University of Georgia Press, 2020).

52. Jerry R. Kline, Carl F. Jordan, and George E. Drewry, *Rain Forest Project Annual Report FY-1967* (Rio Piedras: Puerto Rico Nuclear Center, 1967).

53. The award citation mentions two articles: Carl F. Jordan and Jerry R. Kline, "Mineral Cycling: Some Basic Concepts and Their Application in a Tropical Rain Forest," *Annual Review of Ecology and Systematics* 3 (1972): 33–50, and Carl F. Jordan, Jerry R. Kline, and Donald S. Sasscer, "Relative Stability of Mineral Cycles in Forest Ecosystems," *American Naturalist* 106 (1972): 237–53.

54. H. T. Odum, "Introduction to Section F," in *A Tropical Rain Forest: A Study of Irradiation and Ecology at El Verde, Puerto Rico,* ed. Howard T. Odum and Robert F. Pigeon (Oak Ridge, Tenn.: U.S. Atomic Energy Commission, 1970), v. 2, F3–F7, on F5.

55. Jordan and Medina, "Ecosystem Research in the Tropics," 737–45.

56. Elvira Cuevas and Ernesto Medina, "Nutrient Dynamics Within Amazonian Forests. II. Fine Root Growth, Nutrient Availability and Leaf Litter Decomposition," *Oecologia* 76 (1988): 222–35.

57. Nellie M. Stark and Carl F. Jordan, "Nutrient Retention by the Root Mat of an Amazonian Rain Forest," *Ecology* 59 (1978): 434–37, on 434.

58. H. Klinge and R. Herrera, "Phytomass Structure of Natural Plant Communities on Spodosols in Southern Venezuela: The Tall Amazon Caatinga Forest," *Vegetatio* 53, no. 2 (1983): 65–84.

59. Herrera, Jordan, Medina, and Klinge, "How Human Activities Disturb the Nutrient Cycles of a Tropical Rainforest," 111. See also Klinge and Herrera, "Phytomass Structure of Natural Plant Communities."

60. Jordan and Medina, "Ecosystem Research in the Tropics."

61. Stark and Jordan, "Nutrient Retention by the Root Mat," 434.

62. Stark and Jordan, "Nutrient Retention by the Root Mat."

63. Rafael Herrera, email to author, September 11, 2021.

64. My account of the study is from Rafael Herrera, email to author, September 11, 2021.

65. R. Herrera, Tatiana Mérida, Nellie Stark, and C. F. Jordan, "Direct Phosphorus Transfer from Leaf Litter to Roots," *Naturwissenschaften* 65 (1978): 208–9.

66. Bernard Moyersoen, *Ectomicorrizas y Micorrizas Vesiculo-arbusculares en Caatinga Amazonica,* Scientia Guaianae, v. 3 (Caracas: O. Huber, 1993).

67. Joan Arehart-Treichel, "Saving Tropical Forests," *Science News* 112, no. 22 (1977): 362–63, on 362.

68. Carl F. Jordan and Rafael Herrera, "Tropical Rain Forests: Are Nutrients Really Critical?" *American Naturalist* 117 (1981): 167–80.

69. Jordan and Medina, "Ecosystem Research in the Tropics," 743.

70. Carl F. Jordan, *Nutrient Cycling in Tropical Forest Ecosystems: Principles and Their Application in Management and Conservation* (New York: John Wiley and Sons, 1985); Carl F. Jordan, *An Ecosystem Approach to Sustainable Agriculture: Energy Use Efficiency in the American South* (Dordrecht: Springer, 2013).

71. Nellie Stark, "Man, Tropical Forests, and the Biological Life of the Soil," *Biotropica* 10, no. 1 (1978): 1–10.

72. Cuevas and Medina, "Nutrient Dynamics Within Amazonian Forests," 222–35; Ernesto Medina, email to author, July 5, 2021.

73. H. A. Mooney, O. Björkman, A. E. Hall, E. Medina, and P. B. Tomlinson, "The Study of the Physiological Ecology of Tropical Plants: Current Status and Needs," *BioScience* 30, no. 1 (1980): 22–26.

74. Peter M. Vitousek and R. L. Sanford, Jr., "Nutrient Cycling in Moist Tropical Forest," *Annual Review of Ecology and Systematics* 17 (1986): 137–67, on 155. In 1987 Sanford followed up with an important study showing that tree roots could grow upward on the stems of neighboring trees, providing another way for trees to gain nutrients in these nutrient-poor environments; see R. L. Sanford, Jr., "Apogeotropic Roots in the Amazon Rain Forest," *Science* 235 (1987): 1062–64.

75. R. M. May, "Mutualistic Interactions Among Species," *Nature* 296 (April 1982): 803–4, on 803.

76. D. H. Lewis, "Mutualistic Lives," *Nature* 297, no. 5863 (1982): 176.

77. David Smith and David Lewis, "Professor J. L. Harley," *New Phytologist* 119 (1991): 5–7; D. C. Smith and D. H. Lewis, "John Laker Harley, 17 November 1911–13 December 1990," *Biographical Memoirs of Fellows of the Royal Society* 39 (1994): 158–75.

78. J. L. Harley, "Mycorrhiza and Soil Ecology," *Biological Reviews* 23 (1948): 127–58.

79. J. L. Harley, "Fungi in Ecosystems," *Journal of Ecology* 59 (1971): 653–68, on 667.

80. Sheffield has long been an important center for botanical research and in 2004 opened a modern version of a phytotron, a complex of controlled-

environment greenhouses, growth chambers, and laboratories that is a world-leading research center. It was later named the Sir David Read Controlled Environment Facility. A virtual tour of the facility is on YouTube, youtube.com/watch?v=WGoRkXFrXWI, accessed February 8, 2022.

81. J. Whittingham and D. J. Read, "Vesicular-arbuscular Mycorrhiza in Natural Vegetation Systems. III. Nutrient Transfer Between Plants with Mycorrhizal Interconnections," *New Phytologist* 90 (1982): 277–84; R. Francis and D. J. Read, "Direct Transfer of Carbon Between Plants Connected by Vesicular-arbuscular Mycorrhizal Mycelium," *Nature* 307, no. 5946 (1984): 53–56; D. J. Read, R. Francis, and R. D. Finlay, "Mycorrhizal Mycelia and Nutrient Cycling in Plant Communities," in *Ecological Interactions in Soil: Plants, Microbes, and Animals,* ed. A. H. Fitter, D. Atkinson, D. J. Read, and M. B. Usher (Oxford: Blackwell Scientific, 1985), 193–217.

82. Read, Francis, and Finlay, "Mycorrhizal Mycelia and Nutrient Cycling," 193.

83. A. H. Fitter, D. Atkinson, D. J. Read, and M. B. Usher, eds., *Ecological Interactions in Soil: Plants, Microbes, and Animals* (Oxford: Blackwell Scientific, 1985), viii.

84. Amyan Macfadyen, review of *Ecological Interactions in Soil,* edited by A. H. Fitter et al., *Journal of Ecology* 74 (1986): 602–4, on 604.

85. R. Bajwa and D. J. Read, "The Biology of Mycorrhiza in the Ericaceae. IX. Peptides as Nitrogen Sources for the Ericoid Endophyte and for Mycorrhizal and Non-mycorrhizal Plants," *New Phytologist* 101 (1985): 459–67; R. Bajwa, S. Abuarghub, and D. J. Read, "The Biology of Mycorrhiza in the Ericaceae. X. The Utilization of Proteins and the Production of Proteolytic Enzymes by the Mycorrhizal Endophyte and by Mycorrhizal Plants," *New Phytologist* 101 (1985): 469–86.

86. Gary D. Bending and David J. Read, "The Structure and Function of the Vegetative Mycelium of Ectomycorrhizal Plants. V. Foraging Behavior and Translocation of Nutrients from Exploited Litter," *New Phytologist* 130 (1995): 401–9; Gary G. Bending and David J. Read, "The Structure and Function of the Vegetative Mycelium of Ectomycorrhizal Plants. VI. Activities of Nutrient Mobilizing Enzymes in Birch Litter Colonized by *Paxillus involutus* (Fr.) Fr.," *New Phytologist* 130 (1995): 411–17.

87. J. P. Grime, "Competitive Exclusion in Herbaceous Vegetation," *Nature* 242, no. 5396 (1973): 344–47.

88. J. P. Grime, "Vegetation Classification by Reference to Strategies," *Nature* 250, no. 5461 (1974): 26–31; Jason D. Fridley and Simon Pierce, "Professor John Philip Grime, FRS (1935–2021)," *Trends in Ecology and Evolution* 36, no. 8 (2021): 663–64. For discussion of the CSR theory in relation to other

theories, see J. Bastow Wilson and William G. Lee, "C-S-R Triangle Theory: Community-level Predictions, Tests, Evaluation of Criticisms, and Relation to Other Theories," *Oikos* 91 (2000): 77–96.

89. J. P. Grime, *Plant Strategies and Vegetation Processes* (Chichester: John Wiley and Sons, 1979).

90. Grime, *Plant Strategies,* 36–37.

91. J. P. Grime, quoted in "Ecological Inspirations: John Philip Grime," *Journal of Ecology* blog post, December 16, 2013, *Eminent Ecologist* series, jecologyblog.com/2013/12/16/scaling-up-to-communities-and-ecosystems-by-j-p-grime, accessed February 8, 2022.

92. Grime, "Ecological Inspirations."

93. J. P. Grime, J. M. L. Mackey, S. H. Hillier, and D. J. Read, "Floristic Diversity in a Model System Using Experimental Microcosms," *Nature* 328, no. 6129 (1987): 420–22.

94. Grime et al., "Floristic Diversity," 422.

95. B. D. Campbell, J. P. Grime, J. M. L. Mackey, and A. Jalili, "The Quest for a Mechanistic Understanding of Resource Competition in Plant Communities: The Role of Experiments," *Functional Ecology* 5, no. 2 (1991): 241–53.

96. Grime, "Ecological Inspirations."

97. J. P. Grime, *Plant Strategies, Vegetation Processes, and Ecosystem Properties,* 2nd ed. (Chichester: John Wiley and Sons, 2001), 284–86.

98. David J. Read, "Mycorrhizas in Ecosystems," *Experientia* 47 (1991): 376–91, on 388.

99. The symposium's papers were edited and published in D. J. Read, D. H. Lewis, A. H. Fitter, and I. J. Alexander, eds., *Mycorrhizas in Ecosystems* (Wallingford: CAB International, 1992).

100. Read, "Mycorrhizas in Ecosystems."

101. David J. Read and J. Perez-Moreno, "Mycorrhizas and Nutrient Cycling in Ecosystems: A Journey Towards Relevance?" *New Phytologist* 157 (2003): 475–92, on 476. Read was summarizing the argument that he had proposed in his 1991 review article "Mycorrhizas in Ecosystems."

102. Read and Perez-Moreno, "Mycorrhizas and Nutrient Cycling."

103. Read, "Mycorrhizas in Ecosystems," 387.

104. J. Dighton, "Acquisition of Nutrients from Organic Resources by Mycorrhizal Autotrophic Plants," *Experientia* 47 (1991): 362–69, on 363.

105. Dighton, "Acquisition of Nutrients," 367.

106. J. R. Leake and D. J. Read, "Mycorrhizal Fungi in Terrestrial Habitats," in *The Mycota. IV. Environmental and Microbial Relationships,* ed. D. T. Wicklow and B. E. Söderström (Berlin: Springer, 1997), 281–301, on 281.

107. Suzanne W. Simard, David A. Perry, Melanie D. Jones, David D. Myrold, Daniel M. Durall, and Randy Molina, "Net Transfer of Carbon Between Ectomycorrhizal Tree Species in the Field," *Nature* 388, no. 6642 (1997): 579–82. See also Suzanne Simard, *Finding the Mother Tree: Discovering the Wisdom of the Forest* (New York: Alfred A. Knopf, 2021).

108. David Read, "The Ties That Bind," *Nature* 388, no. 6642 (1997): 617.

109. David Robinson and Alastair Fitter, "The Magnitude and Control of Carbon Transfer Between Plants Linked by a Common Mycorrhizal Network," *Journal of Experimental Botany* 50 (1999): 9–13.

110. A. H. Fitter, "Anthony David Bradshaw, 17 January 1926–21 August 2008," *Biographical Memoirs of Fellows of the Royal Society* 56 (2010): 25–39.

111. "Profile: Alastair Fitter," *New Phytologist* 220 (2018): 977–78.

112. A. H. Fitter, "Costs and Benefits of Mycorrhizas: Implications for Functioning Under Natural Conditions," *Experientia* 47 (1991): 350–55.

113. Robinson and Fitter, "Magnitude and Control of Carbon Transfer," 12.

114. A. H. Fitter, J. D. Graves, N. K. Watkins, D. Robinson, and C. Scrimgeour, "Carbon Transfer Between Plants and Its Control in Networks of Arbuscular Mycorrhizas," *Functional Ecology* 12 (1998): 406–12.

115. Philip L. Staddon, "Mycorrhizal Fungi and Environmental Change: The Need for a Mycocentric Approach," *New Phytologist* 167 (2005): 635–37; O. Alberton, W. W. Kuyper, and A. Gorissen, "Taking Mycocentrism Seriously: Mycorrhizal Fungi and Plant Responses to Elevated CO_2," *New Phytologist* 167 (2005): 859–68.

116. Angela Hodge, Colin D. Campbell, and Alastair H. Fitter, "An Arbuscular Mycorrhizal Fungus Accelerates Decomposition and Acquires Nitrogen Directly from Organic Material," *Nature* 413, no. 6853 (2001): 297–99.

117. This research is discussed in Dighton, "Acquisition of Nutrients," 364.

118. David P. Janos, "Vesicular-arbuscular Mycorrhizae Affect Lowland Tropical Rain Forest Plant Growth," *Ecology* 61 (1980): 151–62.

119. David P. Janos, "Mycorrhizae Influence Tropical Succession," *Biotropica* 12 (1980): 56–64; David P. Janos, "Mycorrhizas, Succession and the Rehabilitation of Deforested Lands in the Humid Tropics," in *Fungi and Environmental Change,* ed. J. C. Frankland, N. Magan, and G. M. Gadd, British Mycological Society Symposium, v. 20 (Cambridge: Cambridge University Press, 1996), 129–62.

120. Catalina Aristizábal, Emma Lucia Rivera, and David P. Janos, "Arbuscular Mycorrhizal Fungi Colonize Decomposing Leaves of *Myrica parvifolia, M. pubescens,* and *Paepalanthus* sp.," *Mycorrhiza* 14 (2004): 221–28, on 226–27.

121. Rebecca A. Bunn, Dylan T. Simpson, Lorinda S. Bullington, Ylva Lekberg, and David P. Janos, "Revisiting the 'Direct Mineral Cycling' Hypothesis: Arbuscular Mycorrhizal Fungi Colonize Leaf Litter, But Why?" *ISME Journal* 13 (2019): 1891–98.

122. Bunn et al., "Revisiting the 'Direct Mineral Cycling' Hypothesis," on 1893, 1894, 1895, 1891 respectively.

123. Trappe, "A. B. Frank and Mycorrhizae."

124. Trappe, "A. B. Frank and Mycorrhizae," 280.

125. Michael F. Allen, *The Ecology of Mycorrhizae* (Cambridge: Cambridge University Press, 1991).

126. K. A. Pirozynski and D. W. Malloch, "The Origin of Land Plants: A Matter of Mycotrophism," *BioSystems* 6 (1975): 153–64; D. W. Malloch, K. A. Pirozynski, and P. H. Raven, "Ecological and Evolutionary Significance of Mycorrhizal Symbioses in Vascular Plants (A Review)," *Proceedings of the National Academy of Sciences of the United States of America* 4 (1980): 2113–18; Roy E. Halling, "Ectomycorrhizae: Co-Evolution, Significance, and Biogeography," *Annals of Missouri Botanical Garden* 88 (2001): 5–13.

127. Allen, *Ecology of Mycorrhizae*, 141.

128. Sally E. Smith and David J. Read, *Mycorrhizal Symbiosis*, 3rd ed. (Amsterdam: Academic Press, 2008), 574. Read had nominated Allen to write the book for the Cambridge series.

129. Martha Christensen, "Soil Fungi: A New Perspective?" *Inoculum* 62, no. 5 (October 2011): 2–3; Juliet C. Brown, "Soil Fungi of Some British Sand Dunes in Relation to Soil Types and Succession," *Journal of Ecology* 46 (1958): 641–64.

130. Edith Bach Allen and Michael F. Allen, "Natural Re-establishment of Vesicular-arbuscular Mycorrhizae Following Stripmine Reclamation in Wyoming," *Journal of Applied Ecology* 17 (1980): 139–47.

131. Martha Christensen, "Renaissance in Mycorrhizal Research," *BioScience* 34, no. 2 (1984): 112–13.

132. Jane Lubchenco, Annette M. Olson, Linda B. Brubaker, Stephen R. Carpenter, Marjorie M. Holland, Stephen P. Hubbell, Simon A. Levin, et al., "The Sustainable Biosphere Initiative: An Ecological Research Agenda," *Ecology* 72 (1991): 371–412.

133. C. Coe Klopatek et al., "The Sustainable Biosphere Initiative: A Commentary from the U.S. Soil Ecology Society," *Bulletin of the Ecological Society of America* 73, no. 4 (1992): 223–28.

134. Deborah Schoen, "Primary Productivity: The Link to Global Health," *BioScience* 47, no. 8 (1997): 477–80.

135. Went diary, September 22, 1952.

136. The East Malling laboratory is described in V. Kolesnikov, *The Root System of Fruit Plants,* trans. Ludmilla Aksenova (Moscow: Mir, 1971), 160–65. A blog post discussing the history of East Malling, with a link to Kolesnikov's book, is "The M9 and Rhizotrons," The Gardens Trust, thegardenstrust.blog/ 2020/08/29/the-m9-and-rhizotrons, accessed February 8, 2022. See also W. S. Rogers, "The East Malling Root-Observation Laboratories," in *Proceedings of the Fifteenth Easter School in Agricultural Science, University of Nottingham, 1968,* ed. W. J. Whittington (London: Butterworths, 1969), 361–78; W. H. Lyford and B. F. Wilson, *Controlled Growth of Forest Tree Roots: Technique and Application,* Harvard Forest Paper No. 16 (Petersham, Mass.: Harvard Forest, 1966).

137. Robert Fogel, "Mycorrhizae and Nutrient Cycling in Natural Forest Ecosystems," *New Phytologist* 86 (1980): 199–212.

138. Bobbie L. McMichael and John C. Zak, "The Role of Rhizotrons and Minirhizotrons in Evaluating the Dynamics of Rhizoplane-rhizosphere Microflora," in *Microbial Activity in the Rhizosphere,* ed. K. G. Mukerji, C. Manoharachary, and Jagjit Singh (Berlin: Springer, 2006), 71–87.

139. Robert Fogel and John Lussenhop, "The University of Michigan's Soil Biotron: A Platform for Soil Biology Research in a Natural Forest," in *Plant Root Growth: An Ecological Perspective,* ed. D. Atkinson (Oxford: Blackwell Scientific, 1991), 61–68, on 63. On the soil biotron's facilities, see James A. Teeri, "The Soil Biotron: An Underground Research Laboratory," in *Inventive Minds: Creativity in Technology,* ed. R. J. Weber and D. N. Perkins (Oxford: Oxford University Press, 1992), 142–53.

140. Teeri, "The Soil Biotron," 147.

141. John Lussenhop, Robert Fogel, and Kurt Pregitzer, "A New Dawn for Soil Biology: Video Analysis of Root-Soil-Microbial-Faunal Interactions," *Agriculture, Ecosystems, and Environment* 34 (1991): 235–49; see also John Lussenhop and Robert Fogel, "Observing Soil Biota in situ," *Geoderma* 56 (1993): 25–36.

142. Michael F. Allen, R. Vargas, E. A. Graham, W. Swenson, M. Hamilton, M. Taggart, T. C. Harmon, et al., "Soil Sensor Technology: Life Within a Pixel," *Bioscience* 57, no. 10 (2007): 859–67; Rebecca R. Hernandez and Michael F. Allen, "Diurnal Patterns of Productivity of Arbuscular Mycorrhizal Fungi Revealed with the Soil Ecosystem Observatory," *New Phytologist* 200 (2013): 547–57; Michael F. Allen and Kuni Kitajima, "In situ High-frequency Observations of Mycorrhizas," *New Phytologist* 200 (2013): 222–28.

143. Megan Raby opened up the study of tropical ecology in relation to conservation of biodiversity, but her focus was on U.S. institutions in the Caribbean.

Raby, *American Tropics: The Caribbean Roots of Biodiversity Science* (Chapel Hill: University of North Carolina Press, 2017).

144. Harold A. Mooney, *The Globalization of Ecological Thought* (Oldendorf/ Luhe: Ecology Institute, 1998).

145. Jonathan R. Leake, David Johnson, Damian Donnelly, Gemma Muckle, Lynne Boddy, and David Read, "Networks of Power and Influence: The Role of Mycorrhizal Mycelium in Controlling Plant Communities and Agroecosystem Functioning," *Canadian Journal of Botany* 82 (2004): 1016–45.

Chapter 8. The Drive for Synthesis

1. Julian Huxley, *Evolution: The Modern Synthesis* (New York: Harper, 1942).

2. Ernst Mayr and William B. Provine, eds., *The Evolutionary Synthesis: Perspectives on the Unification of Biology* (Cambridge, Mass.: Harvard University Press, 1980).

3. William M. Hiesey, "Environmental Influence and Transplant Experiments," *Botanical Review* 6 (1940): 181–203.

4. Arne Müntzing, "Göte Wilhelm Turesson," *Taxon* 29, no. 5–6 (1971): 773–75.

5. Göte Turesson, "The Species and the Variety as Ecological Units," *Hereditas* 3 (1922): 100–113.

6. H. M. Hall, "Heredity and Environment—As Illustrated by Transplant Studies," *Scientific Monthly* 35, no. 4 (1932): 289 302. This article appeared posthumously and was based on a lecture Hall had intended to give in December 1931. For more detail on the program's goals and methods, see Joel B. Hagen, "The Development of Experimental Methods in Plant Taxonomy, 1920–1950," *Journal of the History of Biology* 32 (1983): 406–16; Hagen, "Experimentalists and Naturalists in Twentieth-Century Botany: Experimental Taxonomy, 1920–1950," *Journal of the History of Biology* 17 (1984): 249–70.

7. J. Núñez-Farfán and C. D. Schlichting, "Evolution in Changing Environments: The 'Synthetic' Work of Clausen, Keck, and Hiesey," *Quarterly Review of Biology* 76, no. 4 (2001): 433–57.

8. Jens Clausen and William M. Hiesey, *Experimental Studies on the Nature of Species, IV. Genetic Structure of Ecological Races* (Washington, D.C.: Carnegie Institution of Washington, 1958), iii.

9. G. L. Stebbins, "Botany and the Synthetic Theory of Evolution," in *The Evolutionary Synthesis: Perspectives on the Unification of Biology,* ed. Ernst Mayr and William B. Provine (Cambridge, Mass.: Harvard University Press, 1980), 144.

10. V. Betty Smocovitis, "G. Ledyard Stebbins, Jr. and the Evolutionary Synthesis (1924–1950)," *American Journal of Botany* 84, no.12 (1997): 1625–37.

11. Stebbins, "Botany and the Synthetic Theory," 142.

12. Jens Clausen, David D. Keck, and William M. Hiesey, *Experimental Studies on the Nature of Species, II. Plant Evolution Through Amphiploidy and Autoploidy, with Examples from the Madiinae* (Washington, D.C.: Carnegie Institution of Washington, 1945), iii.

13. Patricia Craig, *Centennial History of the Carnegie Institution of Washington,* v. 4, *The Department of Plant Biology* (New York: Cambridge University Press, 2005).

14. Jens Clausen, David D. Keck, and William M. Hiesey, "Experimental Taxonomy," *Carnegie Institution of Washington Year Book No. 39* (Washington, D.C.: Carnegie Institution of Washington, 1940), 158–63.

15. Hiesey, "Environmental Influence and Transplant Experiments," 199.

16. Jens Clausen, David D. Keck, and William M. Hiesey, "Experimental Taxonomy," *Carnegie Institution of Washington Year Book No. 46* (Washington, D.C.: Carnegie Institution of Washington, 1947), 95–104, on 104. The date of the roundtable is not given in this report.

17. Jens Clausen, David D. Keck, and William M. Hiesey, "Experimental Taxonomy," *Carnegie Institution of Washington Year Book No. 49* (Washington, D.C.: Carnegie Institution of Washington, 1950), 101–14; see 102–9.

18. Jens Clausen, David D. Keck, and William H. Hiesey, *Experimental Studies on the Nature of Species, III. Environmental Responses of Climatic Races in Achillea* (Washington, D.C.: Carnegie Institution of Washington, 1948), 1.

19. William M. Hiesey, "Growth Studies Under Controlled Temperatures," *Carnegie Institution of Washington Year Book No. 50* (Washington, D.C.: Carnegie Institution of Washington, 1951), 99–105; William M. Hiesey, "Grasses," in F. W. Went, *Experimental Control of Plant Growth* (Waltham, Mass.: Chronica Botanica, 1957), 153–63; William M. Hiesey and Harold W. Milner, "Comparative Physiology of Ecologic Races," *Carnegie Institution of Washington Yearbook No. 51* (Washington, D.C.: Carnegie Institution of Washington, 1952), 131–32.

20. Hiesey, "Experimental Taxonomy," and "Growth Studies Under Controlled Temperatures," both in *Carnegie Institution Year Book No. 50* (Washington, D.C.: Carnegie Institution of Washington, 1951), 96–122, on 97 and 104.

21. Hiesey and Milner, "Comparative Physiology of Ecologic Races," 132.

22. Harold W. Milner and William M. Hiesey, "Physiology of Climatic Races," *Carnegie Institution of Washington Year Book No. 53* (Washington, D.C.: Carnegie Institution of Washington, 1954), 162; "Experimental Taxonomy," *Carnegie Institution of Washington Year Book No. 54* (Washington, D.C.: Carnegie Institution of Washington, 1955), 170–72; Harold W. Milner, William M. Hiesey,

and Malcolm A. Nobs, "Physiology of Climatic Races," *Carnegie Institution of Washington Year Book No. 57* (Washington, D.C.: Carnegie Institution of Washington, 1958), 266–70.

23. William Hiesey, "Grasses," in F. W. Went, *Experimental Control of Plant Growth*, 153–54.

24. Jens Clausen and William H. Hiesey, *Experimental Studies on the Nature of Species, IV. Genetic Structure of Ecological Races* (Washington, D.C.: Carnegie Institution of Washington, 1958), 267.

25. William H. Hiesey, Harold W. Milner, and Malcolm A. Nobs, "Working Principles for a Physiological Approach to the Study of Climatic Races," *Carnegie Institution of Washington Year Book No. 58* (Washington, D.C.: Carnegie Institution of Washington, 1959), 344 46.

26. William M. Hiesey, Malcolm A. Nobs, and Olle Björkman, *Experimental Studies on the Nature of Species, V. Biosystematics, Genetics, and Physiological Ecology of the Erythranthe Section of Mimulus* (Washington, D.C.: Carnegie Institution of Washington 1971), 91.

27. Hiesey, Nobs, and Björkman, *Experimental Studies, V*, 91.

28. Biographical information is from "A Conversation with Olle Björkman," interview by Krishna K. Niyogi, 2011, *Annual Review of Plant Biology Conversations*, youtube.com/watch?v=kuIUUkjA6uo, accessed February 5, 2022.

29. Dieter von Wettstein, "The Phytotron in Stockholm," *Studia Forestalia Suecica*, no. 44 (Stockholm: Royal College of Forestry, 1967).

30. Joseph Berry, "Introduction," *Photosynthesis Research* 67 (2001): 1 3, on 1. Berry's essay introduced a festschrift honoring Björkman at his retirement.

31. A. A. Benson, "Following the Path of Carbon in Photosynthesis: A Personal Story," *Photosynthesis Research* 73 (2002): 29–49; see also Thomas D. Sharkey, "Discovery of the Canonical Calvin-Benson Cycle," *Photosynthesis Research* 140 (2019): 235 52.

32. K. T. Glasziou, "The David North Research Centre," *Nature* 187, no. 4739 (1960): 745–46.

33. M. D. Hatch and C. R. Slack, "C_4 Photosynthesis: Discovery, Resolution, Recognition, and Significance," in *Discoveries in Plant Biology*, ed. Shain-Dow Kung and Shang-Fa Yang, v. 1 (Singapore: World Scientific, 1998), 175–96; M. D. Hatch, "C_4 Photosynthesis: A Historical Overview," in C_4 *Plant Biology*, ed. Rowan F. Sage and Russell K. Monson (San Diego: Academic Press, 1999), 17–46; Steve Gartner, "The Discovery of C_4 Photosynthesis," CSIROpedia, csiropedia.csiro.au/the-discovery-of-c4-photosynthesis, accessed February 5, 2022; M. D. Hatch, "The Discovery of C_4 Photosynthesis," *Plants in Action : A*

Resource for Teachers and Students of Plant Science, Feature Essay 2.1, rseco.org/content/feature-essay-21-discovery-c4-photosynthesis.html, accessed February 5, 2022; Joseph A. Berry, "There Ought to Be an Equation for That," *Annual Review of Plant Biology* 63 (2012): 1–17.

34. M. D. Hatch and C. R. Slack, "Photosynthesis by Sugar-Cane Leaves," *Biochemical Journal* 101 (1966): 103–11.

35. L. T. Evans, I. F. Wardlaw, and R. W. King, "Two Decades of Research at the Canberra Phytotron," *Botanical Review* 51, no. 2 (1985): 203–72.

36. Harold A. Mooney, "Carbon Dioxide Exchange of Plants in Natural Environments," *Botanical Review* 38, no. 3 (1972): 455–69, on 462.

37. Herbert G. Baker, "Evolution in the Tropics," *Biotropica* 2, no. 2 (1970): 101–11.

38. E. B. Tregunna, G. Krotkov, and C. D. Nelson, "Effect of Oxygen on the Rate of Photorespiration in Detached Tobacco Leaves," *Physiologia Plantarum* 19 (1966): 723–33.

39. Sam G. Wildman, "Along the Trail from Fraction I Protein to Rubisco (*ribulose bis*phosphate *c*arboxylase-*o*xygenase)," *Photosynthesis Research* 73 (2002): 243–50; Sharkey, "Discovery of the Canonical Calvin-Benson Cycle." Sharkey suggests that the acronym should be either rubisco or Rubisco, without further capitalization.

40. Olle Björkman, "The Effect of Oxygen Concentration on Photosynthesis in Higher Plants," *Physiologia Plantarum* 19 (1966): 618–33.

41. Berry, "There Ought to Be an Equation," 4.

42. Berry, "There Ought to Be an Equation," 3.

43. M. D. Hatch, C. B. Osmond, and R. O. Slatyer, eds., *Photosynthesis and Photorespiration* (New York: Wiley-Interscience, 1971). The U.S. funding agency was the National Science Foundation, and the Australian agency was the Commonwealth Department of Education and Science.

44. Marshall D. Hatch, "C_4 Photosynthesis: A Historical Overview," in C_4 *Plant Biology,* ed. Rowan F. Sage and Russell K. Monson (San Diego: Academic Press, 1999), 17–46; see comments on 27 about what was known at the time of the conference.

45. Olle Björkman, "Comparative Photosynthetic CO_2 Exchange in Higher Plants," in *Photosynthesis and Photorespiration,* ed. M. D. Hatch, C. B. Osmond, and R. O. Slatyer (New York: Wiley-Interscience, 1971), 18–32. The group paper was titled "Characteristics of Hybrids Between C_3 and C_4 Species of *Atriplex,*" 105–19 in the same volume.

46. Björkman, "Comparative Photosynthetic CO_2 Exchange," 19.

47. Björkman, "Comparative Photosynthetic CO_2 Exchange," 30.

48. Hiesey, Nobs, and Björkman, *Experimental Studies, V,* 183.

49. Hiesey, Nobs, and Björkman, *Experimental Studies, V,* 182.

50. Hiesey, Nobs, and Björkman, *Experimental Studies, V,* 196.

51. Berry, "There Ought to Be an Equation," 6.

52. Barry Osmond, "Our Eclectic Adventures in the Slower Eras of Photosynthesis: From New England Down Under to Biosphere 2 and Beyond," *Annual Review of Plant Biology* 65 (2014): 1–32.

53. Berry, "Introduction," 1.

54. Beth Azar, "Profile of James R. Ehleringer," *Proceedings of the National Academy of Sciences of the United States of America* 117 (2020): 20348–50.

55. "Conversation with Olle Björkman," youtube.com/watch?v=kuIUUkjA6uo.

56. Berry, "Introduction," 1.

57. Berry, "There Ought to Be an Equation," 6.

58. Olle Björkman, Robert W. Pearcy, A. Tyrone Harrison, and Harold Mooney, "Photosynthetic Adaptation to High Temperatures: A Field Study in Death Valley, California," *Science* 175 (1972): 786–89.

59. H. A. Mooney, "On the Road to Global Ecology," *Annual Review of Energy and the Environment* 24 (1999): 1–24, on 6.

60. P. J. Kramer, H. Hellmers, and R. J. Downs, "SEPEL: New Phytotrons for Environmental Research," *BioScience* 20, no. 22 (1970): 1201–8.

61. Mooney, "Road to Global Ecology," 15–16.

62. Mooney, "Road to Global Ecology," 17.

63. Björkman, Pearcy, Harrison, and Mooney, "Photosynthetic Adaptation to High Temperature."

64. Berry, "Introduction," 2.

65. H. A. Mooney, E. L. Dunn, A. T. Harrison, P. A. Morrow, B. Bartholomew, and R. L. Hays, "A Mobile Laboratory for Gas Exchange Measurements," *Photosynthetica* 5 (1971): 128–32.

66. H. A. Mooney, "Carbon Dioxide Exchange of Plants in Natural Environments," *Botanical Review* 38, no. 3 (1972): 455–69, on 462.

67. "Conversation with Olle Björkman."

68. "Conversation with Olle Björkman."

69. Robert W. Pearcy, Olle Björkman, Martyn M. Caldwell, Jon E. Keeley, Russell K. Monson, and Boyd R. Strain, "Carbon Gain by Plants in Natural Environments," *BioScience* 37, no. 1 (1987): 21–29.

70. H. A. Mooney, R. W. Pearcy, and J. Ehleringer, "Plant Physiological Ecology Today," *BioScience* 37, no. 1 (1987): 18–20, on 19.

71. Rowan F. Sage and Russell K. Monson, eds., C_4 *Plant Biology* (San Diego: Academic Press, 1999), xiii.

72. Rowan F. Sage, "Russ Monson and the Evolution of C_4 Photosynthesis," *Oecologia* (2021): 823–40, on 823, doi.org/10.1007/s00442-021-04883-1. The article was part of a special issue honoring Monson on his retirement.

73. George J. Williams III, "Photosynthetic Adaptation to Temperature in C_3 and C_4 Grasses: A Possible Ecological Role in the Shortgrass Prairie," *Plant Physiology* 54 (1974): 709–11; Paul R. Kemp and George J. Williams III, "A Physiological Basis for Niche Separation Between *Agropyron smithii* (C_3) and *Bouteloua gracilis* (C_4)," *Ecology* 61, no. 4 (1980): 846–58.

74. Russell K. Monson and George J. Williams III, "A Correlation Between Photosynthetic Temperature Adaptation and Seasonal Phenology Patterns in the Shortgrass Prairie," *Oecologia* 54 (1982): 58–62.

75. Sage, "Russ Monson and the Evolution of C_4 Photosynthesis," 825. The reference is to R. K. Monson, M. R. Sackschewsky, and G. J. Williams, "Field Measurements of Photosynthesis, Water-Use Efficiency, and Growth in *Agropyron smithii* (C_3) and *Bouteloua gracilis* (C_4) in the Colorado Shortgrass Steppe," *Oecologia* 68 (1986): 400–409.

76. Sage, "Russ Monson and the Evolution of C_4 Photosynthesis," 826.

77. Russell K. Monson, Gerald E. Edwards, and Maurice S. B. Ku, "C_3-C_4 Intermediate Photosynthesis in Plants," *BioScience* 34, no. 9 (1984): 563–66 and 571–74.

78. Sage, "Russ Monson and the Evolution of C_4 Photosynthesis," 827.

79. Sage, "Russ Monson and the Evolution of C_4 Photosynthesis," 829.

80. Scientists have continued to debate the reasons for the expansion of C_4 grasslands in the Miocene. For a later review of the subject see Colin P. Osborne, "Atmosphere, Ecology, and Evolution: What Drove the Miocene Expansion of C_4 Grasslands?" *Journal of Ecology* 96 (2008): 35–45.

81. Rowan F. Sage, "Why C_4 Synthesis?" in *C_4 Plant Biology,* ed. Rowan F. Sage and Russell K. Monson (San Diego: Academic Press, 1999), 3–16, on 12.

82. Harold A. Mooney, Peter M. Vitousek, and Pamela A. Matson, "Exchange of Materials Between Terrestrial Ecosystems and the Atmosphere," *Science* 238 (1987): 926–32, on 931.

83. Harold A. Mooney, *The Globalization of Ecological Thought* (Oldendorf/Luhe: Ecology Institute, 1998).

84. Pearcy, Björkman, Caldwell, et al., "Carbon Gain by Plants," 28.

85. Winslow Briggs, "The Director's Essay," *Year Book of the Carnegie Institution of Washington, No. 89* (Washington, D.C.: Carnegie Institution of Washington, 1990), 122.

86. James R. Ehleringer and Christopher B. Field, eds., *Scaling Physiological Processes: Leaf to Globe* (San Diego: Academic Press, 1993). See the review by Russell Monson, "Climbing into the Big Picture," *BioScience* 44 (1994): 188–89.

87. Christopher B. Field, "Plant Physiology of the 'Missing' Carbon Sink," *Plant Physiology* 125 (2001): 25–28, on 25.

88. Field, "Plant Physiology," 25.

89. Christopher Field, "From Ecophysiology to Global Ecology," talk given at the Ecological Society of America annual meeting, August 3, 2020.

90. Hans D. Payer, Lutz W. Blank, Christof Bosch, Gerhard Gnatz, Wolfgang Schmolke, and Peter Schramel, "Simultaneous Exposure of Forest Trees to Pollutants and Climatic Stress," *Water, Air, and Soil Pollution* 31 (1986): 485–91; C. R. Rafarel and T. W. Ashenden, "A Facility for the Large-Scale Exposure of Plants to Gaseous Air Pollutants," *New Phytologist* 117, no. 2 (1991): 345–49; C. R. Rafarel, T. W. Ashenden, and T. M. Roberts, "An Improved Solardome System for Exposing Plants to Elevated CO_2 and Temperature," *New Phytologist* 131 (1995): 481–90.

91. William A. Dugas and Paul J. Pinter, Jr., "Introduction to the Free-Air Carbon Dioxide Enrichment (FACE) Cotton Project," *Agricultural and Forest Meteorology* 70 (1994): 1–2; G. R. Hendley and B. A. Kimball, "The FACE Program," *Agricultural and Forest Meteorology* 70 (1994): 3–14; G. R. Hendry, D. S. Ellsworth, K. F. Lewin, and J. Nagy, "A Free-Air Enrichment System for Exposing Tall Forest Vegetation to Elevated Atmospheric CO_2," *Global Change Biology* 5 (1999): 293–309; Richard J. Norby and Donald R. Zak, "Lessons from Free-Air CO_2 Enrichment (FACE) Experiments," *Annual Review of Ecology, Evolution, and Systematics* 42 (2011): 181–203.

92. John H. Lawton, "The Ecotron Facility at Silwood Park: The Value of 'Big Bottle' Experiments," *Ecology* 77, no. 3 (1996): 665–69, on 666.

93. John H. Lawton, S. Naeem, R. M. Woodfin, V. K. Brown, A. Gange, H. J. C. Godfrey, P. A. Heads, S. Lawler, D. Magda, C. D. Thomas, L. T. Thompson, and S. Young, "The Ecotron: A Controlled Environmental Facility for the Investigation of Population and Ecosystem Processes," *Philosophical Transactions of the Royal Society of London* 341, no. 1296 (1993): 181–94.

94. Lawton et al., "The Ecotron," 184.

95. Lawton et al., "The Ecotron," 192.

96. Ernst-Detlev Schulze and Harold A. Mooney, eds., *Biodiversity and Ecosystem Function* (Berlin: Springer, 1993).

97. *Resolutions Adopted by the Conference: Report of the United Nations Conference on Environment and Development, Rio de Janeiro, 3–14 June, 1992*, v. 1 (New York: United Nations, 1993).

98. "Interview with EICES Director Shahid Naeem," interview by Hari Sridhar, May 25, 2016, Bangalore, Earth Institute Center for Environmental Sustainability, eices.columbia.edu/2016/06/10/interview-with-eices-director-shahid-naeem, accessed February 2, 2022.

99. Shahid Naeem, Lindsey J. Thompson, Sharon P. Lawler, John H. Lawton, and Richard M. Woodfin, "Declining Biodiversity Can Alter the Performance of Ecosystems," *Nature* 368, no. 6473 (1994): 734–37.

100. Shahid Naeem, Lindsey J. Thompson, Sharon P. Lawler, John H. Lawton, and Richard M. Woodfin, "Empirical Evidence That Declining Species Diversity May Alter the Performance of Terrestrial Ecosystems," *Philosophical Transactions of the Royal Society of London* 347 (1321) (1995): 249–62.

101. David Tilman and John A. Downing, "Biodiversity and Stability in Grasslands," *Nature* 367, no. 6461 (1994): 363–65.

102. Mooney, *The Globalization of Ecological Thought*," 88.

103. Yvonne Baskin, "Ecosystem Function of Biodiversity," *BioScience* 44, no. 10 (1994): 657–60.

104. Harold A. Mooney, J. Hall Cushman, Ernesto Medina, Osvaldo E. Sala, and Ernst-Detlef Schulze, eds., *Functional Roles of Biodiversity: A Global Perspective* (Chichester: John Wiley and Sons, 1996), 475.

105. I will not delve into this debate, but interested readers can consult the literature cited here for key texts. See John H. Lawton, "What Do Species Do in Ecosystems?" *Oikos* 71 (1994): 367–74; J. P. Grime, "Biodiversity and Ecosystem Function: The Debate Deepens," *Science* 277 (1997): 1260–61; J. H. Lawton, S. Naeem, L. J. Thompson, A. Hector, and M. J. Crawley, "Biodiversity and Ecosystem Function: Getting the Ecotron Experiment in Its Correct Context," *Functional Ecology* 12, no. 5 (1998): 848–52; David A. Wardle et al., "Biodiversity and Ecosystem Function: An Issue in Ecology," *Bulletin of the Ecological Society of America* 81, no. 3 (2000): 235–39; Michel Loreau, Shahid Naeem, and Pablo Inchausti, eds., *Biodiversity and Ecosystem Functioning* (Oxford: Oxford University Press, 2002); Virginia Gewin, "Biodiversity: Rack and Field," *Nature* 460, no. 7258 (2009): 944–46.

106. Peter Kareiva, "Diversity Begets Productivity," *Nature* 368 (1994): 686–87, on 687.

107. Lawton, "The Ecotron Facility," 666.

108. Lawton, "The Ecotron Facility," 669.

109. Lawton, "The Ecotron Facility," 668.

110. Lawton, "The Ecotron Facility," 668.

111. A. Hector, B. Schmid, C. Beierkuhnlein, M. C. Caldeira, M. Diemer, P. G. Dimitrakopoulos, J. A. Finn, et al., "Plant Diversity and Productivity Experiments in European Grasslands," *Science* 286 (1999): 1123–27.

112. Christopher B. Field, Julie G. Osborn, Laura L. Hoffman, Johanna F. Polsenberg, David D. Ackerly, Joseph A. Berry, Olle Björkman, Alex Held, Pamela A. Matson, and Harold A. Mooney, "Mangrove Biodiversity and Ecosystem Function," *Global Ecology and Biogeography Letters* 7, no. 1 (1998): 3–14.

113. On EcoCELLs, see K. L. Griffin, P. D. Ross, D. A. Sims, Y. Luo, J. R. Seemann, C. A. Fox, and J. T Ball, "EcoCELLs: Tools for Mesocosm Scale Measurements of Gas Exchange," *Plant, Cell and Environment* 19 (1996): 1210–21. A brief tour of the "EcoCells at DRI" can be made via a YouTube video, youtube.com/watch?v=j_nYbGfU2oc, accessed April 7, 2022. On the prairie study, see John A. Arnone III, Paul S. J. Verburg, Dale W. Johnson, Jessica D. Larsen, Richard L. Jasoni, Annmarie J. Lucchesi, Candace M. Batts, et al., "Prolonged Suppression of Ecosystem Carbon Dioxide Uptake After an Anomalously Warm Year," *Nature* 455, no. 7211 (2008): 383–86. The study was reported in the Desert Research Institute *News,* Winter 2008, 2–3.

114. Jacques Roy, Francois Rineau, Hans J. De Boeck, Ivan Nijs, Thomas Pütz, Samuel Abivan, John A. Arnone III, et al., "Ecotrons: Powerful and Versatile Ecosystem Analysers for Ecology, Agronomy, and Environmental Science," *Global Change Biology* 27 (2021): 1387–1407. See the supplementary material included with this article for details about ecotron history and designs.

Epilogue

1. Marvalee H. Wake, "Integrative Biology: Science for the 21st Century," *BioScience* 58, no. 4 (2008): 349–53.

2. Massimo Pugliucci, "From Molecules to Phenotypes? The Promise and Limits of Integrative Biology," *Basic and Applied Ecology* 4 (2003): 297–306.

3. Hans Mohr, "A Life History Between Science and Philosophy," in *Progress in Botany: Genetics, Physiology, Systematics, Ecology,* ed. K. Esser, U. Lüttge, W. Beyschlag, and J. Murata (Heidelberg: Springer, 2005), 3–28, on 26.

4. Sally E. Smith and David J. Read, *Mycorrhizal Symbiosis,* 3rd ed. (Amsterdam: Academic Press, 2008), 4, 115, chap. 3.

5. Joanna R. Freeland, Heather Kirk, and Stephen D. Petersen, *Molecular Ecology,* 2nd ed. (Chichester: Wiley-Blackwell, 2011), 1.

6. Michael Bevan and Sean Walsh, "The *Arabidopsis* Genome: A Foundation for Plant Research," *Genome Research* 15 (2005): 1632–42, doi: 10.1101/gr.3723405.

7. Jim Endersby, *A Guinea Pig's History of Biology* (Cambridge, Mass.: Harvard University Press, 2007), chap. 10; Sabina Leonelli, "*Arabidopsis,* the Botanical *Drosophila:* From Mouse Cress to Model Organism," *Endeavour* 31, no. 1 (2007): 34–38; Leonelli, "Growing Weed, Producing Knowledge: An Epistemic History of *Arabidopsis thaliana,*" *History and Philosophy of the Life Sciences* 29 (2007): 193–223; Leonelli, "Weed for Thought: Using *Arabidopsis thaliana* to Understand Plant Biology" (Ph.D. diss., Vrije Universiteit, 2007); Rachel A. Ankeny and Sabina Leonelli, *Model Organisms* (Cambridge: Cambridge University Press, 2020).

8. See the following accounts Chris Somerville, "Arabidopsis Blooms," *The Plant Cell* 1, no. 12 (1989): 1131–35; Ben Patrusky, "Drosophila Botanica," *Mosaic* 22 (1991): 32–43; Anne A. Moffat, "Gene Research Flowers in *Arabidopsis thaliana*," *Science* 258 (1992): 1580–81; Elliot Meyerowitz and Chris R. Somerville, *Arabidopsis* (Plainsview, N.Y.: Cold Spring Harbor Laboratory Press, 1994), 1–6; Gerald R. Fink, "Anatomy of a Revolution," *Genetics* 149 (1998): 473–77; Virginia Walbot, "A Green Chapter in the Book of Life," *Nature* 408, no. 6814 (2000): 794–95; Elizabeth Pennisi, "*Arabidopsis* Comes of Age," *Science* 290 (2000): 32–35; Elliot Meyerowitz, "Prehistory and History of Arabidopsis Research," *Plant Physiology* 125 (2001): 15–19; Chris Somerville and Maarten Koornneef, "A Fortunate Choice: The History of Arabidopsis as a Model Plant," *Nature Reviews* 3 (2002): 883–89; Maarten Koornneef and David Meinke, "The Development of Arabidopsis as a Model Plant," *The Plant Journal* 61 (2010): 909–21; Maarten Koornneef, "A Central Role for Genetics in Plant Biology," *Annual Review of Plant Biology* 72 (2021): 1–16. See also the interview with Elliot Meyerowitz by Daphne Preuss, 2002, *Conversations in Genetics* series, youtube.com/watch?v=AIIQlHvda9Y, accessed March 21, 2022.

9. Csaba Koncz, "Obituary: George P. Rédei (1921–2008)," *Cereal Research Communications* 37 (2009): 143–47; on *Agrobacterium* see Mary-Dell Chilton, Martin H. Drummond, Donald J. Merio, Daniela Sciaky, Alice L. Montoya, Milton P. Gordon, and Eugene W. Nester, "Stable Incorporation of Plasmid DNA into Higher Plant Cells: The Molecular Basis of Crown Gall Tumorigenesis," *Cell* 11 (1977): 263–71.

10. Especially important were DeLill Nasser, Mary Clutter, and Machi Dilworth.

11. Elliot Meyerowitz and Chris R. Somerville, *Arabidopsis* (Plainsview, N.Y.: Cold Spring Harbor Laboratory Press, 1994).

12. Chris Somerville, "The Director's Introduction," *Carnegie Institution of Washington Year Book No. 93* (Washington, D.C.: Carnegie Institution of Washington, 1994), 51–53.

13. Meyerowitz, "Prehistory and History of Arabidopsis Research," 17.

14. Meyerowitz, "Prehistory and History of Arabidopsis Research," 18.

15. Shauna Somerville, "Molecular Mechanisms of Disease Resistance," *Carnegie Institution of Washington Year Book 93* (Washington, D.C.: Carnegie Institution of Washington, 1994), 54–58.

16. Ute Krämer, "Planting Molecular Functions in an Ecological Context with *Arabidopsis thaliana*," eLife 4 (2015): e06100, doi: 10.75554/eLife.06100.

17. For example, see Christopher W. P. Lyons and Karen-Beth Scholthof, "Watching the Grass Grow: The Emergence of *Brachypodium distachyon* as a Model for

the *Poaceae*," in *New Perspectives on the History of Life Sciences and Agriculture*, ed. Denise Phillips and Sharon Kingsland (Cham: Springer, 2015), 479–501.

18. Gottfried Fraenkel, "The Raison d'Etre of Secondary Plant Substances," *Science* 129 (1959): 1466–70; Paul R. Ehrlich and Peter H. Raven, "Butterflies and Plants: A Study in Coevolution," *Evolution* 18, no. 4 (1964): 586–608.

19. Richard Karban and Ian T. Baldwin, *Induced Responses to Herbivory* (Chicago: University of Chicago Press, 1997); David F. Rhoades, "Reponses of Alder and Willow to Attack by Tent Caterpillars and Webworms: Evidence for Pheromonal Sensitivity of Willows," in *Plant Resistance to Insects*, Symposium Series 208, ed. P. A. Hedin (Washington, D.C.: American Chemical Society, 1983), 55–68, on 67.

20. Ian T. Baldwin and Jack C. Schultz, "Rapid Changes in Tree Leaf Chemistry Induced by Damage: Evidence for Communication Between Plants," *Science* 221 (1983): 277–79.

21. Jack C. Schultz, "Many Factors Influence the Evolution of Herbivore Diets, but Plant Chemistry Is Central," *Ecology* 69 (1988): 896–97, on 897.

22. May R. Berenbaum, "Thomas Eisner, 1929–2011," *Biographical Memoirs of the National Academy of Sciences of the United States of America* (Washington, D.C.: National Academy of Sciences, 2014), nasonline.org/publications/biographical-memoirs/memoir-pdfs/eisner-thomas.pdf.

23. Karban and Baldwin, *Induced Responses to Herbivory*. For a recent review of this field see Richard Karban, "Plant Communication," *Annual Review of Ecology, Evolution, and Systematics* 52 (2021): 1–24.

24. Daniel J. Cosgrove, Simon Gilroy, The-hui Kao, Hong Ma, and Jack C. Schultz, "Plant Signaling 2000: Cross Talk Among Geneticists, Physiologists, and Ecologists," *Plant Physiology* 124, no. 2 (2000): 499–505.

25. Allison Abbott, "Growth Industry," *Nature*, no. 7326 (2010): 886–88. For examples see Danny Kessler, Klaus Gase, and Ian T. Baldwin, "Field Experiments with Transformed Plants Reveal the Sense of Floral Scents," *Science* 321 (2008): 1200–1202; and Ian T. Baldwin, Rayko Halitschke, Anja Paschold, Caroline C. von Dahl, and Catherine A. Preston, "Volatile Signaling in Plant-Plant Interactions: 'Talking Trees' in the Genomics Era," *Science* 311 (2006): 812–15.

26. Jack C. Schultz, "Shared Signals and the Potential for Phylogenetic Espionage Between Plants and Animals," *Integrative and Comparative Biology* 42, no. 3 (2002): 454–62, on 454, 459.

27. "Gene Transfer from Plant to Insect Spotted," *Nature* 592, no. 7852 (2021): 13.

28. Schultz, "Shared Signals and the Potential for Phylogenetic Espionage," 459–60.

29. On plant neurobiology, see Frantisek Baluska and Stefano Mancuso, "Plant Neurobiology as a Paradigm Shift Not Only in the Plant Sciences," *Plant Signaling and Behavior* 2 (July/August 2007): 205–7. For a guide to this literature, pro and con, see Paco Calvo, "The Philosophy of Plant Neurobiology: A Manifesto," *Synthese* 193 (2016): 1323–43; Lincoln Taiz, D. Alkon, A. Draguhn, A. Murphy, M. Blatt, C. Hawes, G. Thiel, and D. G. Robinson, "Plants Neither Possess nor Require Consciousness," *Trends in Plant Science* 24, no. 8 (2019): 677–87; David G. Robinson and Andreas Draguhn, "Plants Have Neither Synapses nor a Nervous System," *Journal of Plant Physiology* 263 (2021): 153467, doi.org/10.1016/j.jplph.2021.153467.

30. Ian T. Baldwin, "Training a New Generation of Biologists: The Genome-Enabled Field Biologists," *Proceedings of the American Philosophical Society* 156, no. 2 (2012): 205–14, on 208, 209.

Index

373